THE ATOMIC COMPLEX

THE ATOMIC COMPLEX

A Worldwide Political History
of Nuclear Energy

Bertrand Goldschmidt

Substantially revised and updated
from the original French edition,
Le Complexe Atomique

AMERICAN NUCLEAR SOCIETY
La Grange Park, Illinois

Library of Congress Cataloging in Publication Data

Goldschmidt, Bertrand.
  The atomic complex.

  Translation of: Le complexe atomique.
  Includes index.
  1. Atomic Energy — History. 2. Nuclear fission —
History. I. Title
QC773.G6313   333.79'24   82-1754
ISBN 0-89448-550-4   AACR2

Translated from the French by Bruce M. Adkins

ISBN: 0-89448-550-4 (hardbound)
0-89448-551-2 (softbound)
Library of Congress Catalog Card Number: 82-70371

American Edition

Printed in the United States of America.

French Edition

CONTENTS

Preface

Nuclear proliferation risks are multiplying. One need only think of Israel, Pakistan, and Iraq to realize what a tangled web the atomic complex is weaving around a destabilized world. All statesmen oppose the spread of weapons. Almost all know little about the complexities involved in the phenomenon of proliferation. All could with great profit read these pages written by a unique person; literally no other man has the experience and perspective of this French scientist, scholar, and statesman. It is not surprising that the U.S. government is beginning to share his views.

In good part, Bertrand Goldschmidt concentrates on relations between the first proliferator, the United States, and the fourth proliferator, France, as he recounts the events of the nuclear age in which he has played such a substantial role.

The events and the man are so closely related that this work is as much a memoir as history. This account is derived not only from deep scholarship, but from personal participation and experience. His exact characterization of nuclear weapons as "monstrous," is not based on hearsay or book learning, but on firsthand observations.

The perception of France's national security interests as requiring that she harness this monstrosity, even while realizing that her security could only be diminished as other nations follow suit, illustrates the deadly dilemma facing all would-be nuclear weapons nations. It faced the United States, which started the proliferation process; it will face all the rest until reason begins to rule over folly in human relations.

The U.S. weapons program was begun during a great war for survival and when it was expected that Nazi Germany might be the first to deploy nuclear weapons. In the end, Germany made little progress toward atomic bombs; and it was Japan, even after it had begun suing for peace, that was the target of these monstrous arms. Now Japan and Germany under the Non-Proliferation Treaty have foresworn the right to possess nuclear weapons — a state of affairs that the United States, the Soviet Union, the United Kingdom, and France especially appreciate. But France still refuses to join the Non-Proliferation Treaty, finding it defective in that it discriminates against nonnuclear weapons nations, a class that France was an unwilling member of for some 15 years.

France declares that she will act "as if" she were a party to this treaty, and this less-than-full participation has marked France's role in a number of other arms control efforts. France backed the Indian position of resistance to international controls in the 1950s; her policy for many years was to leave her chair empty during the United Nations Disarmament Conference meetings at Geneva; she participated reluctantly in the London Supplier talks in the 1970s.

France's shying away from nonproliferation approaches favored by her allies reflects less opposition to their purpose than questioning of their efficacy. France, after many years of exclusion from the Anglo-American nuclear club, has a much keener sense of how nonweapons nations feel about their different status. She believes that aiding such countries to solve their energy problems is more likely to help them resist weapons temptations than denial of supply. It is a close question, about which American policy remains somewhat ambivalent. The war-born Anglo-Saxon effort to monopolize the new energy source gave rise to a deep sense of discrimination in France. The author's resentment on this score drives his narrative, as does his undisguised sense of pride in France's accomplishments in weaponry and in nuclear electric technology where she has become a superpower.

After a generation of France's resisting American initiatives in this field, there seems to be developing a convergence between the positions of the two countries. France has come to recognize the danger presented by the spread of plants that reprocess spent fuel from power reactors — one of whose products is explosive plutonium. It is a long way from the 1963 French offer to supply India with a nuclear power plant without any controls to the stringent conditions that France insisted upon in 1976 for the export of the Iraqi reactor that Israel destroyed in 1981. There is irony in the fact that this preemptive antiproliferation strike was mounted by a country — which France earlier had helped with its reactor program — that is now the source of explosives for a weapons program.

How will nations ever emerge safely from the tangled atomic complex — with its technological, political, and psychological strands? Goldschmidt correctly foresees no end to the spread of nuclear weapons until there is real disarmament by the nuclear weapons superpowers. This, indeed, reflects the basic bargain underlying the Non-Proliferation Treaty.

Only when the weapons powers take the painful steps involved in reversing the weapons buildup of the last generation will there be hope of dampening the lesser rivalries that are generating new weapons programs around the world. May the many readers of this book let this lesson sink in deeply. As Bertrand Goldschmidt writes — the story is not finished.

Gerard C. Smith

January 1982

Foreword

Mankind's understanding and conquest of the tremendous forces in the atomic structure of matter began shortly before World War II. These forces resulted from a purely scientific discovery, that of fission in the nucleus of the uranium atom.

Never has a discovery proved itself so rapidly and completely in the fulfillment of its early promises: the weapon, the submarine engine, and the power station. Never has a discovery been so technically, politically, and psychologically complex; never has the resulting technological progress suffered so many discontinuities; and never has a discovery had so many international implications and consequences.

At a single stroke, the age-old dream of the alchemist became reality: uranium could be transmuted into plutonium, a new element with potential power far exceeding that of the gold that the philosopher's stone was supposed to produce.

But the dream could become a nightmare. The same long series of industrial developments and realizations could lead in two directions: the one toward a new energy source able to meet the needs of the industrial world from now on, and then, from the turn of the century, to help overcome the problems of underdevelopment; the other toward a devastating weapon — the weapon that brought the last world conflict to an end — and toward the thousand-times more powerful weapons since conceived.

The shadow of a major disaster hovers over every landmark along this road of multiple technical successes. Mankind, having mastered the secret of atomic fire, has not succeeded in banning atomic weapons.

Five countries in succession have managed to acquire these weapons of unprecedented destructive power, which now menace every other advance of our civilization. These countries are the five great Allied Powers of World War II — the United States, the Soviet Union, Great Britain, France, and China — those five powers to which the United Nations

Charter of 1945 gave the duty of keeping the peace of the world and the right of veto in the U.N. Security Council.

In acquiring atomic armament, all but the first of these five countries had to overcome obstacles placed in their way by the country or countries that had preceded them. With the exception of China, each rallied in its turn to the joint effort designed to prevent — or at least delay — acquisition of the weapon by further countries.

But just as the uranium atom is divided by fission, its military applications have divided the nations of the world. Thus it has contributed successively to increasing rivalry between the United States and the Soviet Union and then to reinforcing the present division of the world into two camps. It has also created a new and deep discrimination between those states possessing the new weapons and those without them.

Nevertheless, for the first time in history, nations have unilaterally renounced the manufacture of the most powerful armament known, even though the manufacturing process was within their technical and financial capacity. For the first time also, they have agreed to guarantee the peaceful character of certain of their industrial installations — those based on fission — and even to submit those installations, which are indispensable to them, to international control.

It is a fact that nuclear-produced electricity is the only new energy source that can be expected to make a major contribution to world consumption between now and the beginning of the next century. It is technically and economically proven and ready for the precise moment when civilization is faced with its first fundamental energy crisis. But nuclear-produced electricity has also arrived at the very moment when man is becoming seriously concerned over the problems of his environment and of his consumer society. He is no less conscious of the terrible menace of the nuclear arms race, in the face of which he finds himself powerless, so that his anxieties are now turned toward the risks associated with world expansion of civil nuclear activity.

Everything technically possible is being done to ensure that the nuclear industry, despite its complexity, can develop with minimal risk for both population and environment. Everything politically possible is being done to ensure that no country is tempted to make military use of a nuclear installation whose peaceful purpose has been guaranteed.

Yet in spite of this, under the growing shadow of a world energy war, the road toward production of indispensable power from uranium is by no means clear. Artificial obstacles have been created — particularly in many of the Western powers — because of the fear of public opinion concerned with the radioactivity in nuclear power plants and their by-products or as a result of political rules and regulations designed to prevent more countries from acquiring atomic weapons by diverting their civil programs to this "forbidden objective."

Thus mankind, in its concern for the protection of its health and environment, in its pursuit of peace, or for reasons of mistrust or domination, has added its own self-created difficulties and specific obstacles to those already imposed by nature in the exploitation of the atom. If mankind is to succeed in this exploitation, whether for civil or military purposes, these obstacles must be overcome or bypassed.

The story of this exploitation, of these difficulties and obstacles and their consequences, is the underlying thread running through the political history of the atomic adventure, from the discovery of fission to the present day — a present day that finds our world overloaded with megatons of nuclear explosives and desperately short of megawatts of nuclear power. This is the story that the following pages have to tell.

It must be recognized from the beginning that there is a fundamental dissymmetry between the civil and military aspects of the story. On one hand, the development of nuclear weapons preceded that of nuclear power plants and ship propulsion systems by some 10 to 15 years. Thus, weapons development was (and remains) practically independent of any influences from civil nuclear programs. On the other hand, nuclear electricity and the associated nuclear fuel cycles have necessarily grown from technical and industrial roots in the military field.

Had the fission process proved useful only for making a devastating weapon whose vast energy could not be tamed technically or economically, the military side of the atomic story would hardly have been different. But if, by chance, the weapon had proved unrealizable, and only the production of useful energy possible, the development of the new resource would clearly have followed a very different course.

For this reason, rather than considering both aspects in parallel it seemed preferable to me, in the interests of continuity and clarity, to treat separately the political aspects of these two faces of the atomic complex. This book is, therefore, in two parts — and tells two stories. The first is devoted to ''The Explosion'' while the second, ''The Power,'' follows naturally from the first and deals with the recovery of usable energy. In each part, I have tried to give an overall picture of the extraordinary adventure through which I have lived, with (naturally) particular emphasis on French aspects.

For the period from World War II until 1952 I have derived much information and inspiration from the reports written, on the basis of official documents, by the historians of atomic energy in the United States[1] and Great Britain.[2]

It is a pleasure to acknowledge the encouragement of André Giraud and the Commissariat à l'Energie Atomique in my writing this book. My thanks go to all friends and colleagues who have helped in its preparation, particularly Octave Du Temple, Mike Diekman, and Lorretta Palagi of the American Nuclear Society, which has made this American publication possible, and Bruce Adkins who has done the translation so well.

January 1982

Bertrand Goldschmidt

[1]Hewlett, Richard G. and Oscar E. Anderson, Jr. *A History of the United States Atomic Energy Commission, Vol. 1: The New World, 1939/1946.* University Park: Pennsylvania State University Press, 1962.

Hewlett, Richard G. and Francis Duncan. *A History of the United States Atomic Energy Commission, Vol. 2: Atomic Shield, 1947/1952.* University Park: Pennsylvania State University Press, 1969.

[2]Gowing, Margaret. *Britain and Atomic Energy, 1939-1945.* London: Macmillan & Co. Ltd., 1964.

Gowing, Margaret. *Independence and Deterrence — Britain & Atomic Energy, 1945-1952.* 2 vols. London: Macmillan & Co. Ltd., 1974.

Part One

THE EXPLOSION

The first part of this book is devoted to the story of man's accession to the mastery of atomic explosions.

The story is one of unfolding drama in which the actors — that is, the various states acquiring or seeking to acquire nuclear weapons — come in sequence onto a stage constantly haunted by the growth of these armaments.

The first act covers the war period and is dominated by its concluding episode: the decision to use the bomb and the drama of Hiroshima and Nagasaki. Throughout this period the two principal actors, the United States and the United Kingdom, played a deep and complicated game. Their statesmen vied in advance for the new power they hoped to acquire as their scientists unlocked nature's secrets along the road to both explosive and controlled atomic power.

The second act, from 1945 to 1964, saw in 1949 the arrival on the scene of the Soviet Union, whose development work the Anglo-Saxon allies had vainly hoped to delay by keeping their technological data secret in order to prolong their own monopoly. In 1952 the United Kingdom mastered the bomb; France, unaided and in the face of external opposition, followed in 1960. Finally, with China's demonstration in 1964 that it too possessed the bomb, the door to the "Club" of atomic weapons states closed behind this last permanent member of the U.N. Security Council.

The third act begins in 1963, following the Cuban crisis, when the Soviet Union switched from opposing to supporting the international nuclear policies of the United States. This change made possible the progressive establishment of international controls over the civil development of atomic energy and the conclusion of the nonproliferation agreements, fundamental elements in international atomic relations.

This third act, like the whole of this remarkable story, is as yet unfinished. It has been prolonged by the Indian atomic test of 1974, by the negotiations between the Soviet Union and the United States on limiting their strategic super-weapons, and finally by the importance of the unsolved problems of nonproliferation in a world that has shown itself incapable of disarmament and is at the same time divided by an oil crisis, the expansionist policies of the Soviet Union, and the urgent needs of the Third World.

Act One

THE ALLIANCE
1939 to 1945

1. The Fateful Decision

The Death of Roosevelt

The greatest conflict our world has so far experienced — World War II — was also the first war in history to be brought to an end by the decision to use a hitherto unknown weapon.

At the outbreak of hostilities, less than five years before that weapon was used, even its fundamental principle would have seemed pure science fiction. Its development had been undertaken in the strictest secrecy, on the basis of extraordinarily complex theories of pure physics, and had involved an unequaled and unprecedented effort. To the very last moment there had been neither certainty of success nor guarantee of effectiveness.

Both in Britain and the United States, the essential decision to develop the atomic bomb was made without the knowledge of the people, the parliamentary authorities, or even the ministerial services concerned. The decision was made by a tiny group of heads of state and ministers assisted by a small number of their closest aides. The same was true of the nuclear relations between the Allies, and indeed of the supreme decision of the entire affair: the decision that the new weapon would be used in the war.

These heads of state and ministers had put their trust in the theories of their scientists, theories they could not understand and whose validity could be confirmed or disproved only by the final result of this astonishing technical/ industrial undertaking.

On March 15, 1945, President Franklin D. Roosevelt, in a long conversation with Secretary of War Henry L. Stimson, discussed the atomic bomb and its possible military use. For Roosevelt this was to be his last discussion of the matter.

Just back from the Yalta Conference, they had first to take account of the unresolved difficulties with the Soviet Union over the formation of a new Polish government, which according to the Teheran Agreement must result from free elections and must at the same time be acceptable to the Soviet Union — two requirements that seemed incompatible at that time and still are today!

Roosevelt, contrary to the advice of some of his aides, had not told Soviet Premier Joseph V. Stalin about the new weapon. On the other hand, he had secured a Soviet promise to join the war against Japan within two to three months of a Nazi surrender in Europe, which seemed to be near-at-hand.

Stimson was able to report satisfactorily to the president on the progress of work on the bomb, of which two different types would be ready in the summer. These would depend on two "special materials" of which, on the basis of milligram quantities isolated in U.S. laboratories in 1941 and 1942, a few kilograms of each would soon be available. The two materials were: uranium-235, found in naturally occurring uranium in the proportion of 7 parts per 1000 and extraordinarily difficult to separate from its more abundant partner — physically and chemically its twin — the "isotope" uranium-238; the other was the new element plutonium, which man himself had made by the transmutation of uranium-238. The few kilograms of these substances were the result of an immense industrial enterprise that had been created from scratch in only three years and was equal in size to the entire American automobile industry at that time. If the scientists were right, those few kilograms could be used to make explosives of hitherto un-known power.

Stimson raised with the president the question of controlling this remarkable power after the war was over. Here, there were two schools of thought: those who wanted to keep the technology as an Anglo-American secret (and thus give Britain and the United States a monopoly in its exploitation) and those who wanted a solution within the framework of the United Nations.

Roosevelt, exhausted after his long voyage, made no decision. On

April 12th, four weeks after his conversation with Stimson, he died and so disappeared the statesman who would otherwise have had the duty to decide whether or not the atomic weapon should be used. It is probable that, had he lived to make that decision, he would have paid particular attention to the several scientists who had declared themselves opposed to using the bomb at such a late stage in the war.

Responsibility for use and control of the bomb passed from Roosevelt to Harry S. Truman, who — in the office of president for only three months — had not previously known of its existence.

The Discovery of Fission

It all began with the discovery at the end of 1938 that one of the fundamental particles of matter, the neutron, could cause fission in the uranium atom, each fission being accompanied by the emission of energy and the formation of radioactive matter.

The metal uranium, known since the end of the 18th century, is the heaviest naturally occurring element on earth. It is the element that enabled Antoine Henri Becquerel to discover radioactivity in 1896. Two years later, Pierre and Marie Curie isolated its radioactive descendant, radium, which is always present in uranium ores in the proportion of one gram of radium to three tons of uranium.

Before the war, uranium was a virtually unused by-product from the production of radium. Relatively widely distributed in nature, it is generally found in low concentrated ores. At that time, the only rich deposits known were those of the Haut Katanga in the Belgian Congo, discovered in 1913, and those found in 1930 in the Canadian far north.

The phenomenon of fission was discovered in Berlin by Otto Hahn, Lise Meitner, and Fritz Strassmann. The discovery resulted from five years of competition and cooperation between teams of physicists and chemists working in Rome, Berlin, and Paris. The starting point for this international effort had been the discovery of artificial radioactivity in 1934. The discovery was made by the husband-and-wife team of Frédéric and Irène Joliot-Curie at the Radium Institute in Paris, a few months before the death of the laboratory's founder, Marie Curie.

Joliot-Curie said in his Nobel address in 1935: "If we look at past scientific progress, pursued with ever increasing speed, we may reasonably expect future research workers, breaking down or building up atoms at will, to be able to achieve explosive nuclear chain reactions. If such transmutations can be propagated in matter, we can envisage the liberation of enormous quantities of usable energy."

At that time already, there seemed to be two possible processes for the production of energy from the atomic nucleus, one involving the lightest elements found in nature, the other the heaviest.

The first process involved the "condensation" of very light atomic nuclei to produce nuclei of heavier elements (this is the reaction that takes place in the stars and is the source of solar energy); the second involved the splitting of the nuclei of the heaviest atoms into those of lighter elements.

Joliot-Curie's prophecy was confirmed four years later by the discovery of fission in uranium, nature's heaviest atom. And it was again Joliot-Curie who was the first to demonstrate that the splitting of a uranium nucleus by a single neutron, besides producing two lighter radioactive elements (called "fission products") and a release of energy, was also accompanied by the emission of several further or "secondary" neutrons. This is the primordial phenomenon that makes possible the propagation of an "atomic fire": As soon as the secondary neutrons are produced, they migrate into the neighboring uranium nuclei where they cause new fissions.

Identification of neutrons as the "propagating agent" in spreading the fission process paved the way to the liberation of atomic energy from measurable quantities of matter. This access to the immense reservoir of energy in the atomic nucleus could provide, weight for weight, a new fuel some three million times more powerful than coal — a "quantum jump" of a dimension rarely seen in the history of science.

Forty years after the Curie's discovery of radium, atomic physics had ceased overnight to be just a matter of fundamental research or just the prerogative of the detached scientist. A new elite was being born that was to have an ever-growing influence in the life of the major nations of the world: an elite comprised of scientists aware of their moral and political responsibilities.

At the beginning of 1939, even the world press became briefly engrossed in the subject. Every sort of development was envisaged, from an electric power station to a bomb, and including a nuclear-powered submarine. But this new domain was beyond the grasp of public opinion, which was incapable of judging whether these predictions were merely science fiction; however, very soon all the work was covered by a blanket of secrecy.

The Hungarian Expert

Among those worried over the military and political consequences of the discovery, the most concerned was Leo Szilard, a Hungarian-born physicist who had in the past worked with Albert Einstein and who, throughout this extraordinary adventure, was possessed of quite remarkable foresight.

Following the anti-Semitic persecutions in Germany, Szilard had taken refuge in New York from where, in February 1939, he began to make contacts with colleagues in the countries likely to be allies in the coming war that he thought was inevitable. To each he proposed that they should agree together to cease any further publication about nuclear fission. Thus, for the first time in the history of fundamental physics, a policy of secrecy and refusal to exchange scientific knowledge was born.

In a letter to Joliot-Curie, Szilard wrote: ''Obviously, if more than one neutron were liberated, a sort of chain reaction would be possible. In certain circumstances, this might lead to the construction of bombs which would be extremely dangerous in general and particularly in the hands of certain governments.'' Thus, from the very beginning, today's very real problem of the necessity of achieving nonproliferation was defined.

Although this initiative of Szilard's was neither fully understood nor fully accepted, in less than a month — and well before the outbreak of war — each country began independently to impose secrecy on the results of its uranium research.

At the same moment, Szilard was trying to persuade official circles in the United States to take an interest in the matter, for he was haunted by the fear that Nazi Germany might win the atomic race. Finding no response to his first approaches to the Navy Department, and considering the urgency of the problem, he decided he must go higher, for the European war could already be seen on the horizon. In the end, with the help of Einstein, author of the Theory of Relativity, and of Alexander Sachs, an economist friend of Roosevelt's, he established contact with the president when in October 1939 Sachs passed on a letter from Einstein to which a report by Szilard was annexed. This report, among other things, noted that the current work in France was probably more advanced than any other.

Einstein's letter to the president painted a stark and startling picture of the possible effects of an atomic bomb, which it was then believed would have to be very large. ''A single bomb of this type'' he wrote, ''carried by boat and exploded in a port, might very well destroy the whole port together with some of the surrounding territory. However, such bombs might well prove to be too heavy for transportation by air.''

Roosevelt immediately decided to set up a committee.

Five and one-half years after this first alert, from which the whole U.S. nuclear effort began, the same Hungarian scientist, once more helped by a letter from Einstein, approached the president again, this time to warn him of the international consequences should the bomb eventually be used.

However, Roosevelt died without learning of this new approach or of the memorandum that Szilard had prepared. Szilard's letter had set out, with the same clarity as before, the consequences of using the bomb for future U.S. international policies — in particular in relation to the Soviet Union. He pointed out the dangers of the nuclear arms race that would inevitably follow, and in which the United States must eventually lose its lead. Among other things, Szilard wrote: "If there should be great progress in the development of rockets after this war, it is conceivable that it will become possible to drop atomic bombs on the cities of the United States from very great distances by means of rockets. The weakness of the position of the United States will largely be due to the very high concentration of its manufacturing capacity and of its population in cities. This concentration is so pronounced that the destruction of the cities may easily mean the end of our ability to resist."

In his conclusion, Szilard explained the importance of an agreement with the Russians: "In discussing our postwar situation the greatest attention was given in this memorandum to the role that Russia might play. This was not done because it was assumed that Russia may have aggressive intentions but rather because it was assumed that if an agreement can be reached with Russia it will be possible to extend the system of controls to every country in the world."

It was to prove necessary to await the end of the 1960s, and the negotiation of the Non-Proliferation Treaty, before such a world policy could be adopted. In the interval, five countries, led by the United States and the Soviet Union, equipped themselves with atomic weapons under national control.

The Early Soviet Effort

The decision to use the bomb was bound to bring into question relations between the Anglo-Saxons and the Soviet Union. That country, like the Free French, had not been kept officially informed of the Anglo-Saxon atomic research.

It is true that a draft agreement had been worked out early in 1943 for a complete collaboration between British and Soviet scientists engaged in military research. Roosevelt had opposed the scheme, which was therefore abandoned. As a result, there was no real wartime collaboration between Anglo-American and Soviet scientific research in any domain, including of course that of atomic weapons.

But Soviet scientists were clearly not unaware of the uranium question. In the spring of 1940, a special committee on the subject had been set up under the auspices of the Soviet Union's Academy of Sciences, and funds were allocated for uranium prospecting. From reports prepared in 1941 at the Soviet academy, which reached the United States the following year, it was clear that the very far-reaching consequences of the discovery of fission were fully appreciated by a number of Soviet physicists, who were calling on their government to launch immediate studies of the question so as to avoid being overtaken by other countries.

In 1942 the physicist Igor Kurchatov was asked to investigate the possibility of making a bomb, but his work was interrupted by the German invasion of the Soviet Union, which caused the Soviet's specialist teams to be dispersed and assigned to more urgent research. Work in the nuclear physics laboratories was not resumed until 1943, after the Stalingrad victory, and even then the program remained on a rather small scale until the end of hostilities.

However, as we shall see later, the Russians were kept informed of the Anglo-American work by Communist sympathizers — an American mechanic working on the bomb mechanism and two physicists working for the British, one a specialist on weapon theory, the other on controlled chain reactions. The Soviet leaders were thus well aware of what was going on and of its objectives.

The German Failure

Efforts in Germany, initially directed toward nuclear power supplies, were also dispersed following the 1941 attack on the Soviet Union.

The Germans had indeed considered the possibility of isolating and concentrating uranium-235 to make an explosive, but had thought it too difficult and had abandoned the idea. They had concentrated their efforts on the development of a "uranium machine," based on a controlled chain reaction, which they believed could be used in a power-producing motor. They also believed, incorrectly, that this could be used as an explosive if the reaction were sped up. They did not know of the formation of plutonium.

Thus the Germans had decided that a true atomic bomb could not be developed within only a few years. It is to be emphasized that they did not renounce the idea for any moral reason, as has since sometimes been claimed. Being convinced of their scientific prowess, they had no fears that the Allies might be ahead of them.

In Berlin during 1939, the scientists had founded a "uranium society," which included all the great German physicists of the time, in particular the most celebrated of all, the Nobel Prize winner Werner Heisenberg. On the declaration of war, many of these scientists were mobilized; others who opposed the regime, like Otto Hahn, principal author of the discovery of fission, stopped their work on uranium to which, in any case, the government had not given the necessary priority.

In 1942 Marshal Hermann Goering took over responsibility for the work. However, despite the scientific direction being entrusted to a highly competent physicist, competition and lack of liaison between the various research teams continued to impede efficient organization. The total effort involved scarcely 100 scientists and engineers working in small groups with an overall budget that, at the end of the war, amounted to about $10 million, or one-half of one percent of the American investment.

The German scientists followed the same line of research as the French, whose work up to 1940 was known to them. For their "machine," they turned to a system based on uranium and heavy water. This heavy water was provided by a Norwegian plant that was destroyed twice by the Allies; the first time it was sabotaged by a parachute commando working with Norwegian patriots, and the second time, after it had been rebuilt, by British air bombardment. Following these two setbacks, production was abandoned until the end of the war.

The rebuilding of the heavy water plant after the first sabotage was taken by the Allied intelligence services as proof of the importance of the Nazi effort. The Germans, on the other hand, warned by their own espionage networks of the magnitude of the American work, did not take it seriously. Convinced by their own scientists that to make a bomb quickly was impossible, the Germans believed the Americans were trying to acquire industrial and commercial advantages for the postwar period. Thus, the conclusions of the intelligence services of each side, wrong in both cases, had opposite effects on the work underway — stimulating the Americans to greater efforts and reassuring the Germans that such efforts were unnecessary.

It was not until 1945, when at last the Allies occupied enemy territory, that they found final proof that contrary to all their fears the Germans were several years behind them and had not even reached the stage of the first chain reaction, which was achieved in the United States at the end of 1942.

Adolf Hitler had, on the other hand, given his V-1 and V-2 missiles the priority he had thought unnecessary for work on uranium. Intercontinental ballistic missiles and cruise missiles are the direct results of this work. So did the German scientists and engineers contribute, involuntarily, to the atomic weapons of the future by providing a means of transport that today makes these weapons virtually invulnerable.

The Danish Scientist

In the autumn of 1943, the Danish scientist Niels Bohr, the father of modern atomic theory, escaped from Denmark to Sweden. From there, the British persuaded him to go to England. He was the first to bring the Allies reassurance that work in Germany was not yet very advanced. Informed of the Allies' own progress in the military field, at the beginning of 1944 he was sent to Los Alamos, New Mexico, where the new weapon was being developed.

Bohr very quickly perceived the political revolution to be expected from this new force in the world. It seemed clear to him that the only safe course would be to inform the Soviet Union before the bomb was used, and to organize the necessary international controls with that country. He believed such action to be essential if a nuclear arms race between the Soviet Union and the allied Western powers was to be avoided.

Bohr made continual efforts to have his views conveyed to Roosevelt, who let the Danish physicist know that he shared his concern and even encouraged him to raise the matter in England. Returning to London in the spring of 1944, Bohr was able to convince the British minister responsible, Sir John Anderson, that he should be allowed to raise the matter with British Prime Minister Winston Churchill.

A meeting with the prime minister took place in May, less than three weeks before the Normandy landings. It was a disaster. Churchill, preoccupied and impatient, could follow neither the language of the great scientist (well known for his lack of clarity) nor his reasoning.

Despite this setback Bohr, once more in the United States, sought an interview with Roosevelt. The meeting was arranged at the end of August and this time it seemed to be a success. Roosevelt promised that when next he met Churchill they would together examine the consequences of a possible disclosure to the Russians.

The fateful meeting took place the following month at Roosevelt's estate at Hyde Park. Churchill's opinion prevailed and the result, recorded in an *aide-mémoire* of his conversation with the president, was the opposite of what Bohr had sought.

The Hyde Park Aide-Mémoire of September 18, 1944, put an end to Bohr's courageous attempt to create a climate of mutual confidence with the Soviet Union. In the words of the memorandum, "The suggestion that the world should be informed regarding Tube Alloys,* with a view to an international agreement regarding its control and use, is not accepted. The matter should continue to be regarded as of the utmost secrecy; but when a 'bomb' is finally available, it might perhaps, after mature consideration, be used against the Japanese, who should be warned that this bombardment will be repeated until they surrender."

The last paragraph of the memorandum read: "Enquiries should be made regarding the activities of Professor Bohr and steps taken to ensure that he is responsible for no leakage of information, particularly to the Russians."

Churchill wanted to go further and put the scientist under house arrest, remarking that his indiscretions in this matter were "very near the edge of mortal crimes." The prime minister's advisers managed to convince him that this would be going too far, and the matter stopped there, although Bohr continued to try — in vain — to reestablish contact with Roosevelt.

The two principal scientific administrators responsible for the overall American military research effort and for its atomic branch were, respectively, the mathematician Vannevar Bush and the chemist James B. Conant, former president of Harvard. Both were opposed to postwar attempts to maintain an Anglo-American atomic monopoly, both were in favor of an open policy in which the Soviet Union could take part. To this effect they wrote, on September 30, 1944, a memorandum addressed to Secretary of War Stimson. In it they recalled that the bomb, probably equivalent to thousands of tons of classical explosives, should be ready toward the beginning of 1945, and that it was certain to be followed some years later by a hydrogen bomb perhaps a thousand times more powerful still. They also considered that the lead then enjoyed by the United States and the United Kingdom was certain to disappear as other advanced industrial countries caught up.

Bush and Conant thought it desirable to arrange a "preliminary demonstration" of the weapon, either in the United States or in Japan, to be followed by a "real" military application if Japan failed to surrender after this first show of strength. In addition, they proposed that, following the demonstration, all atomic knowledge should be communicated to an international body, except for the detailed mechanisms of the bomb. They believed such a procedure would offer considerable advantages to the Soviet Union and so reduce the chances of a new world conflict.

The memorandum set out for Stimson all the various problems related

---

*"Tube Alloys" was the British code name for the atomic project.

to the decision to use the bomb and to controlling it after the war, as well as the two stages of its development that they foresaw: the probable development of a weapon with a power equivalent to thousands of tons — or "kilotons" — of classical explosives, and the hypothetical development of a weapon with a thousand times more power measured in millions of tons — or "megatons."

Stimson was unable to raise with Roosevelt the international aspects of the possible use of the weapon and of its influence on relations with the Soviet Union until three months after he received the memorandum. Although he had no illusions as to the chances of being able to preserve the atomic secrets for very long, the war secretary nevertheless thought the moment had not yet come to share information about the bomb with the Russians who, as allies, were becoming more difficult daily. Roosevelt agreed.

Stimson was to see the president again for the last time on March 15, 1945, on his return from the Yalta Conference. As we have already seen, no decision was made.

Truman's Initiation

On April 25, 1945, the opening day of the San Francisco conference that was being held to set up the United Nations (U.N.) Organization, President Truman was visited by Secretary of War Stimson and Brigadier General Leslie R. Groves, who had been responsible for the American atomic effort during the past three years. The two men gave the president full details of the remarkable complex of laboratories and industrial plants that had been built in the United States and from which, by an astonishing time coincidence between programs, the first kilograms of two different atomic explosives would soon be available — uranium-235 and plutonium.

To help explain the processes involved in an atomic explosion, they may well have recalled the story of the inventor of the game of chess who persuaded the emperor of India to give him, for his reward, the amount of wheat obtained by placing one grain on the first square of the chessboard, two on the second square, four on the third, and so on. Well before the sixty-fourth and last square had been reached, there would not be enough wheat in all the world to meet the contract. Considering now a quantity of pure uranium-235 (or plutonium), the fission of one nucleus by a neutron releases at least two new neutrons, which in turn cause two further nuclei to split, this time releasing four second-generation neutrons; this leads to eight at the third generation, one thousand at the tenth generation, one billion at the thirtieth generation, and so on until around the eightieth generation there are enough ''secondary neutrons'' to cause fission in the immense number of atoms comprised in several kilograms of uranium-235.

In pure fissile material, this chain reaction takes place in an exceedingly short time, about one ten-millionth of a second, as the interval between two successive generations is about one billionth of a second. Given a sufficient quantity of fissile material, this almost instantaneous release of fission energy will be enormous and explosive.

If the neutron multiplication process takes place in only a small quantity of fissile material — for example, a small sphere of uranium-235 — there will be many neutrons that, on reaching the surface of the sphere, will escape and be lost from any further development of the process. Increasing the radius of the sphere increases its volume more rapidly than its surface, so that the proportion of neutrons lost in relation to those causing further fission decreases. Hence, we arrive at the notion of ''critical size'': below this, too many neutrons are lost so that any chain reaction stops before becoming explosive; above the critical size, more neutrons are formed in the mass than are lost from it, hence, an explosion is possible.

Three months after Stimson and Groves briefed Truman, two types of air-transportable bombs were to become available. With the first, which used uranium-235 and was nicknamed ''Little Boy,'' preliminary testing did not seem necessary. This was not the case with the second, a more complicated plutonium bomb, which was nicknamed ''Fat Man.''

Neither the American nor the British leaders had ever envisaged not using the atomic weapon, provided it could be successfully developed before the end of the war. From the start of work in 1942, the first bomb, expected within three years, had been destined for Japan. This avenging of Pearl Harbor would no doubt have been followed by a breathing space to give Hitler time to surrender before the new weapon was used against Germany, if in fact Germany were still in the war.

The Anglo-American military leaders saw in the bomb a means for successfully creating in the two enemy powers the psychological shock that was expected to produce the ''unconditional surrender'' that — however debatable — had been pursued as an aim of principle by Roosevelt and Churchill from the time of their summit meeting in Casablanca at the beginning of 1943.

Army attitudes toward the new bomb were by no means uniform. In 1942 there were, in fact, two types of atomic weapons being studied: the bomb and radioactive poisons. These latter, produced by a nuclear chain reaction, could be dispersed over an industrial area or a city, which would thus become uninhabitable due to the risk of receiving a fatal radiation dose in only a few hours.

The military authorities finally decided to develop only the bomb, a weapon at least as terrifying, since radioactive poisons seemed more likely to fall within the provisions of the Geneva Protocol on Gas Warfare of 1925. A consideration analogous to that which, a third of a century later, was to animate the arguments over the "neutron bomb," a weapon designed to have greater radiation effects as opposed to explosive results.

In view of the wide range of the weapon's murderous capabilities, these scruples do seem rather surprising. While the explosive effects are essentially confined to the instant of detonation (being related to the blast pressure wave, the intense heat from the fireball, and the radiations from the nuclear reaction), other effects, no less deadly, follow later from the fallout of radioactive fission products over the area attacked.

It was certainly not easy for President Truman, during this first discussion on April 25, to appreciate fully the gravity of the matter. However, Secretary of War Stimson, a mature and profoundly respectable man of the highest morality, did his best to explain. He impressed on the new president the decisive influence that the bomb would have on the development of future international relations between the United States and the rest of the world, getting him to accept the idea of establishing a committee to work out detailed policies to govern use of the weapon. Known as the "Interim Committee," it was to play an important role during the crucial weeks that followed.

Five days after this discussion, Hitler committed suicide. One week later, Germany surrendered unconditionally. Several of its greatest cities, including Dresden, Hamburg, and Berlin, had been appallingly devastated by bombardment with classical weapons — just as Warsaw, Rotterdam, and Coventry had been devastated by German air forces in 1939 and 1940.

If Hitler had acquired the first atomic weapon as he had the V-1 and V-2 missiles, he would certainly have used it not only to win the war but, very probably, he would have continued with its use to gain superiority over the whole world. In the United Kingdom and the United States, fear of this domination was the greatest factor that obsessed and drove forward the scientists — many of whom had fled from fascist persecution — in their efforts to win the race for the atomic bomb.

Consulting the Scientists

The Interim Committee met for the first time on May 9, 1945, the day following the defeat of Germany. Chaired by Truman's future secretary of state, James F. Byrnes, it included representatives of the War and Navy

Departments, the State Department, and the two men responsible for the atomic program, Bush and Conant. It set up a Scientific Panel comprised of four scientists of unchallenged authority whose role in the enterprise had been prominent: three Nobel Prize winners — Arthur H. Compton, Enrico Fermi, and Ernest O. Lawrence — together with J. Robert Oppenheimer who was in charge of the Los Alamos Scientific Laboratory in New Mexico where the weapon itself was being developed.

The committee was informed of the state of the war against Japan and of the further military operations planned. Aerial attacks with incendiary bombs had already inflicted terrible destruction; in Tokyo, an attack on March 9 by 300 aircraft had left 84,000 dead and 200,000 homes destroyed in an area of 40 square kilometres. The American airmen were even hoping to force Japan to surrender without having to invade its territory.

Nevertheless, a landing was planned for the beginning of November, in the face of which it was feared the Japanese would defend their native soil fanatically. The advocates of the operation were hoping that the predicted American losses — some 600,000 men — would be considerably reduced due to the Soviet Union's entry into the war, expected within two to three months, which would probably immobilize a part of the Japanese army on the Asian continent.

The committee rejected immediately any plan that would not have made use of the bomb, considering the earliest possible ending of the war to be the most essential objective. It then discussed whether there should be a preliminary demonstration or immediate military use, arriving, at the end of May, at the unanimous conclusion: ''The bomb should be used as soon as possible against Japan without preliminary warning, and against a target combining high population density with military importance, in order to obtain the maximum psychological effect possible.''

The Scientific Panel was consulted concerning a possible demonstration of the weapon. The idea was certainly attractive, for it would have spared human lives; but it was rejected because it would have required warning Japan in advance, and designating an uninhabited area where the demonstration, which might in fact prove inconclusive, would be carried out.

The trouble was that, although the bomb was certainly revolutionary, it had never been tried in the field and the reliability of its trigger system, therefore, could not be fully guaranteed, even after a successful first test. For these reasons, the four scientists unanimously rejected the idea of a preliminary demonstration, which they believed could not be arranged in a sufficiently convincing manner to bring about an immediate end to the war.

The idea of a preliminary demonstration had been widely discussed among the scientists at the atomic project in Chicago (one of the most important parts of the American enterprise) who were strongly influenced by Szilard. Szilard had, moreover, finally been able to pass to the Interim Committee's Chairman Byrnes, who had been director of war mobilization, the memorandum originally intended for Roosevelt.

The meeting between Szilard and Byrnes, like the one between Bohr and Churchill the previous year, proved a failure. The scientist, possessed of a strong foreign accent, made a bad impression on the future secretary of state who, ill-disposed toward the Soviet Union because of its attitude over the East European countries, failed to see why Szilard should wish it to be brought into the picture. Instead, Byrnes hoped that the use of the bomb against Japan would impress the Russians and make them less intransigent.

In a parallel development, at the beginning of June, a group of senior scientists at the Chicago atomic project, led by the physicist James Franck, a Nobel laureate of German origin, completed a report restating Szilard's arguments and proposing to unveil the bomb to the world by a demonstration over an uninhabited zone and in the presence of U.N. representatives. The group was particularly concerned that the United States avoid any possible loss of prestige which might result from the use of such a destructive weapon; they feared the United States might thus find itself ill-placed to stand as champion of international control over the new power it had unleashed.

The initiative of Franck and his colleagues was unsuccessful. The Scientific Panel did, in fact, reconvene to discuss this new type of preliminary demonstration, but the four top scientists once again rejected the idea and abided by their recommendation in favor of direct military action. Nevertheless, the panel proposed that the use of the bomb should be preceded by consultations on the problems of control in which not only the United Kingdom but also the three other Great Powers — the Soviet Union, France, and China — should take part after having been informed about the future weapon.

The Interim Committee examined this proposition on June 21. The only suggestion it adopted was that of informing the Soviet Union in advance. The committee recommended that the president tell Soviet Premier Stalin of the existence of the atomic project, of the decision to use the bomb in the war against Japan, and, finally, of the hope that there could be subsequent discussions on ways of restricting the uses of the new power to peaceful purposes.

The First Explosion

Meanwhile, the pace of external events had quickened. A summit conference, to be held at Potsdam in the heart of conquered Germany, was arranged.

Not wishing to go to this conference before the bomb had been tested, Truman, with Churchill's agreement, delayed this meeting with Stalin for as long as possible. It was finally fixed for mid-July, the moment when the experimental test was scheduled to take place.

On the day of the test, July 16, 1945, about 100 scientists collected at Alamogordo in the New Mexico desert. Some idea of what was afoot must have leaked out at the nearby atomic establishment at Los Alamos where the bomb (a plutonium type) had been assembled, for at 11 p.m. the night before several hundred participants in the project began surreptitiously climbing the nearby hills to watch the explosion from afar. It was planned to take place in the desert at 4 a.m.

The test was delayed by bad weather and by 5 a.m., having seen nothing, the curious began returning to the township, discouraged and convinced that the mountains in labor had given birth to a mouse. Suddenly the whole sky lit up: They had missed the spectacle, but they knew they had not been shut off from the rest of the world for two years for nothing.

The brilliant light flash was visible for more than 60 miles. A woman driving her car at that distance arrived in a village and began knocking on the doors to wake the inhabitants. "I've got to tell you what I just saw," she said. "It's unbelievable I know, but I just saw the sun *rise* in the *west,* then immediately set again!"

Secret agents had been stationed in this village and in all nearby villages to observe the reactions of the population. It was not easy for them to restore calm; the people could not be told that they had just witnessed the first atomic explosion, the creation of a tiny star that had melted half a square mile of desert sand.

As the bomb passed from being a probable scientific hypothesis to a concrete reality, Japan was beginning to show clear signs of weariness. At the beginning of July, Emperor Hirohito informed his ambassador in Moscow of his profound wish to see a speedy end to the war, and that he wanted to send a special envoy to the Kremlin to learn whether the Soviet Union would be willing to act as mediator in obtaining better terms for Japan than the unconditional surrender demanded by the Allies.

Although the two countries had concluded a neutrality pact that ran until April 1946, Japan's choice of the Soviet Union as mediator was particularly unfortunate because, by an irony of history, no country at that time could have wished less to see Japan lay down its arms, for the Soviet Union was on the point of declaring war itself in order to secure a share of the expected booty.

For this reason, the Japanese ambassador, who was receiving telegrams of ever-increasing urgency from Tokyo, could not be received — despite continued requests — by Soviet Foreign Minister Vyacheslav M. Molotov before the Potsdam Conference. His audience with Molotov did not take place until August 8, 1945, when he was presented with the Soviet Union's declaration of war.

Washington had been aware of the telegrams from Tokyo to the Japanese ambassador in Moscow before the opening of the Potsdam Conference, which would bring the Big Three together on the territory of their common enemy. The United States, knowing the Japanese codes, had intercepted all of the messages. Their content was yet another reason in support of that section of the State Department that for two months had been vainly trying to persuade Stimson and the White House to secure an immediate surrender of Japan — already bleeding to death — in return for a formal declaration that the emperor and his dynasty would be maintained.

The Potsdam Conference

On July 16, 1945, on the eve of the first session of the Potsdam Conference, Truman learned of the complete success of the test with the plutonium bomb. He hastened to inform Churchill who at the end of June had willingly given his consent to using the weapon — consent that was necessary under an Anglo-American agreement concluded in Quebec in 1943.

The numerical results of the test, which had produced a power close to the maximum forecast of 20,000 tons of trinitrotoluene or TNT, convinced Truman and Churchill that they no longer had any need for the Soviet Union to join the war. Rather, they felt just the opposite, but the decision had finally been made and the Allies had no power to stop it.

On July 23, the heads of state approved the Potsdam Declaration, warning Japan that the war would be pursued with the utmost vigor until the end of all resistance. There was mention of a future "peacefully inclined and responsible government," to be set up in accordance with the freely expressed wishes of the Japanese people, but there was no mention of maintaining the emperor and his dynasty. Refusal to accept the terms would result in prompt and complete destruction of Japan; there was, however, no reference to the new weapon.

The following day, the last day of the conference, at the end of the session and alone with Stalin except for his interpreter, Truman decided to inform the Soviet head of state of the existence of the bomb. He recorded what happened in the following words: "On 24th July I mentioned briefly to Stalin that we had a new weapon of unusually destructive power. The Soviet head of state showed no special interest. All he said was that he was glad to hear it and he hoped we would make good use of the weapon against the Japanese."

The Allies had found in Stalin the reaction they were seeking. There had been no mention of the word atomic, but that Stalin had well understood what was going on is evidenced by his remark to Molotov and to Marshal Georgi K. Zhukov that very evening. According to Zhukov's memoirs, Stalin said, "They want to raise the stakes, we must put more pressure on Kurchatov." Kurchatov was responsible for the Soviet nuclear project.

Under the influence of the military rulers in Japan, the Potsdam Declaration was rejected by Tokyo on July 28. From that moment, an American directive dated July 25 became operational. By this directive, as soon as weather conditions were favorable the special bomb was to be used against one of four selected targets, each a city making an important contribution to the war effort. The three other targets were to be attacked also as further special bombs became available.

The Nuclear Sin

The inevitable now followed. Hiroshima was destroyed on August 6, 1945, by the first uranium-235 explosion. President Truman, from aboard the warship in which he was returning home from Germany, announced to the whole world that an atomic bomb had been used.

On August 8, the Soviet Union declared war on Japan and Russian troops entered Manchuria. On August 9, Nagasaki was destroyed by a second plutonium bomb of the same type as that used for the experiment three weeks earlier in the New Mexico desert. The same day in Tokyo, against the advice of his military leaders, the emperor gave the order accepting unconditional surrender. Japanese nuclear physicists had confirmed to him that an atomic weapon was being used. They had themselves envisaged the military application of fission, but had completely lacked the means to achieve it.

The following day, August 10, Truman gave the order halting the dispatch to the Pacific of the third bomb, due to be ready around August 15.

Washington had in fact just learned from the Swiss that Japan was suing for peace, accepting the conditions of the Potsdam Declaration.

So ended, on August 14, 1945, the Second World War and its frightening train of destruction after destruction, the last two due to the new weapon.

Several kilograms of matter in a single device dropped from a single aircraft had produced as many deaths as thousands of tons of ordinary explosives and incendiary bombs dropped from hundreds of bombers. That was the measure of the revolution achieved in the science of destruction and in the history of humanity.

Modern technology could now produce disasters of the same magnitude as the biggest natural catastrophies, such as the destruction of Saint-Pierre in Martinique in the French West Indies. In 1902 the town was destroyed instantly along with its 28,000 inhabitants by the intense temperature rise beneath a scorching steam cloud from the Pelée volcano, leaving as sole survivor a prisoner in a cellar. The destruction was remarkably similar to the annihilation of a city by an atomic explosion. Likewise, the greatest cataclysm of the early part of the century, which took place in Japan, was the 1924 earthquake that caused 140,000 deaths, about the same number as the two atomic bombs.

The decision to use the bomb — a decision that was subsequently frequently and emotionally criticized — was, if not indispensable in ending the war, at least inevitable in the circumstances of the spring and early summer of 1945. It is difficult to blame the politicians who made that decision for failing to foresee a future world — at a time when their bomb was still unproven — haunted by the existence of thousands of weapons, each some thousand times more powerful than the one they currently possessed.

It would have required extraordinarily penetrating and powerful foresight, such as perhaps that of Roosevelt at the height of his intellectual prowess, to be satisfied with a public demonstration, as recommended by a few enlightened scientists but against the advice of their most celebrated colleagues and of those in charge of the American atomic enterprise, and to halt the "natural" outcome of that enterprise. Moreover, during these initial stages of its development, the bomb was not expected to produce more deaths than conventional aerial bombardments, which turned out to be the case.

The decision to use the bomb is explained above all by the wish to end the war without delay so as to reduce the loss of both Japanese and American lives (certainly in greater numbers than those lost at Hiroshima and Nagasaki) that would have resulted from the continuation of incendiary air strikes against Japanese cities and from the landings planned for the autumn.

To these reasons must also be added the circumstances that made it desirable to secure a Japanese surrender before the Soviet war machine had become fully engaged in the East. This would reduce the role the Soviet Union could play in the Asian peace arrangements and, in particular, avoid having Soviet occupation forces in Japan. From this viewpoint, the use of the bomb could be considered one of the first acts in the Cold War.

The United States' acquisition of atomic weapons represented for the Soviet Union a violent disruption in the balance of power, giving the Americans undisputed military supremacy during the years following the world conflict. However, this imbalance between the United States and the Soviet Union would have been the same if the bomb, rather than being used in actual war, had been revealed to the world in a public demonstration as recommended by Szilard and the Chicago scientists. The atomic arms race would no doubt have been just as inevitable and the first international negotiations on controlling the new power would have been no less condemned to failure, as long as the Russians felt themselves in too much of an inferior position compared to the Americans.

On the other hand, the victims of Hiroshima and Nagasaki, in describing their horrible experience, contributed to the creation of an extreme and irreversible world public opinion against the bomb. This adverse public reaction would not have resulted from a demonstration of the bomb.

This public aversion, like the awareness among the political leaders of the Great Powers of the terrible potential of the new weapon, laid the foundations for the balance of terror, which no doubt is to be regretted in principle, but nevertheless is probably responsible for civilization having avoided any nuclear conflict for a third of a century.

If the bomb had not been used against Japan, there would have been a much greater risk of it being used later on during the course of — not just to end — a war such as that in Korea. It may have been used at a time when the two greatest powers both possessed atomic weapons in much greater quantities and with much greater power than the few that were available in 1945.

The descriptions of Hiroshima and Nagasaki have also played a determining role in the gradual creation of a worldwide antinuclear consciousness. This trend lends support to the atomic weapon nonproliferation policies as well as, in a public aware of the inevitable overlap between civil and military aspects, to the campaigns against the development of civil applications of fission.

For this reason, if the idea of a public demonstration had prevailed over that of immediate military use, and if atomic energy had not burst on the world at Hiroshima and Nagasaki, it would (paradoxically) have been the development of civil energy and perhaps even the spread of the weapon that would have been mainly encouraged, because the antinuclear aversion would then no doubt have been less pronounced. There would certainly have been less pressure from world public opinion to delay production of nuclear energy and to resist the spread of a weapon whose martyrs would not yet have existed, for the ''nuclear sin'' would not have been committed.

2. The Pioneers

On August 6, 1945, President Harry S. Truman announced to the world at large the annihilation of Hiroshima and the existence of the atomic weapon. Three days later in another address, the president explained how, for fear that the Germans were ahead of them, the Allies had undertaken the immense enterprise. He justified the use of the weapon on the grounds that it had saved hundreds of thousands of human lives, which would have been lost had the conflict continued.

He then described the responsibility about to fall on the United States in terms of puritanism from which future American policy could already be divined. "We must constitute ourselves," he said, "trustees of this new force, to prevent its misuse and to turn it into the channels of service to mankind. It is an awful responsibility which has come to us. We thank God that it has come to us, instead of our enemies and we pray that He may guide us to use it in His way and for His purposes."

The philosophy of American nonproliferation policy is to be found in these words. During the war, this policy was already evident in the surprisingly difficult atomic relationships between the United States and the

United Kingdom, and in the problems arising from the presence of several French scientists taking part in the Anglo-Saxon enterprise.

To study these relationships and these problems, we must go back to 1939 and 1940 to learn of the pioneering role played by the French at the start of the adventure, of the influence they had on the British, and on the British contribution to the launching of the American enterprise.

In France

In March 1939 at the Collège de France in Paris, Joliot-Curie and his two co-workers, Hans von Halban and Lew Kowarski, showed that the splitting of a uranium nucleus by a single neutron gave rise to the emission of several further neutrons. This phenomenon, predicted by Leo Szilard in his letter to Joliot-Curie the previous month, was the trigger behind the spread of an "atomic fire."

In the following month of April, Joliot-Curie's team, soon to be joined by Francis Perrin, showed the number of secondary neutrons produced in the fission of a single uranium nucleus to be on the order of three. However, in natural uranium only the rare fraction, uranium-235, will easily undergo fission while the 140-times more abundant uranium-238, acting in the mixture rather like the water in damp gunpowder, tends to absorb neutrons before they have had time to fulfill their role.

Despite this, the French scientists had concluded that it should probably be possible in certain conditions to achieve a chain reaction with uranium-235 even when "diluted" in this way in natural uranium.

Experiments have shown that secondary neutrons, which at birth have very high velocities, become better able to cause fission in uranium-235 if they can be slowed down. And slowing down fast neutrons is rather easy. It is enough to place in their paths substances containing light atomic nuclei. When neutrons hit these light atomic nuclei, the neutrons, in successive collisions, lose part of their momentum in the same way that a billiard ball is slowed down when it collides with other balls of the same size so that it gradually loses its kinetic energy. Thus it is necessary to mix with the uranium a material that will slow down the neutrons in this manner but will not absorb too many of them.

In this way, Joliot-Curie and his colleagues had described the conditions under which they hoped, correctly, that an energy-producing machine could be built and controlled. The prediction was brilliant and, had their country not been overrun a year later, they no doubt would have been the first to assemble an atomic pile.

They also gave thought to explosive reactions. Here, however, their predictions were less sure, although they advanced a number of ideas for devices to trigger an explosion after critical size had been obtained (by bringing together two subcritical masses or by compression). These ideas were later translated into standard methods.

Perrin even thought that a chain reaction might be achieved with about 40 tons of natural uranium, which was subsequently found not to be possible. There was a proposal at the time for a secret test in the Sahara — more than 20 years before such a test was in fact carried out by France. Also, when the French test was eventually performed, it was under conditions and with a material very different from that envisaged in those early days.

Between May 1 and 4, 1939, three secret patents were registered in the name of the Caisse Nationale de la Recherche Scientifique (CNRS), the national fund for scientific research that had given considerable support for the experiments. The first two patents concerned energy production, the third concerned explosives. Thus Joliot-Curie embarked upon the path of secrecy that he had rejected a few months earlier following Szilard's appeal to him.

These experiments required large quantities of uranium. Joliot-Curie turned to the Belgian company Union Minière du Haut Katanga, which owned the largest resources of uranium then known in the world in the Belgian Congo. A draft agreement was prepared between the CNRS, holder of the patents, and Union Minière, under which the latter undertook to supply the 50 tons thought necessary for the experiments and even to contribute financially to them. If the experiments proved successful, a joint enterprise was to be set up by the two parties to exploit the results worldwide.

Initialled on May 13, 1939, by Joliot-Curie for the CNRS and by an administrator of the Belgian company, this unusual agreement between a state body and a private foreign trust was never signed or brought into force, no doubt due to the war.

At the outbreak of hostilities in September 1939, Joliot-Curie's work received encouragement in the form of strong support from the minister for armaments, Raoul Dautry. The objective sought was the construction of a power generator, then believed to be more easily accomplished than a bomb. This was an error of judgment that, fortunately, the Germans were also to make. On the other hand, the British and then the Americans took the opposite view and judged it easier to make a weapon based on an explosive chain reaction than to recover energy from a controlled chain reaction. The French had not fully realized the difficulty of this latter objective and believed that an engine for submarine propulsion, which would offer an enormous advantage in that it would not need a supply of oxygen, might be achieved in a few years.

The first technical problem to be solved toward the chain reaction was to find a proper "moderator" for slowing down neutrons. The first substance to be tried was hydrogen, but work with a mixture of ordinary water and uranium quickly showed that, unlike oxygen, hydrogen absorbed neutrons too readily to be suitable. Eventually Joliot-Curie's team concluded that there were two substances that could probably be successful in practice: carbon in the form of graphite, and heavy water. This last is composed of oxygen together with a hydrogen isotope of mass two, known as deuterium, which is found in naturally occurring hydrogen in the proportion of 1 part per 6,000, and which absorbs neutrons far less readily than "ordinary" hydrogen of mass one.

Heavy water is very hard to separate from ordinary water, in which it occurs in the proportion of only 160 milligrams per litre, and its extraction calls for complex processes consuming very large amounts of energy.

Heavy hydrogen was discovered in 1932 by the American scientist Harold C. Urey who, following a considerable number of successive distillations, had identified a denser fraction in water. Until the war, heavy water, which was worth about a half dollar per gram, had no uses other than in scientific research. Nevertheless, a Norwegian enterprise, Norsk Hydro, in which the French held a majority of the capital and which was mainly a producer of synthetic ammonia, had begun small-scale heavy water production despite the absence of any immediately apparent market. Profiting from the low cost of hydroelectricity and a system of fractional electrolysis linked with the main ammonia-production process, by 1940 a total of about 200 kilograms of heavy water had been made.

Joliot-Curie suggested the purchase of this entire stock to Dautry, but it seemed hardly possible to rapidly find the $120,000 required to buy it. In his report Joliot-Curie, therefore, suggested the heavy water might be obtained on loan, specifying in his conclusion: "If our experiments are successful, that is, if we achieve a massive release of energy, even though our materials are likely to be destroyed the loss would be negligible compared with the industrial consequences of such a success. If we fail, then all the materials used will be completely recoverable . . . ."

News from Norway, suggesting that the important German company I.G. Farben Industries was also interested in buying the stock of heavy water, hastened the departure of the French mission to buy the entire stock. Dautry put the secret enterprise under the direction of Jacques Allier, an engineer who was already in contact with Norsk Hydro and who had been recruited into the "Deuxième Bureau" (military intelligence). With the help of that organization, Allier arrived in Oslo at the beginning of March 1940. His mission was a complete success, preference being given to the French government. The whole available stock of 185 kilograms was loaned to France for the duration of hostilities. In addition, Norsk Hydro undertook to accelerate future production, all of which would be reserved for France.

The mission returned, with 26 precious cans, on March 9, 1940, exactly one month before the German invasion of Norway.

Throughout this vital mission to buy the heavy water, French military intelligence had insisted on absolute secrecy and expressed concern that Joliot-Curie's two principal co-workers, Halban and Kowarski, only recently naturalized as French citizens, were of Austrian and Russian origin, respectively. To guarantee that, in case there should be a security leak, neither Halban nor Kowarski could be suspected, Joliot-Curie asked each of them to spend the duration of the mission under surveillance in an isolated residence of their choice. Their compulsory holidays on the two islands of Porquerolles in the south of France and Belle Ile in Bretagne ended on March 16 with the arrival without incident of the heavy water in the cellars of the Collège de France. From being "security risks," Halban and Kowarski became once again indispensable scientists!

The two following months were devoted to setting up the decisive experiment but, in spite of strenuous efforts, this was still not ready when the German advance began to threaten the capital. The laboratory was evacuated to Clermont-Ferrand, France, and the heavy water stored in the safety of the prison at Riom (which soon afterward became renowned, sadly, as the internment center for personalities of the Third Republic detained by the Vichy government). With the continuing German advance, Dautry asked Joliot-Curie, Halban, and Kowarski to go to Bordeaux and on to England.

On June 17, 1940, Halban, Kowarski, and the 26 cans of heavy water boarded ship with a travel authorization signed by Dautry's principal private secretary, Jean Bichelonne, future industry minister in the Vichy government and subsequently a fanatic supporter of collaboration with Hitler. Their mission order read: "They are required to continue in England, in absolute secrecy, the research begun at the Collège de France. They are to present themselves in London to the French Mission led by Colonel René Mayer." Mayer later became prime minister in the Fourth Republic.

So, three months after being under a form of house arrest during the heavy water purchase in Norway, Halban and Kowarski found themselves charged with the great responsibility of trying to preserve France's position, initially so promising, in the atomic race.

Joliot-Curie decided to stay on in France so as to remain in charge of his laboratory, to the establishment of which he had given so much personal devotion. His lack of knowledge of the English language and his fear of not receiving the recognition he merited and hence being denied essential technical support probably prevented his venturing into the unknown and crossing to England. Personal reasons must also have played a part, for his wife was seriously ill and his family had dispersed to different parts of France.

Joliot-Curie did not, in fact, realize the considerable role he might have played. In reality, his world-recognized authority would certainly have contributed to convincing the American and British governments of the vital importance of the uranium enterprise. His sister-in-law, Eve Curie, a distinguished pianist and author of a biography of her mother that had achieved great success in the United States, could have opened the doors of the White House for Joliot-Curie, for she was very close to the Roosevelt family.

Joliot-Curie's prestige would have greatly added to the French participation in the war effort. Had he devoted himself then to the Free French, as he did later to the Resistance and to the Communist party, he could have set up a proper team with himself as the undisputed leader. Without him, the few French scientists were only able to offer individual contributions to the Allied atomic achievements. His participation might even have reserved for France a place in the "Atomic Club," which grew up during the war without her.

In Britain

Halban and Kowarski, with their heavy water and other projects, were received in England with an interest that was all the greater because so far the British had given little thought to the recovery of useful energy. The "uranium business" had, in fact, taken an altogether different course from that in France.

The beginnings had been relatively slow, for the great British nuclear physics laboratories had played little part between 1934 and 1939 in the work leading to the discovery of fission. All the same, from the spring of 1939, those responsible for military scientific research had been aware of what was going on, though they remained very skeptical over the possibility of making an atomic bomb. To a lesser extent, this was also true of

Churchill's scientific adviser and éminence grise Frederick Lindemann, who later became Lord Cherwell. A physicist by training, he explained to Churchill in the summer of 1939 how very difficult the construction of such a bomb would be, if indeed it proved possible at all.

Contacts were nonetheless made with the management of Union Minière du Haut Katanga, but did not lead to any notable uranium purchases. At the same time, some theoretical and practical work was begun in university laboratories. But widespread skepticism persisted and at the outbreak of war all the best British physicists were assigned to other work of the greatest importance, namely, top secret research on ultrashort waves, from which came the radar that played so important a role in the Battle of Britain.

Some months after the start of the war, British scientific interest was abruptly reawakened by a secret theoretical memorandum written by two refugee German physicists, Rudolf E. Peierls and Otto R. Frisch. Frisch had been the first person to produce physical proof of fission, in Copenhagen in January 1939. The memorandum, prepared just before the beginning of the war, concluded with the assertion that a single kilogram of pure uranium-235 would be enough to make a bomb of extraordinary power. (In fact, 10 to 20 times this amount is needed). The authors described a possible method for separating the 235 isotope, set out the principles of a mechanism for the weapon, and calculated its effects. It was a remarkable piece of work, no less basically complete as a first specification for an atomic bomb than were the French patents for future power stations.

In the face of such important conclusions, it was decided that British work should be restarted, to be coordinated by a committee reporting to the minister for aircraft production. The committee was code-named the "Maud Committee."* It included all the top British physicists and from spring 1940 did everything necessary to promote the various lines of work called for.

Soon after the arrival of Halban and Kowarski in June, the Maud Committee heard of their various projects. Convinced that these were of great importance, it invited the two scientists to remain in England and continue their research at the Cavendish Laboratory in Cambridge, where they were already provisionally installed. It was here that they at last completed, in December 1940, the crucial experiment they had planned

---

*The origin of "Maud" is unexpected. At the beginning of the war, Niels Bohr had sent a telegram to one of his British colleagues asking for news of Maud. This colleague decided the word must be an abbreviation for "Military Applications of Uranium Disintegration." In fact, it was the Christian name of an English governess of the Danish scientist's children.

with Joliot-Curie, showing for the first time in the world that there was no doubting the possibility of making "machines" to produce energy from natural uranium. This vital result fully confirmed the promising situation foreseen by Joliot-Curie at the end of 1939 — even though the available quantities of uranium and heavy water were as yet far from sufficient to achieve criticality. In fact, something like 10 times the amount was needed: rather than the few hundred kilograms they then possessed, they needed tons.

Materials were not the only things they lacked. Halban, who had assumed the direction of the work, was constantly asking for additional personnel. He had initially been given a small number of Jewish, German, and Austrian physicists and chemists, some of whom were released from internment camps to take part in this most secret work. The responsible authorities did not, in fact, believe uranium research could lead to anything of practical value in the conflict underway.

In July 1941 the Maud Committee reported distinctly positive conclusions both in the explosives field and in that of energy production. It declared its belief that in less than three years it would become possible to separate uranium-235 on an industrial scale, and that by using about 10 kilograms of uranium-235 a remarkably powerful bomb could be made. It recommended that further work should have the highest priority in Britain or, for certain operations, in North America should this prove necessary.

The report also referred to the possibility of building a "boiler," using uranium and heavy water to generate a controlled heat output. However, this was judged to be less important for the pursuance of the war.

In an annex to the Maud Committee's report, the company Imperial Chemical Industries asserted that work on the boiler could have great interest for the British Empire, and offered to undertake such work on its own account.

Finally, the report emphasized the quite exceptional technical and financial effort that would be required to develop a process and build a plant for isotopic separation of uranium. The process already envisaged was that of diffusion of uranium in gaseous form through a porous membrane; this process is still used today by the main atomic powers.

The British government found itself confronted by a declaration from the country's greatest scientists that a decisive weapon based on known scientific principles could be produced. The difficulty was that neither the proposed bomb nor the boiler could be tested experimentally on a small scale, and that industrial pursuit would involve building uranium and heavy water production plants and an isotopic separation plant — all requiring very substantial expenditures. Moreover, these plants would demand industrial effort particularly difficult to find in a country whose factories, still subjected to incessant bombings, were entirely devoted to immediate objectives such as building fighter aircraft.

Possible U.S. participation in this immense effort was envisaged in the report of the Maud Committee. At the very moment when this possibility and its political consequences were being discussed by the British government, across the Atlantic, the same report and its conclusions were making a considerable impression on the American authorities and on the scientists concerned.

In the United States

Developments on the other side of the ocean had followed a course similar to those in France and Britain with refugee scientists from other countries again playing a major role.

Following the establishment of the Advisory Committee on Uranium by Roosevelt, who was influenced by Szilard, research in 1940 and early 1941 proceeded — though only at the relatively leisurely pace of university laboratories and with little coordination and little support from the federal authorities. Nevertheless, there were some significant advances during this period.

First and foremost, Enrico Fermi, who had fled from fascist Italy after receiving his 1938 Nobel Prize and who was without doubt the most brilliant physicist of his generation, was pressing forward with Szilard on important work at Columbia University in New York.

Early in 1939 they had demonstrated, as had the team at the Collège de France around the same time, the existence of secondary neutrons. Later they proved that carbon, in the form of very pure graphite, could act as moderator and could probably bring about a chain reaction in natural uranium, provided only very refined uranium were used (of which several tens of tons were needed) together with several hundred tons of the graphite. At this time they were seeking in vain for official (or even private) financial support to enable them to attempt an industrial-scale test.

In parallel with this effort, at the end of 1940 at the University of California, the young American chemist Glenn T. Seaborg, using the most powerful cyclotron then in existence, showed that uranium-238 under neutron bombardment was transmuted into a new element previously unknown on earth. He succeeded in isolating some thousandths of a milligram of this, which he baptized "plutonium." He found that the new element, plutonium-239, was fissile under neutron bombardment (like uranium-235). He concluded that if a chain reaction could be established in natural uranium — which Fermi and Szilard were trying to achieve — simultaneously with the conversion of the rare isotope uranium-235 by fission into energy and radioactive products, some of the extra neutrons released would transmute a similar quantity of the abundant isotope uranium-238 into plutonium-239. Thus, the chain reaction would provide the necessary alchemy for the production of a material capable of serving, like uranium-235, as an explosive.

Plutonium does not exist in nature because of its radioactivity: Its atoms are unstable and have one chance in two of disintegrating "naturally" in 24,000 years. Natural uranium is also unstable but the atoms of its two components, the 235 and 238 isotopes (the heaviest naturally occurring atoms) have disintegration times so long that a substantial proportion still remains of what was present when our planet was formed.

In 1941 theoretical physicist J. Robert Oppenheimer reached the same conclusions as the British as to the possible explosive use of uranium-235. Making similar calculations for plutonium, he arrived at a critical mass for the new element of one-third that for uranium-235. (The mass is in fact between 5 and 10 kilograms).

In June 1941 President Roosevelt, as part of his country's preparations for a possible war, decided to set up an organization to deal with all scientific questions that had military significance for the nation. To head this organization, which he made directly responsible to himself, he chose Vannevar Bush. Within it, a committee on defense questions was set up under James B. Conant. The overall structure included a section, code-named S-1, to deal with matters concerning uranium.

All of this represented a substantial and general reorganization of research, which was to have a profound influence on atomic development for which Bush and Conant retained the overall responsibility until the end of the war. Throughout this period, both had direct access to the president at any crucial moment.

The task of the S-1 section was to discover if fission could be successfully exploited and, if so, whether as an explosive or radiation-generating weapon, as a means for submarine propulsion, or for controlled energy production. When work began, American research on uranium-235 separation and on the bomb was thought to be much less advanced than that of the British.

This fact was confirmed in July 1941 when the conclusions of the Maud Committee report became known in the United States as part of the Anglo-American exchanges on the uranium problem. The report gave Bush and Conant exactly what they were seeking: Assurance that results of military significance could be expected before the end of the war and, better still, a program for achieving these results.

3. Strange Partnership

Missed Opportunity

In August 1941, one month after the conclusions of the Maud Committee report became known, Vannevar Bush and James B. Conant, the two responsible for the S-1 uranium section, on their own initiative sent a letter to the British minister, Sir John Anderson, lord president of the council, who had recently been put in charge of his country's atomic affairs. The letter proposed that the development of the bomb should be pursued as a joint undertaking.

On October 9, 1941, two months before the United States entered the war, Bush was received by President Roosevelt and reported to him on British progress and conclusions, as well as on the encouraging American results with plutonium. But he insisted that a great deal more experimental work would be needed before the probability of success could become a certainty. The president asked him to give highest priority to continuation of the research work, but not to pass to the industrial stage until further orders were given. The whole matter was to proceed with the strictest secrecy.

The Americans have recognized that the British work, itself partly derived from that of the French, was of the greatest importance in the final decision to go ahead. On this same day of October 9, Roosevelt wrote to Prime Minister Churchill — as Bush and Conant had written to Anderson two months earlier — proposing that the two national enterprises should not only be coordinated but should be run as a joint project.

Churchill had been informed of the conclusions of the Maud Committee in a favorable memorandum from his intimate adviser Lord Cherwell. He had agreed that atomic matters should have the highest priority, although not without noting on Cherwell's memorandum: "Although personally I am quite content with existing explosives, I feel we must not stand in the path of improvement, and I therefore think that action should be taken in the sense proposed by Lord Cherwell.

Lord Cherwell, like others among the prime minister's advisers, was well aware of the lead then held by the British. For that reason, he was not particularly keen on a joint undertaking with the Americans, or even on building a British isotopic separation plant for uranium in the United States, for he believed that the country in possession of such a plant would be able to dictate conditions to the rest of the world. "However much I may trust my neighbour, and depend on him," he wrote, "I am very much averse to put myself completely at his mercy, and would therefore not press the Americans to undertake this work. I would just continue exchanging information and get into production over here without raising the question whether they should do it or not."

And so Churchill, satisfied with a simple continuation of scientific interchanges, allowed Roosevelt's letter of October 9 to lie unanswered for two months. Then he sent an assurance, in general terms, of the British support for collaboration in the field with the American administration.

The independent proposition of Bush and Conant for a joint undertaking suffered the same fate; in March 1942, seven months after the suggestion had been made, Anderson sent an evasive reply, apologizing for his delay but expressing satisfaction over the state of the cooperation. He could hardly have foreseen that, less than a year later, Bush and Conant would have become (and would remain) convinced opponents of the close collaboration with the British that they had originally backed. Thus did the British government fail to take advantage of the crucial American offers.

This was to be the first demonstration of the fragility of a policy based on the refusal of transfer of knowledge in an attempt to retain supremacy, albeit in this case very ephemeral indeed.

The months following the exchange of letters between Roosevelt and Churchill involved a period of reorganization on both sides of the Atlantic. However, while work in the United States was surging forward, the United Kingdom was marking time and awaiting fundamental decisions necessary for the start of an industrial phase.

On December 6, 1941, on the very eve of U.S. entry into the war, Bush set up the organization of the recently created S-1 uranium section. He called on three of the most celebrated American physicists, Arthur H. Compton, Ernest O. Lawrence, and Harold C. Urey, to undertake assessment of the different technical approaches then envisaged for achieving a bomb.

Meanwhile, in the United Kingdom, it had at last been decided to place the whole matter in the hands of the state organization for scientific research — the Department of Scientific and Industrial Research — where a special secret section was established to work on uranium. This section was given the code name "Directorate of Tube Alloys."

The entry of the United States into the war, and the resulting availability of the resources of the greatest industrial nation in the world to the British atomic enterprise, reinforced the position in the United Kingdom of the advocates who wanted to transfer part of the British development work to America. Thus it was decided, at the first "Tube Alloys" meeting, that Halban's work on the nuclear boiler should be continued in the United States. However, the transfer of his team was to be conditional on its remaining an independent unit under direct control from London.

At the beginning of 1942, the Americans, by simultaneously exploring a great many avenues of research, had virtually caught up with the British and in some areas had even gone ahead, thanks to much larger human and material resources than those available to the British scientists.

So it came about that when the question was raised of transferring Halban's team as an independent unit — despite the fact that this was comprised of no more than six qualified phsyicists, all of German or Central European origin with one exception — Bush, knowing well the reticence of the American security services toward the employment of foreigners in this project, declared himself completely opposed to the idea. The most he would agree to was that Halban and one or two of his co-workers should be integrated into the American team at Chicago. This proposition was unacceptable both to Halban and to the British authorities, who during the summer of 1942, at the very time when the balance of advance was suddenly and rapidly swinging in favor of the United States, persisted in their illusion that their Cambridge team remained in the lead.

Summer 1942 in Chicago

When the British turned down Bush's proposition they undoubtedly renounced their last chance of having full access to the American project, as I was able to see for myself at that time. As a representative of the British, as well as the only Frenchman to take part in the U.S. program, though only for the few months from July to October 1942, I had the good fortune to follow its spectacular development in Chicago.

My particular field was the chemistry of radioactive elements as during the five years preceding the war I had worked in the Radium Institute, where Marie Curie had engaged me in 1933 (the year before her death) as her personal assistant.

After being dismissed at the end of 1940 as assistant at the Paris Faculty of Science (under the anti-Semitic laws then introduced by the Vichy government that among other things prohibited Jews from teaching), I had succeeded in leaving France for the United States in the spring of 1941.

Shortly after my arrival in New York, Enrico Fermi and Leo Szilard called me and suggested that I join their team at Columbia University, where I would devote myself to the problems of producing very pure uranium.

Throughout the summer I awaited my assignment to Columbia; Szilard always assuring me that it was imminent, the delay being due to no more than the formalities of security clearance. But when the decision finally came in October, it was negative.

Fermi and Szilard broke the news to me. They had themselves managed, at last, to obtain greater government support for their work, but on the condition that they recruit no more foreigners. They were thus forced to abandon their offer for me to join them, an offer made still more difficult because Washington at that time did not recognize the Free French forces. The Free French, therefore, proposed my services to British scientific research so that I could join the Cambridge group led by Hans Halban and Lew Kowarski.

While awaiting the outcome of this proposal, I worked for several months at the New York Cancer Hospital, where the first tests were being made of internal radiotherapy using artificial radioisotopes. From there I went to Canada as a consultant at the radium and uranium extraction plant at Port Hope, Ontario.

Then, at the end of June 1942, I was called to Washington by the British embassy, where I learned that instead of going as I had expected to join Halban's team in Cambridge, I was to be sent, on behalf of the British, to Chicago to learn about the chemistry of the new element, plutonium.

A group had been set up at the University of Chicago since the spring of 1942, under the direction of the physicist Compton and was known by the code name "Metallurgical Project." Compton had been asked to bring into the group, among others, the New York team of Fermi and Szilard and the unit recruited by Glenn Seaborg, who had been responsible for the discovery of plutonium. The new combined group was assigned a double task: to determine whether a chain reaction could be achieved with natural uranium and graphite, and at the same time, to try to develop a chemical method for extraction of the plutonium produced in such a reaction.

I arrived in Chicago in July 1942, where I was to spend nearly four fascinating months. I was received by Compton himself, one of the most respected scientists in the States. He explained that, as a representative of the British team, I would find all doors open to me. However, he asked me to voluntarily restrict my work to the area of my special competence, namely, chemistry. Then, to my great surprise, he told me that among the many secrets I would learn, the chemical behavior of plutonium was no less important than the actual discovery of that element had been. In fact, it was then correctly believed that the Germans had no cyclotron powerful enough to enable them to discover and isolate the new element, which had been found by Seaborg to have chemical properties different from what might have been expected.

Fermi and Szilard also gave me a most friendly welcome. They were very amused to find me at last inside the inner sanctum of the American project, under the dual banner of the Free French forces and British scientific research.

At the time of my arrival in Chicago, there were already more than 100 scientists at work in various laboratories of the university. There was an excellent atmosphere within this young and enthusiastic group; they knew their objective was a weapon that, if successful, would have a destructive power beyond all comparison with armaments of the past. Moral scruples had been overcome by the fascination of the research and by the haunting fear that the Germans were following the same path, and were even possibly ahead. The timetable then envisaged, which was miraculously adhered to, foresaw that a bomb would be achieved within three years.

Fermi was responsible for producing the chain reaction. In the greatest secrecy, beneath the stands of the university football stadium, a structure of graphite and natural uranium was being assembled. Because it was built by piling up tens of thousands of graphite blocks, in some of which cavities had been made and filled with uranium oxide or metal, it became known as an "atomic pile." (The term "reactor" came later.)

The chemists were from time to time allowed into this mysterious place, all shining with graphite powder. There they saw a bizarre cubic structure, black and glistening and several yards in each dimension, being built by men covered from head to foot in black powder. The sight of the experiment was highly moving, for we knew that the outcome of the war, and therefore the destiny of the world, could perhaps depend on it.

According to calculations, when the structure reached about a 7-metre cube, critical size would be achieved and the chain reaction would progressively begin — slowly, because the graphite would slow down the neutrons and increase the time between succeeding generations of fissions and because "safety rods" of cadmium, a material that absorbs neutrons, had been arranged so that they could be pushed into the pile as required to prevent the reaction from getting out of control.

Accordingly, a system of natural uranium and graphite (the essential elements of the Chicago pile) has a critical mass as in the case of a bomb. But there are two fundamental differences. The first is that, with natural uranium, tons are needed before the critical mass is obtained, as opposed to the few kilograms required in the cases of uranium-235 or plutonium. The second difference is that, because the neutrons are slowed down, the production of their successive generations takes place much less rapidly than in the bomb. It is this aspect that makes the pile controllable.

A natural uranium atomic pile is also a true "alchemy machine," for as the uranium-235 is gradually consumed in fission, a quantity of plutonium is produced approximately equivalent to that of the uranium-235 that has been used up.*

Now, while the separation of the two isotopes of uranium is extremely difficult, that of plutonium from uranium is relatively easy, except that the latter separation is complicated in practice by the intense radioactivity of the fission products that are present.

This complex development of a chemical process for plutonium extraction was the objective of Seaborg's team to which I was assigned. We all worked in a single large room, previously used by chemistry students for their laboratory training. Some 10 small groups, each comprised of two or three scientists, were studying various possible methods of separation, using trace quantities of plutonium that were insufficient to be weighed but detectable by their radioactivity. Working with Seaborg's deputy Isadore Perlman, my job was to identify the principal long-lived fission elements that must be separated from plutonium during its extraction. It was the golden age of this new chemistry and we quickly discovered some new radioelements — isotopes of then unfamiliar elements — that were to

---

*For each gram of uranium-235 consumed, slightly less than a gram of plutonium is produced, together with some 20,000 kilowatt-hours of energy.

represent an important part of the radioactive wastes from the operation of future nuclear power reactors.

On August 20, 1942, during one of the weekly meetings of the Metallurgical Project researchers (meetings at which the numbers of participants grew at a rate worthy of a chain reaction), Seaborg got up to announce that on that very day a substance transmuted by man had actually been seen for the first time: a miniscule quantity — a few micrograms — of a pink-colored plutonium compound. Edward Teller, another brilliant scientist of Hungarian origin who headed the theoretical physics group and was later to become "father of the H-bomb," asked what the compound was. Seaborg replied that he was not allowed to say, so strictly compartmented was our knowledge as a precaution against leakage.

The chemical data obtained from the separation of a quarter of a milligram of the element in September were the basis of a most remarkable feat of technical and industrial achievement. Less than three years later, kilogram quantities of plutonium were to be extracted in a fascinating plant, remotely controlled through thick concrete walls protecting the operators from lethal irradiations.

A few months after this first fraction of a milligram of plutonium had been isolated, another vitally important event took place in Chicago. On December 2, 1942, a truly historic date in the atomic era, Fermi's pile reached its critical size and the chain reaction was initiated in this strange structure of uranium (6 tons of metal and 50 tons of oxide) and graphite (400 tons).

Slowly withdrawing the cadmium rods caused a growing rise in radioactivity and in the neutron flux in the pile. Within a few minutes, after an energy release of less than one watt, the reaction had to be slowed down to keep the level of radioactivity from becoming dangerous for the operators.

Following several days of running at powers of a fraction of a watt, Fermi's pile had to be shut down; otherwise the radiation would have become dangerous even for passers-by and for people living across the road from the football stadium. This first pile, built by successive approximations and for purely experimental purposes, had not been equipped with protective shielding. It had been by no means certain that it would work, and in any case it had never been envisaged as a permanent, or even semipermanent, installation.

Nearly four years after Joliot-Curie and his team had beaten that of Fermi by a single week in the discovery of secondary neutrons from fission, the Italian scientist took his brilliant revenge. But it is probable, had it not been for the invasions of France and Norway, that Joliot-Curie and his co-workers would have won this second race also, achieving a chain reaction with uranium and heavy water.

Compton, who was present throughout the whole of the December 2 operation — termed the "divergence" of this first atomic pile — telephoned Conant in Washington to report the success. He said simply "you will be interested to learn that the Italian navigator has landed in the New World." At the same moment when the German advances had for the first time been halted, before Alexandria in the Egyptian desert and before Stalingrad in the icy Russian steppes, a successful experiment had opened for mankind the door to a new world full of hopes and threats, the world of modern alchemy.

Thirty years later, in the laboratories of the French Atomic Energy Commission, a detailed analysis by French scientists of uranium deposits from the country of Gabon in Africa was to prove beyond a doubt that a chain reaction had already taken place in nature.

Because uranium-235 is a radioactive isotope more unstable than uranium-238, as one goes back in prehistory it is found that natural uranium then was more "enriched" than it is today. When the mineral deposits in the Gabonese mine at Oklo were being formed two billion years ago, the amount of uranium-235 in the mixture at the time was five to six times greater than today, the concentration being about the same as that now used in ordinary (light) water-moderated nuclear power reactors, the most common type in current operation.

At that time, our planet was about two-thirds its present age, the African and American continents were joined together, and the most advanced living organisms on the earth's surface were the monocellular blue algae. Thus, conditions were physically favorable, given an appropriate concentration of uranium (as at Oklo), for a chain reaction to start in the sedimentary deposits each time there was an incursion of water. Evidence has been found of several natural piles of this type in the prehistory of the area, all within a few miles of each other. They must have been active over thousands of years, their operating cycles "controlled" by water evaporation due to the heat they generated, leading to an interruption of the nuclear reaction until a renewed penetration of water.

And now, following this short excursion into prehistory, and my technical reminiscences of the golden age of atomic research in Chicago in summer 1942, it is time to take up again the thread of the developing American project and its rivalries with its British counterpart.

Breakdown in Anglo-American Relations

That same summer in the United States, a momentous political decision had been made: to launch into full-scale industrial development of atomic weapons with the whole enterprise under army responsibility and control.

The resulting military organization set up its headquarters in New York, leading to the project being known as the "Manhattan Engineer District." Its establishment provoked serious conflict with most of the scientists, who feared that a military atmosphere would be harmful to the freedom they considered essential for the proper growth of scientific research. This fear was not unfounded, for the secrecy regulations were strengthened considerably, and compartmentalization became so strict that it often put the scientists in severe difficulty.

Nevertheless, there is no disputing the contribution of the army's weight and influence in obtaining priorities vital for the rapid advance of these important technical undertakings.

The Manhattan Engineer District was ably led by General Leslie R. Groves, a young and dynamic general of the Engineer Corps. He had confidence in the scientists despite their hostility toward him, and although he could not follow all their complex theories and calculations, he believed them when they declared themselves convinced that an atomic bomb was feasible.

On the British side, the rapid progress of the American enterprise and the new role of the army finally convinced the British that they could not continue on their own. But Halban remained unable to accept the absorption of his little team into the American complex in Chicago. Instead he sought another solution, that of transferring his team to Canada with the idea of setting up an Anglo-Canadian project under his own direction. He saw this as allowing him to remain independent of the United States while nonetheless having the advantage of being close to Chicago and the American resources. He hoped in this way to be able to obtain the first tons of heavy water expected from an American-owned plant situated in Canada.

At the end of July 1942, the British minister Anderson proposed to Churchill that Halban's team should indeed be transferred to Canada, and that a pilot uranium isotopic separation plant based on British development work should be built in the United States. This installation, costly but indispensable for a definitive evaluation of the British process, was never built during the war.

Churchill approved the recommendations of Anderson, who then wrote to Bush, the research leader, to inform him of the transfer of Halban's group to Canada and to ask him to add the British pilot separation plant to the American program with the same priorities as the S-1 section projects. On the assumption that this would be acceptable, which he expected to be the case, he proposed that British representatives should be included in the committee directing the American enterprise.

It had taken nine months for the British to accept Roosevelt's proposal for a partnership; but in the meantime the balance of power between the organizations concerned had shifted strongly in favor of the Americans — they no longer needed the help of their allies.

With Britain fighting for its life and with twilight about to engulf the once supreme British Empire, whatever the outcome of the war, Churchill and his advisers did not realize that their hopes of winning the race for atomic power were illusory, for the lead held by their scientists over those of the United States would disappear as soon as the industrial stage was reached.

Having previously hoped to secure the atomic monopoly, the British were now to be denied even an equal share of it with the Americans, for whom it was to become an essential element of their postwar nuclear supremacy.

The proposals to install the Cambridge team in Canada, as well as for the establishment of an Anglo-Canadian project, were eagerly accepted by the Canadian government. An agreement, concluded without difficulty at the end of October, specified the conditions for setting up the new enterprise, which would be under the aegis of the Canadian National Research Council, would be installed in Montreal with Halban as director, and would concentrate on developing the uranium heavy water system.

Meanwhile, in the United States the outlook for British affairs was by no means as promising. In the very same month of October 1942, Secretary of War Henry L. Stimson, in a discussion with Roosevelt, reported on progress of the S-1 enterprise. It seemed certain that the Chicago pile was going to be a success; the problems of extracting plutonium were well on the way to solution; and several methods for separating the uranium isotopes seemed highly promising.

Stimson also raised with the president the subject of collaboration with the United Kingdom. Underlining that 90 percent of the work was now being done in the United States, he suggested that future exchanges with the British should be kept to a minimum. Roosevelt agreed, considering that his obligations here vis-à-vis Churchill were extremely vague.

With the backing of Roosevelt, the scientific administrator Bush left the British no hope. In his reply to Anderson, he refused the request to build a British pilot plant for isotopic separation in the United States on the grounds that the existing American program was already overloaded.

The British protested and it was decided to put the matter to Roosevelt again. This was done in December, in the week following the startup of Fermi's pile, during a meeting of the S-1 Steering Committee. The president confirmed the decision to launch an industrial effort with the aim of producing the first bombs for the beginning of 1945. Exactly one year after the attack on Pearl Harbor, the powerful American industrial machine was put into gear in pursuit of the most technically formidable military project of all time.

As for the British, Roosevelt decided on a solution limiting exchanges to nothing but information that the recipient country might require for immediate military purposes.

At the beginning of January 1943, the Americans sent a memorandum to the British and the Canadians setting down the conditions that would apply to future information exchanges between themselves and their partners. The outspoken wording made it virtually a declaration of disassociation. It was specified that: "Since neither the Canadian or the British governments are in a position to produce the elements '49' " (code name for plutonium) "and '25' " (code name for uranium-235) "our interchange has been correspondingly restricted by order from the top." The restriction affected nearly all areas. The only doors left open were those concerned with exchanges of fundamental scientific data or, in the case of practical developments, those where the Americans thought they might still gain something from the British.

The United States was taking its first steps along the road to nonproliferation. In contrast to their previous liberal attitude, they were adopting a policy of nontransfer of technology in the two areas that are today considered the most sensitive — the chemical reprocessing of irradiated fuels for the extraction of plutonium and the separation of uranium-235 for fuel enrichment.

For the Anglo-Canadian team, having just settled in Canada and expecting, in particular, to profit from being at an easy distance from the Chicago laboratory and from many of the American resources, the decision was a terrible blow. Until then there had been complete interchange of information between British and American scientists. My own experiences in Chicago, where all doors had been wide open to me, were proof of this.

When Anderson received the text of the American memorandum in London in February 1943, he immediately warned Churchill, who had just left for the Casablanca summit meeting. He asked Churchill to alert

Roosevelt of these unjust propositions which he believed, incorrectly, to have been made without the president's knowledge. At Casablanca, Churchill evidently did not raise the matter with Roosevelt but with his special assistant, Harry L. Hopkins, who promised to look into it on his return to Washington. When no news came from Washington, Churchill expressed his growing agitation in a series of telegrams, the last of which, in April 1943, read: ''I am much concerned at not hearing from you about 'Tube Alloys.' That we should each proceed separately would be a sombre decision.''

In the confrontation that followed, the British still believed they held a winning card, that of Canada's possession of the second most important uranium mine in the world. That card had, in fact, already been trumped.

Late in 1940, Edgar Sengier, managing director of Union Minière du Haut Katanga, had with great foresight decided to send from Africa to New York a considerable quantity of ores very rich in uranium — some 1,200 tons or about as much as the uranium reserves held in Belgium at the beginning of the war. This stock was later purchased by the Manhattan Project, together with other quantities that had already been mined but were still at the pithead in the Congo. These purchases were vital for the success of the American effort since extraction of the mineral had come to an end in Africa as of 1937 and was not restarted until 1944.

For this reason, the Canadian mine became a primordial bargaining card for the British Empire, the Canadian government being informed to this effect in 1942. It was following this that Ottawa, in December 1942 and without informing the British, authorized the Canadian mining company to sign a sales contract with the American government covering its entire production to the end of 1946. The British did not learn of this contract until the following spring (thanks to some information conveyed privately to me personally). They were then told that Clarence Howe, the Canadian minister of munitions and supply had been unaware of the importance of the transaction when he countersigned the contract and, as Churchill is reputed to have said, ''sold the British Empire down the river.''

Thus the 100-strong Montreal team led by Halban, denied access to uranium, heavy water, and American technical help, found themselves condemned to inaction almost before they had begun work. The scientists became extremely irritable and demoralized; every one of us was more interested in the results of a visit to Washington by a British personality, or of a summit conversation, than in the possible results of scientific experiments that we could not carry out. There was a feeling of helplessness, mixed with a fear of seeing the United States taking over the monopoly of the new source of energy, and some of us even wondered whether the other members of the Grand Alliance should not be informed. Today, in retrospect, it seems possible that these difficulties of 1943 may have subsequently encouraged Alan Nunn May, the first British physicist to join Halban's team, to transmit secret information to the Russians (see p. 90).

In the month of May 1943 in Washington, Churchill again met Roosevelt and once more raised the uranium question. He upheld his conviction that collaboration should exist in the atomic field just as in all other fields of the common effort, whatever might be the relative proportions of the contributions from the two partners. That was the position in aircraft construction, where the Americans were responsible for the bombers and the British for the fighters. Roosevelt again gave completely satisfactory pledges which, again, were followed by no practical action.

Luckily, in July 1943, there was an abrupt change in this situation. The American secretary of war, Stimson, was in London with Bush to discuss, among other things, problems of antisubmarine warfare. Churchill sent for Bush on July 15 and did not conceal from him his extreme anger over the way the uranium problem had developed: Every time he saw Roosevelt, Roosevelt gave his word that the matter would be settled satisfactorily, but later there was always someone at a lower level who stopped it. Bush retorted that the British were concerned above all about postwar commercial aspects, the proof being their interest in technical data of commercial and industrial significance, as well as the role they had let the private firm Imperial Chemical Industries take in the enterprise. He also added that the presence of too many foreigners among the scientific leaders of the British team was a security risk.

Churchill replied bluntly that none of these considerations interested him — the only things that mattered for him were the prosecution of the war and its outcome.

One week later, the prime minister received Stimson and Bush at Downing Street, in the presence of Anderson and Cherwell. After bitterly recalling the history of the immediate past, Churchill, who recognized in Stimson a convinced supporter of the Anglo-American alliance, confirmed his total disinterest in postwar commercial applications by declaring himself ready to accept any proposition concerning future industrial aspects that the president of the United States considered equitable. He asked Anderson to prepare a draft agreement to this effect that would be judged acceptable by the Americans.

From that moment on, all problems were quickly resolved. Anderson went to Washington and had no difficulty in developing a final text for a collaboration agreement, which two weeks later was presented to the heads of state at the Quebec summit conference.

Churchill's fierce energy had finally won through in one of the most painful episodes in the history of Anglo-American collaboration during the Second World War. The United Kingdom came out relatively well of an affair generated by mutual misunderstandings between the Americans and the British. What is more, atomic energy appeared for the first time on the agenda for a summit meeting. From then on, it was never to be totally absent.

The Quebec Agreement

On August 19, 1943, in Quebec, an agreement was signed "Governing Collaboration between the Authorities of the U.S.A. and the U.K. in the matter of Tube Alloys." Canada, though not a signatory, was kept continuously informed of the negotiations, for it was to be associated with the implementation of the agreement.

This first international atomic treaty was to influence world politics well beyond the end of the war. Not only did it lay down the conditions of military nuclear collaboration between the two allies, it was also a true nonproliferation agreement, for it gave to each of the two signatories, through a right of veto on the communication of secret information to a third party, the power to prevent any transfer of knowledge.

The treaty established between its two signatories the principle of free exchange of information, but only where that information could serve the war effort. For data that could be used with a view to postwar industrial and commercial advantages of atomic energy, it was specified that because the burden of production would fall upon the United States, the British prime minister would leave it to the U.S. president to decide, in all equity and after the war, what part of the advantages should accrue to the United Kingdom. The completely unique provisions of this clause, however, were never applied in practice.

In the military field, the agreement included a mutual engagement by the two signatories never to use the bomb against one another, as well as a right of veto over the use of the weapon against a third party.

The Quebec Agreement was to operate in three different areas: procurement of nuclear materials, participation of British scientists in work undertaken in the United States, and the Anglo-Canadian project.

Concerning the supplies of uranium, the United States and the United

Kingdom established an understanding, which remained operative until 1961 when an abundance of this source material rendered it unnecessary. The practical result of the understanding was the creation, in 1944 of a commercially oriented joint supply agency — the "Combined Development Trust" — in the operation of which Canada also was associated. The agreement thus gave the Anglo-Saxons a veritable monopoly for many years on the Western World's main uranium resources. The role of the agency was to purchase all uranium that became available and divide it between the Americans and the British, thereby avoiding competition between them in foreign markets.

The first task of this agency was a delicate negotiation with the Belgian government in exile. This led, at the end of 1944, to a 10-year agreement under which Belgium gave the American and British governments first option to purchase all the uranium ores produced in the Congo. In practice, however, during the war the total Belgian and Canadian production went to the United States.

The success of the Anglo-American scientific collaboration under the Quebec Agreement was assured by the excellent relations that developed between General Groves and the most celebrated British nuclear scientist of the time, the Nobel laureate James Chadwick, who was chosen by the British to maintain liaison with the Americans.

Chadwick, who had discovered the neutron in 1932, impressed Groves with his simplicity and the solidarity of his scientific judgment. He believed it would be better for the United Kingdom to abandon its proposed industrial developments and to try to assign as many as possible of the best scientists and engineers from their laboratories to work in every phase of the American effort. This was partially achieved in the field of isotopic separation of uranium and was a great success in the development of the bomb itself. The only areas not penetrated were those of graphite piles and plutonium extraction.

X, Y, and W

At the end of 1942, three large American nuclear centers had been established, intially known only by their code names X, Y, and W. Even their locations were secret.

Nuclear center X was in the Tennessee Valley, at Oak Ridge. It was built between 1943 and 1945 to accommodate two uranium isotopic separation plants, each based on a different process, as well as the first experimental air-cooled graphite pile.

Hanford, Washington, was the site of W, where three large plutonium-producing graphite piles were built.

Finally, Y was the "Nobel Prize Winners' Concentration Camp," at Los Alamos in New Mexico, where design and construction of the bomb were carried out in a most beautiful setting, but where every worker had signed a guarantee not to leave until six months after the end of the war.

At all of these sites, particularly at Oak Ridge and Los Alamos, giant scientific establishments were created, bringing together for the first time thousands of technicians, scientists, and engineers of all disciplines — from pure mathematics to biology and including physics, chemistry, electronics, and metallurgy. The engineers in charge of building the big industrial installations also took part in the laboratory research, so as to cope better with the transition in scale. Whole battalions were thus able to attack the development of a given scientific field on a broader front than anything known hitherto.

Throughout the United States, a total of nearly 150,000 workers was engaged in different aspects of the enterprise, which for this reason became well known in scientific and industrial circles, at least for its size if not for its exact purpose. Its very substantial cost, totaling $2 billion at the time, was completely exempt from congressional control.

One of the most difficult technical problems to be resolved was that of cooling the Hanford piles, where several hundred grams of plutonium were produced each day. The heat released was not recovered but was dissipated by circulating the very pure water of the Columbia River through the annular aluminum cladding that enclosed the uranium rods. The water was circulated under pressure and at very high speeds.

To avoid external release of harmful radiation, these piles were enclosed within thick walls of concrete and lead. Irradiated uranium rods were periodically withdrawn by remote control mechanisms and sent to the chemical plant where their plutonium was extracted. This plant was built as a sort of man-made canyon, a corridor several hundred yards long with concrete walls several yards thick between which, in a series of cycles of remotely controlled chemical operations, the plutonium was finally separated from the uranium and fission products. During the commissioning of the pilot installations of such a plant, when a certain unit broke down, 150 men were needed to repair it; each worked on the equipment for only half a minute, the health service having forbidden any longer exposure.

Having extracted the plutonium, it was necessary to transform it into very pure metal and then cast it in the shape needed for the bomb — very delicate and dangerous operations if strict precautions were not observed, for plutonium, like radium, is extremely toxic if ingested because of its radioactivity.

At Oak Ridge, huge isotopic separation plants were being built to isolate uranium-235 from its more abundant isotope uranium-238. Because of the near-identity of the physicochemical properties of the two, it is necessary to use methods involving many successive separation stages, which slowly enrich a fraction with the rarer isotope leaving an excess of the more abundant one.

Several methods — based on known principles that since 1941 had been used to produce, with great difficulty, some fractions of a milligram of uranium-235 — were adapted in America to a sufficiently large scale to permit kilogram quantities to be produced by 1945.

One of these methods, that of gaseous diffusion, was carried out in a building with a perimeter of some two miles and a similarly enormous internal complexity. It involved 4,000 diffusion stages or passages of a gaseous uranium compound (hexafluoride) through porous membranes comprised of many extremely small pores. Another method, electromagnetic separation, required an amount of steel on the same order as that needed to build a fleet of warships, not to mention an important fraction of the silver reserves of the U.S. Treasury, used because there wasn't enough copper available.

Los Alamos was stranger still, situated on a magnificent isolated plateau at a 6,000-ft. altitude in New Mexico, a sort of paradise/prison where more than 1,000 scientists lived with their families. It was there that the outcome of the whole enterprise was to be decided.

The young physicist J. Robert Oppenheimer was in charge. A remarkable leader of men, he had assembled about him a galaxy of mathematicians, physicists, and specialists in micrometallurgy and ballistics. His ideas inspired the theoretical and practical studies that led to the solution of one of the most difficult problems, the solution to which the details, if not the principles, remain relatively secret even today — the precise mechanism of the bomb.

The difficulty lies in the requirement that subcritical masses or volumes of fissile material be brought together to form a super-critical mass in a time sufficiently short to avoid too many premature fissions and, consequently, a premature explosion.

The operation is even more difficult with plutonium than with uranium-235 for, from the moment of its production in a pile, plutonium-239 is slowly transformed by the neutron flux into another plutonium isotope of mass 240. This isotope, because of its tendency to fission spontaneously, emits neutrons that increase the chances of a premature explosion of the weapon and, therefore, acts somewhat as a poison hampering the military use of plutonium-239.

In both uranium-235 and plutonium bombs, the critical mass must be attained in an extremely short time. This is achieved by the use of very powerful chemical explosives. In one system, the two subcritical masses are thrown together in a tube similar to a conventional gun barrel. One of the masses is projected by an explosive charge into the other mass, which is situated at the far end of the tube and hollowed out to receive the first mass. Another method makes use of the phenomenon called "implosion," obtained by using shaped charges that enable the combined effects of an assembly of these explosive materials to concentrate on a single focal point. This is used to crush a hollow sphere of, for example, plutonium, which initially has a subcritical geometrical configuration, but which when crushed very rapidly becomes a solid super-critical sphere.

To set off the explosive reaction at the precise moment when the super-critical mass is attained, it is necessary to have a trigger, a device to produce a sudden jet or "puff" of neutrons. Finally, the super-critical masses must be held together for as long as possible — about a millionth of a second. This is achieved by placing them inside a very heavy tamper, which by its inertia prevents the fissile mass from expanding too rapidly.

It can be easily understood that the complexity of these mechanisms makes the atomic bomb, even today, an extremely delicate device — practically impossible to produce by makeshift methods. Its efficiency, i.e., the ratio of the mass of fissile material that actually undergoes fission to the initial mass, can vary between extremely wide limits for any given system.

The success of Oppenheimer's team lay not only in solving these problems for uranium-235 and for plutonium, but also in reducing the size of the devices so that they could be carried in a Super-Fortress, the largest bomber in existence at the end of the war.

Restricted Collaboration

At Los Alamos, some 20 U.K. scientists took part in the final phase of development of the bomb, as well as in the extremely complex work related to its construction and — most delicate of all — the perfection of its detonating mechanisms. Among these scientists were the great British expert on explosives, William G. Penney, and a brilliant theoretical physicist of German origin, Klaus Fuchs. Twenty years later, the former was to find himself directing his country's atomic energy program, the latter in charge of an atomic laboratory in East Germany, after serving 10 years in an English prison for spying on behalf of the Soviet Union.

The British contribution to the American project over the last two years of the war was beyond anything the team's numerical strength could have suggested. It is quite impossible to evaluate quantitatively the time saved as a result of this contribution; nonetheless, it is very likely that without it the first atomic bombs would never have been ready in time to end the war.

Thus was the British collaboration reestablished following the Quebec Agreement. Nevertheless, mistrust had persisted, and the Anglo-Canadian enterprise in Montreal was not without renewed difficulties.

During the first week of January 1944, an official delegation from Montreal, of which I was a member, went to Chicago to reestablish collaboration. A meeting, chaired by Groves and Chadwick, was attended by 24 experts from two laboratories. It was characteristic of the international composition of the teams that 13 different nationalities were present — witness to the generous and intelligent attitude, before the war, of the Anglo-Saxons toward the refugees from Nazism. The success of the atomic venture was their reward.

After Groves had welcomed us and Chadwick had replied, to our astonishment the two leaders started a heated *sotto voce* argument, the subject of which was soon revealed to us. Plutonium chemistry and extraction methods were to be excluded from the renewed collaboration on heavy water piles.

The American attitude was as follows. The collaboration was to be restricted to fields where it could benefit the war effort; therefore, since the design of the American plant at Hanford for extracting plutonium from irradiated uranium — an incredibly complex and entirely remotely controlled process — was already finalized, it could hardly be modified, let alone improved, as a result of possible British participation. Hence, British participation was rejected.

Such singularly strict interpretation of the Quebec Agreement, among allies, was tantamount to blocking technological transfers in one of the key sectors of the vital forthcoming issue of nonproliferation, namely, the ''reprocessing'' of irradiated fuels.

In spring 1944, it was finally decided to build a large heavy water pile in Canada. This, destined to be the main result of the collaboration between the three Quebec Agreement participants, gave the Montreal team, after a year of uncertainty, a purpose as well as interesting work to do — useful if not for the war (for the decision came very late) at least for the future.

The Americans further wanted this Anglo-Canadian project to have a British director. The choice fell on John D. Cockcroft, already well known as a nuclear physicist, and one of the team that had developed radar, and a future Nobel laureate. Quiet and efficient, he replaced Halban, the original founder and subsequent director of the Montreal laboratory, whose leadership had not been free of friction with the members of his team.

The large Canadian heavy water pile was preceded on the same site by a much smaller one, built under the direction of Kowarski. Completed in September 1945, the latter was the first pile in the world to operate outside of the United States, where a first heavy water pile had already been constructed in May 1944 in Chicago. The larger Canadian pile contained some 10 tons of heavy water and about the same amount of uranium metal. Materials were supplied by the United States and the pile was built in the two years from 1945 to 1947 at a magnificent isolated site at Chalk River, on the banks of the Ottawa River in Canada.

The British had tried in vain to persuade the Americans to change their decision on plutonium, but had succeeded in obtaining one concession. The Americans were willing to supply to the Canadian laboratory, though without any relevant technical information, a few uranium rods that had been irradiated in their Oak Ridge experimental pile and therefore contained several milligrams of the precious element. We were left free to develop a method for its extraction on the basis of the data we already had from the early work of the Seaborg group in Chicago, a group in which I had been the only one of the Anglo-Canadian team to take part before the breakdown of the collaboration.

Given this problem, I could offer a choice of two solutions: Either we could try to reinvent the method used by the Americans, of which I knew the main outline from my last mission to Chicago, which was before the February 1943 breakdown; or we could try a different way, based on organic solvents, which Seaborg had always thought the most promising although, pressed for time, he had not been able to develop it.

Lord Cherwell, during a visit to Montreal at the end of 1944, favored this second solution which he thought more original. Standing before the great man known as "The Professor," who was so close to Churchill, I had felt like a young student as I explained on the blackboard in my laboratory the chances of success with the two possible methods. Throughout my explanation he had remained sombre and silent.

A systematic program of research was therefore undertaken to choose the most promising among some 300 organic solvents that were commercially available in the United States. A half-dozen of these were selected and, in 1945 in Montreal, we were able to use one of them as the basis for a completely new plutonium extraction process.

Four years later a more efficient solvent than all the others was discovered in the United States, and it is on this that all of today's large plutonium extraction plants are based. This solvent was not commercially available when we were making our choice in Montreal, underlining the chancy nature of such applied research. Our work in this case had, nevertheless, enabled us to demonstrate during the war the relative ineffectiveness of policies of secrecy, even in such a specifically sensitive field as the reprocessing of irradiated fuels.

The work of the Anglo-Canadian project thus provided a very useful complement to the knowledge acquired by the British scientists working in American laboratories. The Canadian input to the British effort was extremely important, not only in the technical contributions of numerous very able experts, but also in the politically unparalleled help in material, administrative, and financial fields.

The British wartime financial investment in atomic projects was only about one hundredth of the American, which amounted to $2 billion. Therefore, although they were treated rather like poor relations, in fact, the British did quite well. When hostilities ended, they were in possession of nearly all technical data concerning the American atomic work, which certainly facilitated their acquisition of nuclear weapons after the war.

In this respect, it is interesting to note that the restrictions imposed by the United States on information transferred to the British, particularly in the fields of uranium enrichment and plutonium production, owed at least as much to concern for commercial and industrial competition as to anxiety over nonproliferation. This is clearly shown by the fact that the British had access to Los Alamos and the bomb.

Despite this, a new approach to Roosevelt by Churchill, during the September 1944 summit meeting at Hyde Park, gave him at least some hope for continuing and close postwar collaboration with the United States. This was recorded in the Hyde Park Aide-Mémoire that put an end to Bohr's attempt to have the Russians informed. There it was written: ''Full collaboration between the United States and the British government in developing Tube Alloys for military and commercial purposes should continue after the defeat of Japan unless and until terminated by joint agreement.'' As we shall see later, this engagement remained a dead letter.

The United Kingdom did in fact have to pay for the knowledge it acquired from the Americans by renouncing under the Quebec Agreement its right to communicate atomic information to a third party without

American consent, even when the information had been discovered by British scientists in Britain.

The Quebec Agreement thus forced the United Kingdom to observe the rules the Americans later imposed on themselves, and to restrict, on a number of occasions, the United Kingdom's freedom of action in the international atomic field vis-à-vis Europe and the Commonwealth. Even before the end of the war, these restrictions were already affecting relations with the French.

Informing General de Gaulle

Besides Halban and Kowarski there were three other Frenchmen — Pierre Auger, Jules Guéron, and myself — who worked for the British atomic project during the war, being detached on an individual basis by the Free French forces.

Auger was in charge of the physics division in the Montreal laboratory from its establishment until France's liberation in the summer of 1944, when he left to return to France as the director of higher education. Guéron had been part of the Cambridge group from December 1941; he played a leading role in the chemistry division of the Montreal project, while I was responsible for the work on plutonium.

Although the total French contribution was important and out of proportion to our tiny number, it could never represent a real political asset for France for we were not grouped in a coherent unit with a recognized leader who could have negotiated with the British on our behalf. Had Joliot-Curie gone to England, he would naturally have assumed this role.

Despite our own work being confined to atomic piles and plutonium extraction, we knew the formidable importance of the weapon that was being developed. It was for this reason that Auger, Guéron, and I took the initiative of warning General Charles de Gaulle, during a visit of a few hours that he made to Ottawa on July 11, 1944, of the consequences of this new element in world politics. This was a year before the bomb was used, and the Free French Committee had never been informed officially of the enterprise or its progress.

The matter was delicate, for we could not tell the British authorities of our intentions, and we did not wish to inform anyone other than the general himself. As a first step, while not divulging the reason, we had to persuade the Free French delegate in Canada, Gabriel Bonneau, to seek an interview for us with the general on a top secret matter of the highest importance, during the 15 minutes or so that he was expected to spend with the French delegation in Ottawa. Bonneau, trusting in our good faith, agreed to ask the leader of Fighting France to receive just one of us in private. Thus it was Guéron, whom de Gaulle already knew, who gave him our information — in a secluded little room at the end of a corridor where the general, forewarned, allowed us three precious minutes of his time.

We believed very sincerely, in these still difficult days for our country, that it was imperative to inform the general of the importance of the project that, in a reference in his memoirs to our disclosures, he called an "apocalyptic undertaking." We wanted him to be aware of the very considerable advantage that possession of the new weapon would represent for the United States, to be ready to relaunch atomic research in France as soon as practically possible, and finally to know of the existence of uranium resources in Madagascar. At that time we believed these resources to be very much larger than they turned out to be.

A few minutes after his meeting with Guéron, all three of us were officially presented to the general, with other Free French representatives in Canada. When it was my turn he said simply — giving me for the first time in my life the title "Monsieur le Professeur ": "Thank you; I have understood very well."

Halban and the Patents

Before the end of that same year, 1944, French participation in the project was the cause of a major incident between the British and the Americans in connection with the early patents taken out in France.

When Halban first arrived in England he had been very conscious of the French lead in the chain reaction field. As he saw it, this lead was confirmed by the existence of five basic patents, taken out in 1939 and 1940 by the *Caisse Nationale de la Recherche Scientifique*, the national fund for scientific research, on behalf of the group at the Collège de France, for whom they had also been registered in Britain. The British government had subsequently become increasingly interested in these patents, in step with its financing of the research at Cambridge since 1940; this research had enabled Halban and Kowarski to register new and more advanced patents, in particular concerning the use of heavy water in the uranium chain reaction.

When the Department of Scientific and Industrial Research took over the direction of the British enterprise, long and complex negotiations began with Halban over the patents, which were not completed until summer 1942. Halban declared that he had verbally received from Joliot-Curie the authority to negotiate to the best possible advantage all the rights in the French patents. As part of his British contract, he had endeavored to secure from the first French government after the liberation a somewhat unbalanced agreement giving the British all rights in the French patents throughout the British Empire and the rest of the non-French world, and giving the French similar rights to exploit, in France and the French Union, all the subsequent new patents taken out in Britain.

The British not only hoped that this agreement would yield substantial financial benefits, but also thought that possession of the rights in the patents would give them the monopoly and the control of the new source of energy throughout the world. This idea, which would also have necessitated an agreement over the patents between the United States and the United Kingdom, was very badly received in Washington in 1942, where the American officials concerned were obsessed with their antitrust legislation. The British attitude was indeed one of the reasons behind later accusations by the Americans that the United Kingdom was only interested in future industrial and commercial exploitation.

The patents problem thus contributed to the deterioration of Anglo-American nuclear relations. It also greatly complicated relations between the Anglo-Saxons and the French in the last year of the war, all the more regrettable since in the end the financial and political value of these ''certificates of invention'' proved very minor, particularly because, despite years of negotiation, the Americans refused to recognize them — partially, perhaps, because of the importance of their subject matter.

In November 1944, Halban asked to go to England, then to recently liberated France, to see Joliot-Curie and submit to him his patents agreement with the British. Learning of the proposal, Chadwick and Cherwell, who happened to be in Washington, told the minister concerned in London, Sir John Anderson, that they considered the visit totally undesirable. In their view, it would risk starting up a flow of secret information into France, a partner about whom they were now even less sure since Joliot-Curie had recently become a member of the Communist party. However, Anderson, himself interested in the patents question and encouraged by Halban, let Halban proceed after referring the matter to Churchill.

As soon as he reached London, Halban explained that it would be impossible for him to discuss the patents question with Joliot-Curie without giving him some glimpse of the evolution of research since 1940, in particular the successful achievement of a chain reaction, and of the existence and fissile properties of plutonium. Anderson did not object, but asked the Americans for their agreement. The proposals for the visit and for the communication of the information were therefore submitted to the

American ambassador in London, John G. Winant. Despite a telegram from General Groves officially blocking the visit, Anderson put pressure on the ambassador who, in the absence of a reply from Washington to a request for further instructions, gave his consent.

When Groves saw the text of the communication to Joliot-Curie that Winant, without any scientific knowledge, had approved on his own responsibility, his fury was unbounded. The matter should have been submitted to the Combined Policy Committee set up by the Quebec Agreement but Anderson's haste, fanned by threats from Halban that he was going back to Canada, had prevented the affair from following a normal course. To crown it all, Halban's visit, as well as a meeting between Anderson and Joliot-Curie in Paris, had a completely different result from that sought; for Joliot-Curie, not being convinced that Halban's personal policy was the best, refused to discuss any cession of rights in the French patents other than in the framework of a wider collaboration agreement.

The Problem of the French

In this affair, the Canadians backed the Americans, so bringing Anglo-American atomic relations once more to the breaking point. The United States was particularly reproachful of the United Kingdom for not having revealed its ties with France at the time the Quebec Agreement was signed.

Secretary of War Stimson was so alarmed that he decided to put the whole matter before Roosevelt. This he did on December 30, 1944, at the same time as he assured the president that the first bomb would be ready at the beginning of August 1945. Stimson was against making any concessions to France. He believed the French had contributed nothing to the American enterprise and he was afraid, he said, that if the French were given an inch they would use the opportunity to make continued and perhaps successful efforts to get a mile — in the form of information and/or participation.

Roosevelt, who made no secret of his reserves vis-à-vis the Free French, or of his hostility toward their leader, was categorically against any engagement with France that could make her a fourth atomic partner, and he insisted that the British should put off any negotiations with Joliot-Curie until the end of hostilities.

Following this drama over Halban's visit to Joliot-Curie, Anderson, more aware than anybody of the British debt to the French, felt he could not leave the problem unresolved. He therefore put the whole question to the prime minister so that it might be discussed shortly afterward with the American president at the Yalta Conference. Churchill was adamant: In his view, the only thing that counted was the agreement made with Roosevelt at the Quebec Conference. Anderson tried to reinforce his argument by suggesting that if the British promises to the French were not honored, there was a risk that Joliot-Curie might advise de Gaulle to seek atomic collaboration with the Soviet Union. "If there is any real danger of this," replied Churchill, "Joliot should be forcibly but comfortably detained for some months."

In March 1945, Anderson made a final attempt through British Foreign Secretary Anthony Eden, who also feared a possible Franco-Soviet atomic agreement and therefore favored offering France a degree of participation in postwar Anglo-American work. Once again the British prime minister's reaction was negative, and this put an end to Anderson's efforts to resolve the problem of French participation.

After the war was over, in October 1945, Joliot-Curie and Auger, who were setting up the Commissariat à l'Energie Atomique, the French Atomic Energy Commission, informed the three of us who were still in the Anglo-Canadian group (Halban had been obliged to resign at the beginning of the year) that they wished us to return to France. Kowarski and Guéron went back but Cockcroft, with Joliot-Curie's agreement, asked me to remain during 1946 as leader of the chemistry division of the Canadian enterprise so as to perfect the plutonium extraction process we had developed independently of the United States.

However, because of my official connections with the newly created Commissariat à l'Energie Atomique (CEA), the United States insisted that the British and Canadian governments should agree to my departure. Thus I had to quit the project just two weeks after having signed a 1-year contract with the British.

Before leaving North America at the end of January 1946, I was received, at my own request, by General Groves in Washington. He explained to me that, because the war was now over and the United States no longer had atomic relations with France, the employment of a French specialist detached by his government could no longer be justified. But he added, to show that no personal reflection was implied, that if I could agree to bind myself to the enterprise in Canada, the United States, or the United Kingdom for the next four or five years he would — despite my nationality — be perfectly happy for me to stay on. He added moreover that I would not be the only foreigner to stay in such circumstances, for he had been pleased to accept the continued participation at Chalk River of the brilliant Italian physicist Bruno Pontecorvo, who did not have official links with his government. General Groves could hardly have suspected that, a few years later, other far more dangerous links would cause him to regret not having made Pontecorvo share the same fate as myself — removal from the project. For the scientist, a member of the Communist party, was to defect to the Soviet Union in 1950, abruptly quitting the England that he had recently chosen as his adopted country when he had been naturalized as a British citizen.

We who returned to France were in no way formally released from our promises of secrecy, which were required of everyone in the Allied enterprise. Although Cockcroft had given us, on our departure, a letter authorizing communication of our knowledge to our government, shortly afterward he asked us to return it saying there had been a misunderstanding and that we would receive another, more official communication. This did in fact happen. Three months later each of us received a new letter from Minister Anderson, informing us that we must continue to respect our pledges of secrecy, from which he could not release us — and from which we were never formally released.

I had also raised this delicate matter of secrecy during my last visit to General Groves; his attitude had been more flexible. It was tacitly understood that we could use our knowledge to benefit France by giving information to our research teams, but without publishing it and only to the extent necessary for the progress of our work. That was a reasonable compromise. We applied it during the first years of the CEA, with the agreement of all those in charge and without complaint from any of the interested parties.

So ended the French participation in the wartime nuclear enterprise. We had, despite our small numbers, undoubtedly made our contribution. Without Halban, Kowarski, and their heavy water, it is possible that during the conflict the British would not have pursued their researches toward plutonium production and the industrial uses of the new source of energy. Without Halban, Kowarski, and their heavy water, it is certain that the Montreal project could not have been born, and that the Canadian program, true descendant and heir of the work at the Collège de France, could never have achieved today's leadership in the development of heavy water power reactors.

Act Two

THE CLUB
1945 to 1964

1. The Last Chance

The end of 1945 left humanity with few, very few, years in which to try to reestablish a world free from atomic weapons. It is doubtful that the leaders of the main nations really appreciated this. Their successors, a third of a century later, with tens of thousands of far more powerful bombs, gravely and anxiously go over and over the problem of proliferation. This problem could only have found complete and nondiscriminatory solution in the early years after World War II.

Having, in the words of President Harry S. Truman on August 9, 1945, " . . . emerged from this war the most powerful nation in the world — the most powerful nation, perhaps, in all history," the United States attacked this immense problem during their first years of rivalry with the Soviet Union — rivalry then dominated by the American nuclear monopoly.

The simple possession of the atomic bomb, even without any threat to use it, was certainly partially behind the early stages of the Cold War. An example was the blocking by Washington, in 1946, of Russian ambitions in the Iranian province of Azerbaïdjan or in the Turkish Straits of the Dardanelles.

In March 1947 Truman launched his aid program for Greece and Turkey, followed three months later by the Marshall Plan, which Soviet Premier Joseph V. Stalin forbade Czechoslovakia to join. That country, after the Prague coup d'etat of February 1948, was to join Poland and Hungary behind the Iron Curtain. In June 1948 the Berlin blockade began, while elsewhere Yugoslavia was excluded from the Cominform. At the beginning of 1949, Mao Tse-tung founded the People's Republic of China. The North Atlantic Treaty was concluded in April, and in May the Berlin blockade was at last lifted.

In the nuclear field, in parallel with the establishment of national legislation and an organization for the peacetime development of the new power, Washington's international action was directed toward the establishment of world control to prevent the acquisition of atomic weapons by other countries. In the event this action turned out to be unsuccessful, the American lead and monopoly were to be maintained for as long as possible.

The Smyth Report

Nevertheless — and in spite of the initial intention to pursue secrecy — 10 days after Hiroshima, the Department of War in Washington published a document, the "Smyth Report," on the history of the great scientific and technical success of the Manhattan Engineer District project. This report, nearly 250 pages long, made public the broad outline of the American effort, described the principles behind the main lines of development that had been successfully followed, and even pointed out certain technological problems that had been particularly difficult. However, it was also stressed that all items of scientific, technological, and industrial data that had not been revealed should remain secret.

The report had been prepared, at the request of General Leslie R. Groves, by the physicist Henry D. Smyth during the last year of the war without any consultation with the governments of either the British or the Canadian allies.

One of the purposes was that, in case it should finally turn out that the weapon could not be achieved, it would be necessary to justify before a congressional committee of inquiry the enormous enterprise that had developed without Congress being informed.

The details revealed in the Smyth Report were invaluable for any country launching into atomic work; for nothing is more important, when undertaking technical research over a wide field than knowing in advance which lines of approach can or cannot lead to success, even if this knowledge relates only to basic principles.

In view of its repercussions, the decision to publish the report was undoubtedly wise, for the rather general data revealed were becoming more and more difficult to keep secret in peacetime. The data were in any case known to the British as a result of their participation in the work, and no doubt also in large measure to the Russians as a result of wartime unlawful disclosures, which at the time had not yet been uncovered.

The publication was also a practical concession to scientific circles, which were strongly opposed to excessive secrecy and to restrictions — restrictions from which they had suffered during the war — on the freedom of work that even extended to laboratories for fundamental science. In any case, the information revealed was indispensable and probably sufficient as background for those about to attack the political problems raised.

True, it was not yet possible to foresee the full importance of the future contribution of atomic energy toward meeting the electricity requirements of the industrialized countries, nor its more immediate role in naval submarine propulsion. But nuclear science and technology at the end of the war had reached the stage where the peaceful benefits of fission could be seen very clearly: benefits both from applications of artificial radioelements and from the production of usable energy, although as yet questions of competitivity remained undetermined.

The two wartime explosives, uranium-235 and plutonium, were destined to be the concentrated fuels of the future. In this connection it is worth emphasizing that the main methods in use today for isolation and production of these two materials were either already perfected or very well advanced. Therefore, the technical and political solutions proposed at that time for the control of their production have not been outmoded by subsequent progress in nuclear technology.

The scientific education of the American people was accompanied by a veritable crusade to make them aware of the political aspects of atomic weapons and their dangers. A brochure jointly prepared by some 15 leading scientists and journalists sold nearly a million copies. Its title, *One World or None,* was self-explanatory: It demonstrated the impossibility of defense against the new weapon, and a description of the annihilation of Hiroshima was accompanied by terrifying representations of an atomic attack on New York.

The Policy of Secrecy

On October 3, 1945, President Truman sent a message to Congress on the revolutionary character of the methods that would be needed, both nationally and internationally, for the control of the atom. Nationally, he proposed the creation of a commission to supervise all atomic materials and plants, taking care however to minimize interference with research projects and private enterprises. Internationally, he proposed discussions, first of all with the British and Canadian partners and then with other nations, to elaborate an agreement under which atomic cooperation could replace atomic competition.

The proposal for preliminary discussions with the Anglo-Saxon allies was in the form of a request to British Prime Minister Clement R. Attlee, who at the end of the war had replaced Churchill after an electoral landslide to the Labour Party. Attlee had subsequently taken the first steps toward restarting British atomic research and he wished to be sure of American help. In his view, this was an additional reason for arranging a summit meeting between the Anglo-Saxon allies without delay.

The first such meeting, devoted entirely to atomic matters, took place on November 11, 1945, on board the American presidential yacht on the Potomac River. It was attended by Truman, Attlee, and their Canadian colleague W. L. Mackenzie King. They had no difficulty in agreeing to continue the policy of secrecy, as well as maintaining purchases of all available uranium in the Western World, as had been done since 1944 following the Quebec Agreement.

As information and uranium are the two essential elements of any nuclear program, the maintenance of secrecy and the monopoly of uranium were, in practice, the best guarantees against any others acquiring the atomic weapon, or as is said today, the best solution leading to nonproliferation.

The new policy was solemnly announced from the White House on November 15, 1945, by the three heads of government in the following terms:

> We are aware that the only complete protection for the civilized world from the destructive use of scientific knowledge lies in the prevention of war. No system of safeguards that can be devised will of itself provide an effective guarantee against production of atomic weapons by a nation bent on aggression. Nor can we ignore the possibility of the development of other weapons, or of new methods of warfare, which may constitute as great a threat to civilization as the military use of atomic energy . . . .

> We have considered the question of the disclosure of detailed information concerning the practical industrial application of atomic energy. The military exploitation of atomic energy depends, in large part,

upon the same methods and processes as would be required for industrial uses.

We are not convinced that the spreading of the specialized information regarding the practical application of atomic energy, before it is possible to devise effective, reciprocal, and enforceable safeguards acceptable to all nations, would contribute to a constructive solution of the problem of the atomic bomb.

On the contrary we think it might have the opposite effect. We are, however, prepared to share, on a reciprocal basis with others of the United Nations, detailed information concerning the practical industrial application of atomic energy just as soon as effective enforceable safeguards against its use for destructive purposes can be devised.

In order to attain the most effective means of entirely eliminating the use of atomic energy for destructive purposes and promoting its widest use for industrial and humanitarian purposes, we are of the opinion that at the earliest practicable date a Commission should be set up under the United Nations Organization to prepare recommendations for submission to the organization.

A month later, during a conference in Moscow of the foreign ministers of the Big Three, this proposal was accepted by the Russians. Soviet Foreign Minister Vyacheslav M. Molotov was persuaded without too much difficulty that Canada, China, and France should also be invited to join in sponsoring a resolution, for submission the following month to the first General Assembly of the United Nations, by which a commission was to be established for the international control of atomic energy.

The Acheson-Lilienthal Plan

The first General Assembly of the United Nations, held in London in January 1946, set up the U.N. Atomic Energy Commission (AEC) as a dependent body of the U.N. Security Council. Its ambitious program covered the complete elimination of all major weapons adaptable to mass destruction, but it was mainly concerned with atomic energy.

Before leaving for London for this U.N. meeting, American Secretary of State James F. Byrnes appointed his undersecretary Dean G. Acheson to chair a committee to formulate the American position on international control of atomic energy. This committee included, among others, the three leaders of the American wartime project — Vannevar Bush, James B. Conant, and General Groves — as well as Assistant Secretary of War John J. McCloy.

The practical work of the committee was delegated to a group of five advisers, chaired by David E. Lilienthal who was president of the national enterprise for the development of the Tennessee Valley, the Tennessee Valley Authority. Also in the group were three industrialists, one of whom had been involved in the production of plutonium, and the scientist responsible for perfecting the nuclear bomb, J. Robert Oppenheimer, who undertook to educate the group in the subject of its study.

After six weeks of intensive work, including visits to the various American nuclear centers, the group presented the committee with a report of some 60 pages containing some revolutionary conclusions. It proposed, in fact, that the entire business should be internationalized.

The report was accepted, with hardly any modifications, by all the members of the committee, despite the fact that one of them, Groves, was inwardly in favor of using the bomb as a means to prevent its acquisition by other countries. The text, which later became known as the Acheson-Lilienthal Report, was made public in March 1946. It was the first and one of the most important documents relating to nonproliferation.

Largely inspired by Oppenheimer, the report states that no effective security system for atomic weapons can be based on treaties of renunciation (such as the illusory 1928 Kellogg-Briand Pact that proclaimed war to be illegal), nor even on a system of agreements for which the only checks of compliance would be an international inspection.

The report's philosophy is close to that of the opponents of the actual world system of atomic energy safeguards, which is based on a treaty of renunciation — the Non-Proliferation Treaty — and on international verification of the pledges of peaceful utilization. These opponents consider such guarantees inadequate, for they allow nonnuclear weapons states to have installations that can be easily adapted for weapons production. The fear is that, sooner or later, one of these countries is bound to be tempted to use such installations for military ends.

The authors of the Acheson-Lilienthal Report believed that complete protection against a possible nuclear war could not be assured, even if all nations undertook to outlaw atomic weapons and to submit all of their civil nuclear activities to safeguards based on international inspection.

According to Lilienthal and his colleagues, nuclear temptations and rivalries will be inevitable so long as individual nations can indulge in "dangerous" activities, i.e., activities that could be quickly adapted more or less to arms production. They concluded that such activities should be under international management. This is the solution that today is again being advanced as a counter to fears caused by nonnuclear weapons states possessing uranium enrichment installations or fuel reprocessing plants, even if these are subjected to inspection.

The Acheson-Lilienthal Report was of course addressed to a world as yet free from stocks of nuclear arms, whereas today's discriminatory situation must recognize the existence of nuclear weapons states that are not prepared to disarm. Nonproliferation policies today are essentially designed to guarantee the continued abstention of other nations.

Immediately rejecting any idea of forbidding civil applications of atomic energy, the report asserted confidence in the early development of fission as a new source of energy, in addition to coal and oil, for the production of heat or electricity in large power plants. Ship propulsion was not mentioned.

The report defined as "dangerous activities" not only installations for producing uranium-235 and plutonium (enrichment and irradiated fuel reprocessing plants, and reactors* producing energy and plutonium) but also, and more surprisingly, uranium mines and the corresponding refineries. In brief, almost the entire nuclear fuel cycle is classified as dangerous, and for that reason should be removed from national control and placed under an international authority.

The Acheson-Lilienthal Report also recognized the possibility of intentionally "denaturing" fissile materials by adding isotopes that would make them unusable for military purposes while still allowing their use in civil projects. Uranium-235 was to be "diluted" with several times its weight of uranium-238 and plutonium denatured by high content of plutonium-240 (see p. 55). However, this idea was soon abandoned because progress in the military field made it possible to use plutonium, even with a high content of the 240 isotope, as an explosive.

The report pointed out that two types of energy-producing plants were possible: the one producing plutonium during operation, and therefore "dangerous"; the other consuming either uranium or "denatured" plutonium without producing more plutonium. The latter type could, while always subject to inspection, be owned and operated by national organizations. Thus, a given country could have had two kinds of nuclear electricity production, one under the country's own authority, and the other, due to its dangerous character, under "international authority."

It was also proposed, as is currently envisaged today, to spread the

---

*Use of the term "reactor" had now replaced that of "pile."

geographical locations of dangerous installations so as to reduce the risks in case one of them should be illegally seized by a country.

In order to carry out its tasks effectively, the international authority envisaged would need to be in the forefront of nuclear progress and to have the exclusive right to carry out research on nuclear explosives so as to be in the best position to detect activities with military applications. There was to be an ''inspector corps'' specifically for detecting clandestine activities, in particular, the illegal mining of uranium.

The authority would thus have been a supranational organization, having its own large-scale industry that it would have exploited and developed in the name and interest of all nations. In brief, it was to be a demonstration of world government in a business of worldwide importance, which in the long term must doubtlessly either lead to true world government or fail.

A series of transition stages was proposed to ease the path from the national stage to that of international exploitation. Progressively, the United States was to have handed over to the authority first their information, then their installations, later their fissile materials, and finally their military laboratories and weapons. From the political point of view, these transition stages were the most delicate aspects of the operation. The authors of the report also believed that one of the first steps to be taken should have been the compilation of a verified world inventory of uranium mines, implying the admission of foreign inspectors into the Soviet Union.

The publication of the report in March 1946 made a deep impression on the American public, which saw in it the favorable influence of scientists and a possible solution to the frightening problem that had been in every mind since Hiroshima. But it was recognized that a most important step, yet to be attempted, was to persuade the Soviet Union to accept the plan.

To carry out this delicate negotiation, the American secretary of state chose a faithful personal and political friend, the veteran statesman Bernard M. Baruch. Seventy-five years old, already a legendary personality, éminence grise of many presidents and a powerfully rich financier, Baruch enjoyed a high reputation in government circles, although others, among them Acheson and Lilienthal, were less enchanted and, in fact, were severely critical of him.

Baruch would have been excellent as a campaigner to persuade Congress to adopt an international treaty involving renunciations of sovereignty. But it was much more questionable from the viewpoint of the first and principal difficulty to be overcome: that of negotiation with the Soviet Union. Baruch's vanity was not flattered by having to present and defend a plan he had not conceived. Further, as if the dispositions of the plan were not in themselves already sufficiently opposed to the traditional Russian practices of secrecy, one of the cornerstones of Soviet Union self-protection against the rest of the world, the American negotiator decided to add a supplementary political clause withdrawing, for the purposes of atomic energy control, the right of veto enjoyed by the five permanent members of the Security Council under the U.N. charter. This right of veto had been one of the conditions of the Soviet Union's participation in the organization that succeeded the League of Nations, which in the late 1930s had failed to prevent the conflicts in Africa (Abyssinia) and Asia (Manchuria) or the Second World War. This veto clause in the U.N. statutes meant that no Security Council resolution could be adopted as long as any one of the Big Five was opposed to it.

Baruch had thus obtained President Truman's agreement that the American proposal (by now also known as the Baruch Plan) for the creation of an international authority for atomic development, based on the Acheson-Lilienthal plan, should exclude the right of veto, a right that could have been invoked to avoid condemnation by majority vote of any country in breach of its formal promise not to develop or use atomic energy for destructive purposes.

This clause and a certain number of additional provisions favorable to the United States were included in the American proposal officially submitted to the United Nations on June 14, 1946.

Five days later the Soviet delegate Andrei A. Gromyko, then in the early days of his long diplomatic career, proposed an international convention of indefinite duration, open to all nations and calling for an absolute prohibition of the use of atomic weapons, prohibition of their production, and the destruction of all existing weapons within three months of the convention's ratification. Signatory states would undertake, within six months, to enact national legislation severely punishing any breach of the treaty, which would be considered a crime against humanity. Being national legislation, the signatory states would themselves have the duty of ensuring observation of the treaty. In other words, this was to be self-control, the very opposite of international inspection.

These two positions seemed irreconcilable from the start. The Soviet Union was supported only by Poland, whereas the United States was backed by the other 9 countries of the U.N. AEC, which was comprised of the 11 members of the Security Council plus Canada. This was the period when any U.S. proposal in the United Nations was almost automatically adopted by a considerable majority of votes, which did nothing to allay Soviet mistrust.

Certain countries, like France, tried to find some sort of compromise between these opposing points of view, and in late July the French representative Alexandre Parodi proposed that the AEC set up a scientific and technical committee with the mandate of making some form of preliminary analysis to establish whether, in fact, an effective control of atomic energy was possible.

The Bikini Tests

Before proceeding with the story of these U.N. negotiations, it is necessary to make a detour via the Pacific, for some of the experts who were to take part in the negotiations were many thousands of miles from New York on the high seas aboard an American warship. They were on their way to witness a full-scale naval atomic exercise on Bikini Atoll in the Marshall Islands involving the first nuclear bomb explosions since the end of the war.

Altogether, this political Noah's Ark was comprised of 22 foreign representatives (2 from each member country of the U.N. AEC, including the Soviet Union and Poland) invited as observers by the State Department despite a marked lack of enthusiasm by the navy.

The U.S. Navy had no share of the wide praise and publicity given to the army for production of the bomb and to the air force for its delivery. To remedy this, the navy had decided to organize tests to study the effects of the bomb on a fleet made up of various units from the many then waiting to be scrapped.

The decision to hold this first, and last, public demonstration of a nuclear explosion had been made as far back as late 1945, and neither President Truman nor the State Department subsequently raised objections to its being held at the very moment when the first negotiations on international control were beginning in New York.

In addition to the 42,000 men responsible for the organization of these exercises, the tests were watched by press representatives from all over the world, as well as the 22 official foreign observers. It was in this last capacity that I witnessed the tests, the holding of which would have been fiercely opposed by today's public opinion, and which the world press at the time reported in an exaggerated and unpleasant manner.

Two bombs, of comparable power to those used over Japan, were experimentally exploded; one at a height of a few thousand feet, the other about 30 feet below the surface of the sea. The airborne test in July 1946 was comparatively unspectacular and therefore somewhat disappointing, but the underwater explosion, set off by radio from some 20 miles away, was totally different. It produced an immense column of several million tons of water, which lifted, then smashed like a fragile straw, an old battleship lying 400 yards from the center of the explosion. This gigantic column, one and one-half miles high and nearly half a mile in diameter and looking like an enormous oak tree, produced at its base a wave over 100 yards high, swamping the assembled ships and sinking several of them. The whole supernatural and unforgettable scene, lit by brilliant sunshine, lasted for nearly a minute.

On my return to Paris, I described the terrifying effects of the atomic weapon at a meeting that, despite its being chaired by Joliot-Curie, led to my being branded as a bogeyman by the Communist press, which was then trying to play down the American military lead.

Some U.S. leaders were pleased by the coincidence of the tests with the first meeting of the U.N. AEC, seeing in them a means of encouraging the policy of international control by impressing foreign observers and world public opinion. The national defense saw it as an opportunity to display the power of its new weapon, whereas the navy wavered between this point of view and a desire to prove that its role was not too seriously affected by the advent of the bomb. And responsible American admirals had not yet fully realized that fission, thanks to its use in a nuclear submarine engine, would within 10 years bring about a complete revolution in their service, as well as a certain degree of revenge on the army and air force that the Bikini tests had failed to provide.

By an ironic turn of fate, today's only well-known reminder of this giant and unhealthy demonstration is the name of a charming item of feminine apparel.

The U.N. Negotiations

In August I returned to New York, together with several other scientific observers who had witnessed the Bikini tests, to take part in the U.N. AEC's negotiations on controlling the new force whose fearsome power we had seen with our own eyes.

This was also my first experience with a large international organization, a world apart in which it takes some time to get use to the complex rules of play. No less complex is the structure: the public stage with actors delivering successive monologues; in the wings a whole battery of commissions, committees and subcommittees, working groups and groups of experts; the system of resolutions and amendments with such importance attached to their wording and the sense of each word; separate negotiations among groups of countries and votes where one or another power bloc can exert special pressures; not to mention the difficulties of communication across language barriers, which can occasionally defeat the most competent interpreters.

Gradually, one learns to appreciate and value this artificial world. Initial doubts as to effectiveness are replaced by the realization that these contacts between opposing nations — contacts that otherwise would often not take place at all — can only be beneficial.

The early deadlock between the American and Soviet positions was hidden for some time by the work of the experts. In early September, they presented their report, the conclusions of which seemed acceptable to the Russians; there was nothing in the available scientific data to prevent the creation of a technologically feasible and effective system of safeguards. However, the report avoided the question about whether such a system would be politically acceptable, and no recommendation was made as to how satisfactory guarantees might be assured.

Following this report, informal talks began on the controls to be applied to different stages of the fuel cycle. They concluded that while a simple inspection system would be adequate for the early stages of the natural uranium cycle, from mine to fuel, management by an international authority would be necessary for isotopic separation plants, plutonium extraction units, and certain types of reactors. The Soviet expert even contributed to the definition of the international management that the future authority should provide. There is a remarkable similarity between these conclusions and current trends relating to safeguarding sensitive stages of the fuel cycle.

The honeymoon, however, was short. By the end of October, the verbal hostilities of the Cold War had again appeared in the General Assembly in the form of a violent attack by Molotov on the United States. He proposed that the atomic question should be dealt with in the framework of general disarmament negotiations, evidently a way of postponing the affair indefinitely.

Baruch counterattacked. He tried to get the Russians to make their attitude toward the American plan, with its clause abandoning the right of veto, unambiguously clear. A vote was taken on December 30, 1946. The following week Baruch resigned. He had secured 10 favorable votes against 2 abstentions, by the Soviet Union and Poland. Some 20 years later it became known that, just a few days before that vote, the first Soviet reactor had begun operation. The United States had mistakenly believed that, confronted with the power of their bomb, the Soviet Union would accept their proposals. But instead the Russians decided to put their trust in their technicians and equip their army and defense services, the most powerful in Europe, with the same devastating weapon.

The Baruch Plan, by now the "majority plan," was nonetheless studied in detail throughout the whole of 1947, under the ironic eyes of the Soviet representatives who from time to time highlighted an evident fault in the structure being developed by the Western experts. The exercise was clearly hopeless.

But although there was no likelihood of the Soviet Union accepting its conclusions, the task was nevertheless fascinating, generating an absorbing intellectual game. Even within the majority, agreement was sometimes difficult to achieve. For example, many sessions were devoted to deciding whether or not uranium ores still in the ground should belong to the international authority. In the end, as a concession to the nations owning the mines, the draft treaty gave them the right to retain ownership of their ores so long as these were not mined. The international authority was then to be entrusted with deciding, each year and for each country, how much ore could be mined and what quantities of concentrated fissile materials should be produced.

This was nonetheless the first international evaluation of the various stages of the nuclear fuel cycle from the viewpoint of safeguards. Some of its conclusions have lost none of their relevance today; they are given in a report to the Security Council that still makes the most instructive reading in the current situation. This evaluation clearly demonstrates the conflict between the requirements of security and those of a broad development of civil applications of atomic energy. It is particularly obvious in relation to the storage of fissile materials: if prohibited, there is the risk of impeding peaceful developments; if authorized, materials that could be used for military purposes become available to all. The conclusions emphasized that the possibility of "denaturing" fissile materials to make them unsuitable for military applications is far less certain than the Acheson-Lilienthal Report had suggested.

The evaluation deals for the first time with the important role of plutonium in obtaining energy from its alchemic parent, the abundant uranium-238 isotope. This is the process known as "breeding," which is vitally important for the future of atomic energy. The process involves "burning" plutonium in reactors in the presence of uranium-238 (or natural uranium) so that the excess neutrons create fresh plutonium in quantities equivalent to — or even slightly greater than — those consumed. The new plutonium, once it has been recovered by chemically reprocessing the irradiated uranium, can in turn be used under the same conditions. Thus, both the isotopes in natural uranium become potential sources of energy: the uranium-235 directly and the uranium-238, which is 139 times more abundant, after it has been gradually transformed into plutonium by successive stages.

Finally in this report, uranium enrichment and plutonium extraction were considered, as they still are today, the most sensitive operations from the viewpoint of possible diversion.

Other countries in the U.N. AEC were in the same position compared to the United States, the United Kingdom, Canada, and to a lesser extent France, as many of the developing countries are today when similar negotiations are taking place. They had no way of launching into this extraordinary field of activity that the United States and its allies had explored during the war, and the only knowledge they could have of the general technology involved came from what the Americans had chosen to release. The economic and political integration proposed in the "majority plan" did indeed open highly promising prospects for these countries, but only in exchange for relinquishing a part of their sovereignty, an act made only slightly easier by the fact that the Great Powers would have to do the same.

A partial application of this plan to those countries that were prepared to accept it (i.e., those countries that were not under Russian influence at the time) was unthinkable in a plan whose main objective was the complete elimination of atomic weapons through the universal internationalization of nuclear activities. At the same time, the United States could not consider putting their atomic installations, essentially military at this time, under international management unless the Soviet Union agreed to do likewise.

In June 1947 the Russians took a step forward and proposed to the United Nations an international control of all civil nuclear activities from mining to the production of fissile materials and energy, these activities being subjected to periodical international inspections, the application of which could, however, be vetoed in the Security Council.

The Americans considered such a proposal inadequate. Had it been made a year earlier it could have opened definite possibilities for negotiations. In spirit it was hardly different from the Non-Proliferation Treaty negotiated 20 years later. Had it been adopted at this earlier moment and provided the Soviet Union, taken at its word, had not withdrawn, it would have resulted in that country being opened to international safeguards in an ideal world comprised of only nonnuclear weapons states, all of which would have been party to a truly universal nonproliferation treaty.

In spring 1948, after two years of work and more than 200 sessions and on the eve of further American atomic tests in the Pacific, the U.N. AEC finally admitted it had reached a deadlock and stopped work. The first attempt to achieve international nuclear disarmament had failed.

The American and Russian positions had remained irreconcilable. The United States, before giving up their bombs, wanted first to ensure the establishment of an international management system throughout the world and particularly in Russia while the Soviet Union wanted first to see American nuclear disarmament and then to consider international control.

It is understandable that the Soviet Union, in the absence of mutual trust between the two Great Powers, could not accept the American plan, which in Russian eyes would have meant mortgaging their future security. The secrecy covering the location of Soviet industrial centers was at the time the Russians' best protection against possible atomic attack by the Americans, thus they were quite naturally anxious to avoid making these centers easy targets in case a conflict should arise. (Today, primarily as a result of U-2 aircraft flights and more recently of spy satellites, this secrecy no longer exists.)

The failure of the Baruch Plan, despite its utopian aspects, was a severe setback on the road to world peace and the evolution of humanity. It would in fact have been easy, had the American plants been opened to international inspection at that time, to know the precise number of the few dozen bombs manufactured since 1945.

Today this is out of the question, with tens of thousands of far more destructive bombs now present in the nuclear arsenals of the world's two most powerful nations. It is unlikely that there can ever again be complete trust until the world is finally united as one nation. For now, should international nuclear disarmament ever be attempted, there will be no way of checking the accuracy of weapon stock declarations by a country possessing vast numbers of such weapons. A country could, therefore, always keep back a clandestine and undetectable reserve. The last chance for returning to a world free from atomic bombs, and free from discriminatory measures aimed at nonproliferation, has vanished for a very long time.

The thread that might have led from a supranational authority in this sector to a world government was broken. The hope of living in a world free from nuclear weapons must now await the day, no doubt far in the future, when by an opposite approach an already-established world government may lead to complete suppression of the bombs.

Indeed, as long as there are sovereign states liable to engage in armed conflict with one another, no system will be materially able to stop them from making military use of the resources of nuclear or any other science or technology if they feel that their existence or their freedom may depend on their doing so.

2. Monopoly Lost

In the four years following the failure of negotiations in the United Nations on international controls for atomic energy, two countries were to join the Americans as nuclear weapons powers: in 1949, their rival in the Cold War, the Soviet Union, then in 1952 their closest ally, Great Britain.

The first of these developments was a complete surprise for the American government; the second, although expected, was hardly a less traumatic experience. It was a flagrant demonstration that the policy of secrecy had failed. The Americans, to maintain their lead, would now have to develop and produce more powerful bombs in even greater quantities.

The McMahon Act

For American nuclear development, the early postwar years had been a period of hesitation, difficulties, and even deterioration, linked with the industrial demobilization of the country, with the absence of any apparent competition, and with the maintenance in peacetime of the policy of secrecy.

The setting up of a legal regime designed to govern atomic energy in the United States in peacetime had led to a full-scale public political battle, the first of its kind to be experienced.

In October 1945, a draft law, the May-Johnson Bill prepared by the War Department, was submitted to Congress. It envisaged a commission nominated by the president — and in which there could be participation from the military — which would have extensive powers that would effectively give it absolute control of all scientific research in nuclear physics. An accelerated procedure was begun to ensure its rapid adoption.

As soon as the content of this bill became known, the scientific circles concerned rose in arms against it. They saw in it a continuation of army control, an excessive maintenance of secrecy, and loss of freedom in their work all the way back to basic physical research.

The call for revolt came from Samuel K. Allison, one of the directors of the Chicago nuclear center, who declared that atomic scientists would prefer to study the color of butterfly wings rather than continue to work under the conditions of secrecy imposed by the army. The scientists during the war had tolerated — albeit under protest — the measures imposed by General Leslie R. Groves, in particular his favorite method, compartmentalization. Because of this, research scientists had complained they were sometimes led to duplicate work already done, while at other times they were prevented from linking up with relevant research elsewhere.

This dissatisfaction led to the scientists in the various nuclear centers setting up pressure groups, which had soon joined together to form a single association, the Federation of Atomic Scientists. This body acquired a considerable influence over public opinion and among the politicians. For the first time in history, scientists had shown themselves determined, through the spoken and written word, to influence the political consequences of their discoveries.

Their first success was to prevent the rapid adoption of the May-Johnson Bill proposed by the military. Then in late 1945, an external incident was to provide them with appreciable support. The press published accounts of the dismantling by the American army of the five cyclotrons existing in Japanese universities. The component parts of the cyclotrons had been thrown into the sea. One of these instruments of pure research was a superb machine purchased from the United States just before the war. Groves was obliged to admit that he personally had given the order for this stupid act, at the very moment when the Supreme Allied Command had independently authorized the use of the same large cyclotron for biological and medical research in Tokyo.

With support from the scientists' movement, the young senator Brien

McMahon persuaded the Senate to set up a commission, comprised of senators from both parties, to study future atomic legislation.

This commission met from November 1945 to April 1946 in public and secret hearings to gather evidence from all who had held positions of responsibility in the American atomic adventure.

The scientists tried hard to demonstrate that there was no real "atomic secret," either for the bomb or for the production of energy. The basic scientific data were known to physicists throughout the world and the main technical secrets concerned industrial developments that were often delicate, tedious, and costly to implement but would always be within the capability of a large country determined to make the necessary effort. Leo Szilard, in his evidence, cited as an example the success of my own Montreal team, independent of the Americans, in the plutonium extraction field.

Despite the efforts of supporters of army control, the influence and arguments of the scientists won the day, and President Harry S. Truman made it known that he was in favor of a commission comprised of only civilians. This tipped the balance, and the McMahon Act was finally adopted by Congress in late July 1946.

Although this was indeed a victory for civil over military influence, the real winner in the affair was Congress, for the McMahon Act created a Joint Committee for Atomic Energy (JCAE), comprised of 18 members from the two houses of Congress and both political parties, whose task was to oversee the activities of the new U.S. Atomic Energy Commission (U.S. AEC). The JCAE was to have a considerable influence on all future American nuclear policies.

Under the McMahon Act, all questions relating to atomic energy were made the responsibility of the new U.S. AEC, which was to be comprised of five civilian commissioners nominated by the president with Senate approval. Everything concerning atomic energy, from uranium ore to nuclear fuel, was to come under the authority of, and become the property of, the U.S. AEC. Secrecy was to be maintained, however, and the death penalty was prescribed for anyone found guilty of passing information to a foreign power, even in peacetime. A Division of Military Application, directed by a high-ranking army officer, was to be responsible for the production and testing of weapons.

In external relations, the new act led to atomic isolationism. It was in fact specified that the U.S. AEC's policy should be to control the dissemination of classified information in a manner favorable to defense and security. To ensure this, the following guidelines were laid down:

(1) That until Congress declares by joint resolution that effective and enforceable international safeguards against the use of atomic energy for destructive purposes have been established, there shall be no exchange of information with other nations with respect to the use of atomic energy for industrial purposes; and

(2) That the dissemination of scientific and technical information relating to atomic energy should be permitted and encouraged so as to provide that free interchange of ideas and criticisms which is essential to scientific progress.

The American Reorganization

The U.S. AEC took over nuclear duties from the army on January 1, 1947. During the first months of that year David E. Lilienthal, nominated by Truman as chairman of the AEC, underwent a protracted and distressing ordeal in the Senate before that body was prepared to confirm his appointment. His success as administrator of the most important government-owned enterprise in America, the Tennessee Valley Authority, as well as the part he had played in the development of the earlier plan for international management and control of atomic energy, had made him a *bête noire* of those who opposed any form of nationalization in the electricity industry. His confirmation in office was secured only after three months of unpleasant hearings not far short of a witch hunt.

The civil reorganization of the nuclear enterprise took place, not without difficulties, during 1947. Some industrial groups had withdrawn, wishing to return as quickly as possible to peacetime activities, while many scientists who were still on the threshold of their careers (80 percent of the scientists involved in atomic research during the war were under 35) went back to their university work. In peacetime, they had found it difficult to accept the disconcerting aspect of work in which, having devoted extraordinary efforts to wresting from Nature some of her most enigmatic secrets — those of the structure of matter — the successful research results were then locked away in the strongboxes of an artificial, man-made secrecy.

Many leading figures in the atomic venture, such as J. Robert Oppenheimer, Glenn T. Seaborg, Edward Teller, and others, also went back to their university posts, although they all maintained links with the U.S. AEC and their former laboratories, to which they even returned to work, sometimes for several weeks at a time. Thus they remained well-informed advisers on developments in work they had previously initiated and directed.

The annual budget for the whole of the atomic enterprise was sharply reduced in 1946 and 1947, increasing again in 1948. It was not until the end of 1949 that, in the face of Soviet competition, the budget again reached the levels in personnel and investment that it had at the end of the war.

In the early stages of the Cold War, the main objective was to produce atomic weapons despite limited uranium supplies, the problems of producing fissile materials, and the need to improve a rather primitive understanding of the mechanisms of the nuclear explosion itself.

The first three bombs, the one used for the New Mexico test and those exploded over Japan, were the only ones the Americans had at the time of their use. They had been "handmade" at Los Alamos, and no arrangements for production in series had been foreseen. When such production began after the war, it quickly ran into so many snags that for a time it was virtually at a standstill.

By a strange interplay of technology and politics, at the very moment when the Soviet Union was fruitlessly calling for an end to American production of bombs, and others were suggesting this as a goodwill gesture in the negotiations on international control, the veil of secrecy prevented the world from seeing that this production had, in fact, practically stopped.

Nearly two more years were needed to set up, at the Sandia Base near Los Alamos, a research and production establishment to develop and make more advanced weapons in quantity. A test center was built on Eniwetok Atoll in the Pacific and three bombs, the most powerful of which had three times the explosive power of the Nagasaki bomb, were exploded in the spring of 1948 at the same moment when the international situation over the Berlin blockade was getting worse. The knowledge gained from these tests was to make possible the production of more powerful and compact weapons, which were at the same time more economical in their requirements for fissile materials.

The reorganization of production plants for fissile materials progressed in step. For the uranium-235 that had been used in the Hiroshima bomb, an acrobatic combination of three methods of isotopic separation had been necessary and three different installations had respectively contributed to the different stages of the process. Comparison of the economics of the three methods favored the gaseous diffusion process, and the corresponding plant was therefore extended to embrace the entire operation, while the other two installations were closed down.

Production of plutonium had meanwhile been dangerously threatened in 1947 following a process of aging in the basic materials used in reactors (particularly graphite) when subjected to radiations, which led to mechanical deformations. The huge atomic piles came close to being abandoned, but appropriate modifications became possible and, from 1948 onward, the reactors returned to a regular operating schedule.

Another source of plutonium for the American weapons was Canada. Besides directly supplying uranium, the Canadians also made available, for extraction of the contained nuclear explosive, irradiated uranium from their two large heavy water research reactors that they had built with U.S. assistance.

The supply of uranium, despite the Canadian contribution, was another cause of concern for the Americans, for the resources of the Belgian Congo had to be shared with the United Kingdom. The U.S. AEC therefore launched a campaign to encourage home prospecting in Colorado and Utah, while talks with South Africa were also started by the joint Anglo-Saxon supply agency with a view to uranium extraction from gold-bearing sands. These normally have a low uranium content; however, that uranium could be comparatively easily extracted by chemical means.

The problem of controlled energy production was of course already being studied, together with its possible applications in the military field: nuclear propulsion for aircraft and ships. However, during the years immediately following the war, the weapons program retained its priority option on the limited uranium resources available.

Thus it was in 1949, with work started on fundamental atomic reactor prototypes for possible energy production and with increasing numbers of more effective bombs being made, that the world was surprised and startled by the first Soviet atomic explosion.

The Soviet Breakthrough

Four years after Hiroshima, the balance of world power was to undergo the first stages of a fundamental change. In early September 1949, an American meteorological reconnaissance aircraft, specially equipped to analyze atmospheric radioactivity, detected a slight but abnormal increase during a routine flight between Japan and Alaska. The aircraft was part of a long-range detection system recently established for this purpose.

Washington immediately ordered all available military aircraft to the area, as well as stations equipped with meteorological sounding balloons, to take samples. Chemical analysis quickly confirmed what Soviet scientists and industry had achieved; the radioactivity could not have been produced other than by the explosion of a plutonium bomb.

It had taken place on August 29 in Siberia, near Semipalatinsk. The

British, warned of what to look for, detected the radioactive cloud over the Atlantic during its rotation around the earth, confirming the American conclusions.

The most difficult person to convince was Truman who, as he said, could not come to terms with the idea that "these Asiatics" could build something as complex as the bomb. Initially reticent he finally agreed — particularly to avoid an indiscreet leak to the press or a Soviet announcement — to take the first step and make public this new and sensational factor in world politics. He made the official announcement personally on September 23.

The surprise with which this event was greeted was proof of the effectiveness of Soviet secrecy. The American leaders had made two errors of judgment: They had underestimated Russian industrial potential in a field that certainly had the highest priority, and they had overestimated the effectiveness of their own atomic secrecy that they had tried to impose.

Although most of the scientific experts had evaluated the time it would take the Soviets to produce the bomb at three to six years, government authorities had forgotten these forecasts, while many scientists, failing to appreciate the passage of time, had continued to believe that this important event was still several years away.

The Soviet technicians had followed avenues of research similar to those of their American predecessors, all clearly outlined for them in the Smyth Report, which also indicated the pitfalls to be avoided.

As early as 1944, the Soviets had attacked the problems of producing pure uranium and graphite. The American success in 1945 had resulted in the highest priority being given to plutonium production, studies relating to the bomb itself, and the construction of a test site. On Christmas Eve, 1946, the first Soviet reactor had achieved criticality, almost four years to the day after the Fermi reactor, which it closely resembled. This 4-year gap had remained, which meant that the construction of plutonium-producing reactors, and of a plutonium extraction plant, as well as the final development of the bomb, were all achieved at the same remarkable pace as the corresponding American operations.

The gaseous diffusion process for isotopic separation of uranium-235 was to be developed in parallel. Starting late in 1949, the Russians began design studies for the construction of their first atomic power station and several experimental reactors, all using enriched uranium. This indicates that the isotopic separation plant must have been achieved some two to three years after the industrial production of plutonium. The man responsible for the overall project was the redoubtable Interior Minister Lavrenti P. Beria.

In step with each achievement of a further technological success, the

Russian attitude became more and more inflexible in world politics, and indeed in late 1947, Soviet Foreign Minister Vyacheslav M. Molotov had announced in the United Nations that the Soviet Union also possessed the secret of the atomic bomb. Resulting from this, and in answer to Truman's announcement in September 1949, two days later the Tass Agency confirmed that the Soviet Union already had the weapon at their disposal and explained — in an embarrassed fashion — that explosions related to important work in the Soviet Union might be detected beyond their frontiers.

Although some 250 German scientists and technicians (some of indisputable merit) who had been captured during the occupation of their country at the end of the war were employed in the Soviet nuclear enterprise, their participation does not seem to have been a decisive factor in its success. They had been installed at Soukhoumi, on the Soviet Black Sea, an area comparable to the French Riviera, where a large laboratory had been set up for them in an old sanitarium. They were employed on nonsecret projects that were marginal to the main venture — measurement and analysis work and electronics — thus freeing Soviet physicists for more important and secret research. Ten years later, the captured scientists were given permission to return to Germany; only three of them, who had married Russian women, remained in the Soviet Union.

The Russians undoubtedly received a certain amount of help from information passed to them by pro-Communist scientists who had participated in the wartime Anglo-Saxon program. The two men mainly responsible for these leakages were from Britain. The first incident goes back to 1946, when it was discovered that Alan Nunn May, one of the leading British physicists in the Anglo-Canadian outfit, had passed important information to the Soviet military attaché in Ottawa toward the end of the war. The importance of this act was lessened by the publication of the Smyth Report. When arrested, May admitted the offense and was sentenced to 10 years imprisonment for violation of the Official Secrets Act, although what he had done was in favor of a country that was then an ally.

The ineffectiveness of the protection of secrecy became even more obvious when, in February 1950, the theoretical physicist Klaus Fuchs, who was of German origin and one of the leading scientists in the British group, admitted communicating to the Soviet Union all the information he had acquired since 1942. This information included important details concerning the isotopic separation of uranium, and vital data acquired in the United States at Los Alamos in 1944 and 1945, where Fuchs had contributed to the development of the first bomb. He had even taken part in highly secret discussions on the possibility of a hydrogen superbomb.

Son of a Protestant minister in Leipzig, East Germany, Fuchs had gone to England in 1932 at the age of 21 as a refugee from Nazism. The British security services, aware of his Communist affiliations when he was a student in Germany, had failed to warn the Americans when he was transferred to the United States. After the war he became head of the theoretical physics division in the British atomic program.

Fuchs was sentenced to 14 years in prison, after a somewhat unusual trial in which the accused himself provided the only proof of his guilt. Released after serving two-thirds of his sentence, he was allowed to go to East Germany where he became deputy director of the National Institute for Nuclear Physics.

However, there was no U.K. monopoly of atomic espionage for, in 1944 at Los Alamos, David Greenglass, a skilled mechanic, had passed to a Soviet agent (the same one who was Fuchs' correspondent in New Mexico) details of the mechanism used to obtain critical mass in the plutonium bomb. Greenglass was the brother of Ethel Rosenberg, whom he denounced together with her husband Julius, claiming they were responsible for his own act, which had none of the importance of Fuchs' revelations and was probably superfluous in view of them. Greenglass provided the main evidence on which the couple were sentenced to death in 1951 and executed in 1953 for espionage on behalf of the Soviet Union, former ally during the hard years of the fighting war, but now an enemy in the Cold War.

The Rosenbergs' sentence, at a time when McCarthyism in the United States was in full cry, was in singular contrast to the punishment inflicted by the British for similar offenses. It struck down equally a husband and his wife (whose guilt was conceivably not equal) and who protested their innocence to the last, and who refused an offer of pardon in exchange for information on the espionage network used. The affair stimulated world-wide emotional reproaches, and was widely used in anti-American propaganda.

Whatever the significance of these external contributions, the Soviet achievement in producing the bomb in four years was an unquestionable triumph. Such success, in a country then undergoing complete reconstruction, could not have been possible without a very considerable effort and the highest priority at both scientific and industrial levels.

This achievement was a striking demonstration of the ineffectiveness of the policy of atomic secrecy. Still more, that relatively useless policy vis-à-vis the Soviet Union had also impeded the postwar reorientation of the American atomic program and its subsequent development.

This was clearly confirmed by the reaction to the Soviet achievement of the U.S. AEC's own chairman. Lilienthal declared that the news of the

Russian explosion proved it was time to stop the senseless stifling of American effort by the excessive secrecy then being enforced; an excess that was harming both technical progress and national defense.

The policy of nontransfer of information was no more effective vis-à-vis the United Kingdom. In the early postwar years, it was to contribute to a fresh episode in the delicate atomic relations between the two Anglo-Saxon allies. To follow this, we must first take a step back in time to 1945.

Anglo-American Traumatism

Before the end of the tripartite (United States, United Kingdom, and Canada) summit meeting at the White House in November 1945, British Prime Minister Clement R. Attlee proposed to President Truman the establishment of a new framework for future atomic collaboration between the Anglo-Saxon allies. This proposal was brought about because the 1943 Quebec Agreement, which had been the basis of collaboration until then, had been essentially designed to cover the wartime situation, even though certain clauses were intended to apply after the end of hostilities.

There was, of course, the clause relating to the handing over of American industrial data to the United Kingdom after the war, but Churchill had left this matter entirely in Roosevelt's hands. The American president, at Hyde Park in September 1944, had personally promised to the British prime minister a close collaboration after the war between the British and American governments in developing ''Tube Alloys'' for both military and commercial purposes. Now Churchill was no longer prime minister, and Roosevelt was no longer alive. Furthermore, this Hyde Park memorandum was hardly binding on Roosevelt's successor.

Prime Minister Attlee, whose government did not conceal that it now intended to develop the production of fissile materials, obtained complete satisfaction; on November 16, 1945, the three heads of state signed a document approving the principle of extensive cooperation. ''We desire,'' they declared, ''that there should be full and effective collaboration in the field of atomic energy between the United States, the United Kingdom, and Canada.''

This document was not produced without difficulty for once again, in the course of the discussions, General Groves had opposed the principle of full collaboration. Neither was Vannevar Bush, responsible for the American national research program, any more favorable toward the idea, fearing

now that too close a link between the United States and the United Kingdom
(a link that he had been the first to propose, in 1941, and thereafter had tried
to prevent) would decrease the chances of success in future negotiations
with the Soviet Union on the question of worldwide atomic energy control,
which to him was a matter of primordial importance.

On the other hand, the joint purchasing policy for uranium available
outside the partners' territories was extended without argument, since it
was very much to the advantage of the United States. Now that the war was
over, however, the United Kingdom would soon be claiming for its own
purposes its half-share of the production from the Belgian Congo.

This possibility was no more attractive to General Groves than was the
prospect of ''complete and effective'' collaboration. Such an agreement
seemed to him quite unacceptable, for he saw in it the equivalent of a true
military alliance, which would make it easy for Britain to build the necessary installations for producing nuclear explosives. Groves therefore decided to sabotage the project, which was of course ''top secret,'' for at the
time Congress itself was unaware of the existence of the 1943 Quebec
Agreement and the American obligations toward the British.

Grasping at every straw, Groves pointed out to Secretary of State
James F. Byrnes that the proposed agreement could not remain secret since
the U.N. Charter required, under clause 102, that any international treaty
should be published. As a result the whole world would see that the United
States, while pretending to seek the establishment of international controls
on the civil development of the new energy source, was at the same time
helping another state in its ''nonpeaceful'' development.

Such a resort to legal argument was unexpected, particularly from
someone who had never hesitated to treat the Manhattan Engineer District,
for which he bore the extremely heavy responsibility, as a separate entity to
which normal rules did not apply especially in matters of secrecy, which he
had turned into a religion.

But it would be enough if he could obtain a few months delay since
internal legislation on atomic energy then under discussion was soon to
provide a more solid basis for no longer honoring collaboration promises
made by Roosevelt, or later by Truman, to British prime ministers.

The question of collaboration thus remained in suspense, and Attlee
eventually raised it again with Truman in April, asking that the directive of
November 16 be applied. Truman's answer was short and clear: The phrase
''complete and effective collaboration'' approved five months earlier was
in his opinion vague and could be applied only to purely scientific data. He
had certainly never contemplated, even at the time when the allies were

working together to seek a worldwide control agreement, the conclusion of a pact whereby the United Kingdom would be helped to establish the necessary industrial complex for manufacturing atomic bombs.

Thus, three and one-half years after Roosevelt had agreed, in the midst of an all-out war, to the first rupture with the British, his successor had repeated the act, again citing as his reason the distinction between fundamental scientific data and industrial information.

Shortly afterward in July 1946, Congress voted to adopt the McMahon Act, and so provided decisive arguments for the opponents of Anglo-American entente. Internal legislation and the fear of congressional reaction were the excuses put forward to justify breaking earlier engagements on atomic energy. It was not to be the last time in the history of American nuclear policy that this happened.

In fact, Congress on this occasion certainly had reason to complain. Nobody had dared to inform it of agreements signed during the war or of others since envisaged. What is more, an undertaking by an American president is not legally valid unless it has been approved by Congress.

As already mentioned, the new law authorized the exchange of scientific data but prohibited, while awaiting the establishment of an effective international safeguards system, any transfer of data relating to industrial uses. Here again we find the distinction between information of industrial nature and that of scientific and technical nature, which President Truman had invoked to release himself from his earlier engagement toward the British prime minister.

In reality there was no such thing, at this time, as an ''industrial'' use of atomic energy. Only certain industrialized stages existed, such as isotopic separation plants, plutonium-producing reactors, and plutonium extraction plants, all indispensable for making nuclear explosives but only at a later stage, possibly, for the production of usable energy.

The British Decision

While the United States was thus building their legal wall of secrecy that was to be the basis of their nonproliferation policy, and while the Soviet Union was pursuing by forced marches the road to atomic weapons, the United Kingdom, quietly and without fuss and without informing the public, although this would certainly not have caused any difficulties, launched its own program in the same direction.

Under a law of November 1946, the entire responsibility for Britain's civil and military atomic energy development was entrusted to the government, which in fact already held the monopoly. During this same year, the Defense General Staff, together with the scientists responsible, had been making their estimates of the number of bombs required and the time necessary to produce them, using one or possibly two graphite reactors similar to those in the United States. The possibility of awaiting the outcome of international negotiations before launching the program, which in any case would take about five years to yield results, was never considered. The three service chiefs, fully appreciating the dissuasive power the new weapon would provide, although unable to quantify that power, were thinking in terms of a few hundred bombs rather than just several units or even some dozens. They envisaged an initial annual production of about 15 units.

The breakdown of the most recent overtures toward the Americans and the adoption in Congress of the McMahon Act reinforced the determination of the British Labour government under Prime Minister Attlee. In early 1947, without any international control agreement, Attlee made the positive decision to develop, in the greatest secrecy, the production of nuclear weapons.

In passing, it is interesting to note that the British decision to "proliferate" was followed within a month by undertakings to the opposite effect from several other countries. In February 1947, in fact, 21 years before the signing of the Non-Proliferation Treaty, five countries became the first to renounce possession, production, and testing of atomic weapons. True, this was hardly a voluntary renunciation, being a condition imposed by the Allies as part of their peace treaties with Bulgaria, Finland, Hungary, Italy, and Rumania. The 1955 peace treaty with Austria was to include the same clauses. As yet there was no problem with either Germany or Japan, for both were occupied by the victorious armies.

Plans for the British program were soon decided. It was not to follow exactly the same pattern as the Americans, and in particular, rather than cooling the plutonium-producing reactors with water and discharging the heat into a river, it was decided to extract this heat at a higher temperature using air. This was a first step toward recovery of energy and the production of electricity.

Hence, the British atomic program began in an essentially military direction, though in no way forgetting possible civil applications. The scale of the enterprise and the resulting impossibility of concealing it from public opinion made it more and more difficult to maintain secrecy as was originally intended. The government therefore decided on an official announcement, though without publicity, of the existence of a military program. In May 1948 the minister of defense arranged for a question to be asked in parliament as to the state of advancement of production of the most modern weapons. He was then able to reply that all types of weapons, including atomic weapons, were under development.

The following day the London *Times* gave a few lines to this moment-ous announcement, which otherwise went virtually unnoticed. With the start of the Cold War following the Communist takeover in Prague and on the eve of the Berlin blockade, both the British parliament and public opinion found it quite natural that their country should become a nuclear weapons state like the United States. Possession of the bomb was a symbol of prestige and a sign of power that seemed perfectly normal for a nation that as yet had not realized it had lost its empire and its former world status.

It was therefore in a climate totally different from today's psycho-political environment that the British government began its development of atomic armaments. It even believed it was a few years ahead of the Soviet Union, and was not in the least concerned over the prospect of being the second country in the world to possess the bomb, therefore setting the first example of proliferation, making it more difficult to persuade other countries not to follow.

Previously, in 1947, following its firm decision to launch a military program, London made an effort to reopen negotiations with Washington, seeking now to include nuclear collaboration in a general defense agree-ment that would have been compatible with the new American law.

It was now the Americans who were asking for concessions. It was becoming increasingly difficult for them to accept the British right of veto on the use of their principal instrument of defense. Furthermore, their accelerated bomb production, in parallel with the intensified Cold War, had increased their uranium requirements so that the time was approaching when these could not be met unlesss the British gave up their share of the Belgian Congo production, which at that time was stored in England and largely unused.

Any new agreement with the British would of course have to be approved by the recently created congressional JCAE, already well on its way to becoming the U.S. AEC's "supreme commander." Thus it became necessary in the spring of 1947 for the American leaders concerned to pluck up courage and reveal the existing links to the leading members of the JCAE. The latter were both shocked and outraged to learn of their country's double dependence on the United Kingdom, both for nuclear raw material and for utilization of the finished product. They found the Quebec Agree-ment intolerable, and proposed to President Truman that he immediately cancel it, in exchange for the financial aid to be given under the recently announced Marshall Plan. This was of course quite impossible.

The Anglo-American Modus Vivendi

In late 1947 new negotiations were opened between the two wartime allies. The Quebec Agreement was ''adjusted.'' The British right to veto American use of the bomb was abolished but the reciprocal restriction on the transfer of information to third parties was maintained. The joint supply agency (Combined Development Trust) was also maintained, symbolizing the importance of uranium in Anglo-American relations. However the British agreed to let Washington have their 1948 and 1949 shares of the Congo uranium, and if necessary to hand over part of the stock held in the United Kingdom. This important concession was made in view of the joint national defense requirements. In return, various technical areas were to be reopened to the British so that ''inter-ally'' exchanges could be resumed. The technical areas concerned were ones in which the British had themselves made their own contribution, such as that of plutonium extraction, which had not been part of the collaboration agreed on at the Quebec Conference. None of these areas came anywhere near the design of the bomb.

This series of agreements, or rather declarations of intent, was concluded in early 1948 and referred to as a ''modus vivendi,'' which, unlike an international atomic agreement, could be exempted from submission to Congress. Had it been effectively respected, it would certainly have helped the independent British program; but in fact it went the same way as many of those distressing deals painfully negotiated between divorced parents for the care of their children and the division of their resources. The Americans were very soon largely ignoring it, both in letter and in spirit, and it proved extremely disappointing for the British.

However the British, no longer concealing their interest in producing a weapon, now proposed — in the autumn of this presidential election year — that an information exchange on atomic weapons should be set up. The moment was well chosen for with the reelection of President Truman during the middle of the Berlin blockade, combined with a Republican defeat in Congress, the political climate in Washington had become less isolationist and less hostile to such a form of collaboration.

This feeling that some additional form of cooperation should be considered, beyond that of the modus vivendi, was encouraged by the relative vulnerability of the United Kingdom in the event of conflict with the Soviet Union. For this reason a proposal was advanced under which British plants for the production of fissile materials and the manufacture of bombs would as far as possible have been situated in the United States or Canada, with the minimum number of unassembled bombs essential for the requirements of joint defense kept on British soil. Because the United States would thus have assumed the main responsibility for bomb production, they would have been allocated 90 percent of the uranium available over the coming five years, with the remaining 10 percent shared between Canada and the United Kingdom.

These proposals were unacceptable to London. They were also opposed in Congress on the grounds that they were much too favorable toward the British. The British had the support of President Truman, of Secretary of State Dean G. Acheson, and of chairman of the U.S. AEC, Lilienthal. On the other hand, one of the five U.S. AEC commissioners, Admiral Lewis L. Strauss — a financier who had made his fortune on Wall Street before working for the navy during the war — together with the secretary of war and a large number of members of the JCAE were opposed to so complete a collaboration.

The British were in any case determined to go ahead on home territory with their own production of fissile materials. They therefore sought a full exchange of information on production plants for explosives and on the manufacture of ''improved'' weapons. They had no intention of exchanging the real thing for its shadow, and they would certainly not consider giving up their own production of nuclear weapons unless they could be absolutely certain of being supplied with their own bombs, fully under their own control, on British soil.

By late September 1949, negotiations had again reached a deadlock for it was clear that British demands would not be accepted by Congress. The Soviet atomic explosion that same summer led to a brief renewal of the talks, but toward the end of the year they still seemed doomed to failure. Then with the discovery in early 1950 of the very serious Fuchs espionage on behalf of the Soviets, the last British hopes came to an end.

At precisely the same moment, the American government decided to take a further step in the arms race and officially announced the start of research on a new weapon, the hydrogen bomb, far more powerful and far more deadly than the Hiroshima bomb. The British, naturally, were not to take part in this new venture.

The Korean Threat

Toward the end of 1950, the issue of the atomic bomb again reached the top level of the American and British heads of state. For the second time, President Truman was faced with the agonizing decision of whether or not to use the atomic weapon.

The Korean war, which had begun on June 25, 1950, had turned out to be a series of unexpected disasters for the American troops and U.N. contingents. Following the Chinese breakthrough in late November, Truman, prompted by the commander in chief of the U.N. forces, the American general, Douglas MacArthur, gave the impression during a press conference that use of the bomb in Korea was imminent.

The impression was false and was quickly corrected, but nevertheless it provoked fierce opposition from the French, Canadian, and British governments. The Quai d' Orsay (French Foreign Office) issued a declaration that the objectives of the Korean War could not justify the use of the new weapon, while British Prime Minister Attlee, after discussions in London with his French counterpart René Pleven, hurried to Washington on December 5 to see Truman. The meeting ended with a communiqué in which the United States, acknowledging that use of the atomic bomb could lead to a third world war, promised to warn the United Kingdom if the international situation changed sufficiently to warrant serious contemplation of the use of the bomb. This was far from the right of veto that had been established in Quebec and surrendered by the British in 1947.

American public opinion at this time was reflected in the various stands taken in the Senate. Most of the members wanted the bomb to be handed over to the U.N. forces in Korea, its possible use to be preceded by an ultimatum from that international organization. Some senators were totally opposed to any use of the bomb in such a localized conflict, whereas at the other extreme a senator called for an atomic attack on the Soviet Union itself, unless an agreement could be reached with the Soviets "within the next few days." Another member of the Senate wanted to authorize MacArthur to use the bomb at his discretion to prevent further Chinese infiltration. In the end the moderates, who wanted the general dismissed, won the day, at least in the United States, and in April 1951 the government announced General MacArthur's retirement.

The British Explosion

The following year the British Labour government asked Washington to authorize a test explosion of the first British bomb at one of the American sites in the Pacific, so that the British could benefit from the equipment and experience of their allies. This request between allied countries was sensible and logical, if only from the viewpoint of the resulting economy.

Despite the intimacy of Anglo-American military relationships in many fields, some of which were highly secret, American agreement was obtained with so much difficulty and reticence resulting from application of the McMahon Act that the British Defense General Staff finally chose for its test site the island of Monte Bello, off the west coast of Australia in the Indian Ocean (the Australian government having agreed).

Nevertheless a final, though fruitless, effort to renew Anglo-American collaboration was made by Churchill, following the return to power of the Conservatives in the general election of late 1951. The British leader believed, incorrectly, that he would be able to revive both the spirit and the letter of the Quebec Agreement.

Having accordingly failed to renew an extensive nuclear collaboration with the Americans, Churchill did however have the satisfaction of demonstrating to them, thanks to the Monte Bello explosion on October 3, 1952, that British technicians were quite able to do without their help.

Thus ended for a while the checkered story of Anglo-American nuclear collaboration since 1941; a story full of instruction, for it was during this time that the United States' "nuclear denials" atomic policy was defined and applied for the first time.

In practice, the countries that suffered the most from the effects of this restrictive policy were certainly not the Soviet Union or the United Kingdom, who joined the Atomic Club four and seven years, respectively, after the end of the war. The real sufferers were the countries with less nuclear advancement, which nevertheless had launched individual civil programs in the new field (such as France, Norway, and Sweden) together with others that, either from discouragement or from lack of uranium, had simply given up.

As for the British, their dream of a full and intimate nuclear relationship with the United States (which never materialized) denied them the more important role of "nuclear leader" among the West European and Commonwealth countries (other than Canada) — a role that at the time, in view of their technical advance, could easily have been theirs but which they would never again have the chance of playing, since to acquire it they would have had to collaborate closely with the other countries during the early stages of their nuclear programs. To allow this, the British government, in the face of unfulfilled promises, would have needed to renounce the secrecy pledge of Quebec in exchange for the difficult road of independence. But would it have been possible for the British, in their then state of economic and financial subservience vis-á-vis their protector, to have taken such a step?

Despite the establishment in 1958 of privileged relationships with Washington in the fields of nuclear armaments and submarine propulsion and their wider policies aimed at securing — on the basis of Churchill's efforts and of the past alliance — ''special links'' with the United States, the sought-after status as equal partner in the direction of world affairs was denied them by Washington. And in the long run those very policies were to contribute to the British decline, in the nuclear field as elsewhere.

As for the United States their role of sacred guardian of the new power, enshrined in their policy of nonproliferation, had suffered a double setback with the acquisition of nuclear weapons by their main rival and their closest ally. They had never been in a position to stop the British military program, either by means of technical constraints or by political absorption into a truly joint program, two extreme solutions between which they had oscillated back and forth.

3. The Superbomb

The Principle

By late 1942, American physicists working on the first atomic bomb had realized that if they could produce the weapon they might be able to achieve, in its center at the moment of explosion, temperatures never before attained on earth. These temperatures, in the region of several tens of millions of degrees Celsius, could be sufficiently high to trigger reactions such as those occurring in active stars.

Deep inside such stars, the nuclei of light atoms are constantly undergoing transmutation to produce heavier atomic nuclei and neutrons. This ''nuclear condensation'' reaction is accompanied by a loss of mass, resulting in the production of immense amounts of energy, such as takes place in our own sun and by which all life on earth is maintained.

This reaction, known as a ''fusion'' or thermonuclear reaction (because of the role played by the higher temperature) yields 10 times more energy than the fission of uranium from a given initial amount of matter. In addition, once the fusion process has been triggered, the resulting heat output is sufficient to maintain the initial high temperature, thus propagating the reaction rapidly throughout the mass of light atoms (the limiting factor in the process), which are arranged around the ''conventional'' atomic bomb used to trigger the fusion.

From this idea to the concept of an atomic fission bomb setting off a much more powerful weapon, known as a ''hydrogen'' or thermonuclear bomb, was a comparatively simple step for the scientists at Los Alamos. They had in fact already studied the possibility, so as to be sure the first atomic explosions would not set fire to the entire world atmosphere. While the war was in progress, however, research on this hypothetical weapon went forward very slowly, since its outcome was dependent on the success of the atomic bomb necessary to trigger it. Soon after the war, the Los Alamos scientists came to the conclusion that this new development would indeed be possible, but the disorganization following the end of hostilities hindered further valid research in the area.

From the start two possibilities had been considered. One comprised the use of a fission explosion to initiate the fusion of a small quantity of light elements. This latter reaction, and the additional neutrons released, would increase, or "boost," the explosive power obtained from the fission reaction. The other possibility, far more ambitious, depended on using the temperature produced at the center of a fission explosion (in the region of 100 million degrees Celsius) to trigger a thermonuclear reaction in a substantial mass of light elements, thus obtaining an explosion with a power of a completely different order of magnitude, for example, 1000 times greater than that of the triggering explosion. This is what became known as the "superbomb" or "H-bomb."

A major difficulty was the fact that the fusion reaction could not be tested on a small scale, for there was no laboratory method of obtaining the very high temperatures achieved in the atomic bomb. Theoretical calculations showed that the heavy isotopes of hydrogen were the most suitable, but that the reaction would take place less readily with heavy hydrogen of mass two, or deuterium, than with tritium, an unstable hydrogen isotope of mass three that can be produced by bombarding another light element, lithium,* with neutrons.

In spring 1949, research into the "booster" principle was sufficiently advanced for it to be decided to test a boosted bomb during a series of trials scheduled in the Pacific in 1951. Theoretical calculations for the superbomb had also made some progress, thanks in part to the use of the first electronic computers, as well as to a better understanding of the theory of fission explosions. No practical superbomb, however, was yet in sight.

---

*Lithium, which contributes to the general feeling of anxiety linked with the threat of nuclear war, is paradoxically an element that, ingested in quantities of less than one gram, is a preferred drug for the treatment of certain cases of nervous depression.

Truman's Decision

Even before the news of the Soviet Union's first atomic explosion hit Washington like a thunderbolt, the U.S. Atomic Energy Commission (U.S. AEC) had submitted to the president a proposal to extend the production of more effective atomic weapons. However a decision was still to be made on this proposal when, in October 1949, during the month following America's loss of its monopoly, it became necessary to decide what Washington's reaction should be to this event. In particular, should priority now be given to research on the superbomb?

Among the convinced supporters of such a course were Lewis L. Strauss, the most isolationist and anti-Communist member of the U.S. AEC; Senator Brien McMahon, chairman of the Joint Committee for Atomic Energy (JCAE); and in scientific circles the Hungarian-born physicist Edward Teller, one of the leaders of the Los Alamos team, who since 1942 had been convinced that research on thermonuclear weapons had not received the attention it merited.

The next four months were reminiscent of those that had preceded the use of the bomb against Japan. Once again opinions were divided. For some, war looked difficult to avoid; the hydrogen bomb seemed the only way to stop it and to prevent the Russians from invading Europe as soon as they had sufficient conventional atomic bombs at their disposal. For others, the United States had already created a monster and should not now set out to produce a "Super-Frankenstein."

In late October, two sessions of the U.S. AEC's scientific consultative committee, under the chairmanship of J. Robert Oppenheimer, were devoted to the technical and political aspects of the problem. One possible program, which appeared to be free from any great risk, was the accelerated development of improved conventional atomic weapons including, among others, tactical weapons and the bomb boosted with tritium that was to be tested in 1951. A second possible program, aimed at producing the superbomb, would call for considerable extra work and expense in an intensive research program over five years, with a likelihood of success estimated at only slightly more than 50 percent.

All the scientists, who initially held differing views, were finally unanimous in supporting Oppenheimer's opinion and opposing the second venture. They considered the hydrogen bomb to be technically uncertain and, if successful, likely to be too expensive, particularly in comparison with production costs of conventional nuclear explosives. Moreover the production of tritium, then considered essential for the superbomb, would have to have been undertaken at the expense of conventional bomb production.

In the conclusions of the consulting committee's report, the participants also noted that, although their opinions differed as to the technical chances of producing the superbomb, they were unanimous in hoping that one way or another the production of such a weapon could be avoided, for they had no wish to see the United States take the initiative in precipitating its development.

Some members of the committee, including Oppenheimer, went a step further and in a separate report that described the superbomb as a weapon of genocide, urged that it never be produced. They felt that its possession by the United States would have a psychological effect contrary to the nation's interests. Two members, the Nobel Prize winners Enrico Fermi and Isidor I. Rabi, went further still and called on the U.S. president to declare, before the American public and the world at large, that it was contrary to basic ethical principles to develop such a weapon. They asked him to invite all nations to make a solemn pledge to join the United States in renouncing research leading to its production. They added that, should their proposal be accepted, any illicit experimental test by another power could be detected; furthermore, the American nuclear arsenal already included weapons sufficiently powerful for an adequate reaction to any such production or use of the superbomb.

Fermi and Rabi were in fact thinking in terms of a truly universal nonproliferation treaty, limited to an H-bomb whose technical feasibility was still unproven. In their eyes the United States, having committed the nuclear sin — Oppenheimer had a particular and personal sense of responsibility for this — had an obligation to take such an initiative. Some of their arguments were similar to those used later and still today to persuade countries not yet in possession of conventional atomic weapons to renounce weapons development or production forever.

But the United States was in the middle of the Cold War, and the fear of being overtaken by the Soviet Union overruled ethical and moral considerations. The scientists in favor of developing the new weapon, backed by Teller, made direct approaches to the military and to the influential members of the JCAE, whom they easily won over to their point of view. In November 1949, a senator committed an intentional indiscretion on television, which was naturally taken up by the entire press, by saying that progress had been made toward building a bomb a 1000 times more devastating than the one used on Hiroshima.

President Harry S. Truman was thus forced into making a rapid decision and announcing it publicly. He gave the task of assembling data for this decision to a three-man committee comprised of Secretary of State Dean G. Acheson, Secretary of Defense Louis A. Johnson, and the chairman of the U.S. AEC, David E. Lilienthal. Only the last, who was about to resign, opposed production of the H-bomb, being afraid that an even fiercer arms race would result and would reduce the ultimate chances of success for the plan to which he had given his name four years earlier. Lilienthal hoped at least to delay any decision by a few months.

During its last meeting, on January 31, 1950, the committee learned of the arrest of Fuchs in England four days earlier and of the importance of the information he had passed to the Soviets. The gravity of this news merely added last minute emphasis to the cause of the H-bomb protagonists, who had in any case already won the day.

The committee's recommendation was immediately passed to President Truman, who lost no time endorsing it and the same evening announced it officially. As Commander in Chief of the Armed Forces, he said, he had instructed the U.S. AEC to pursue its work on all possible atomic weapons, including the "so-called hydrogen or superbomb."

Secrecy had of course prevented the public from learning of the divergent opinions that had shown up both in scientific circles and in governmental authorities; and, no less evidently, neither the American people nor world opinion had been consulted. Twenty-seven years later, a public debate on the development and production of a considerably less revolutionary and essentially defensive weapon, the "neutron bomb" (see p. 221), was to cause another president, Jimmy Carter, leader of an America still suffering from the trauma of the Vietnam War and seeking détente with the Soviet Union, to pause and reflect.

The Expanding American Enterprise

Truman's decision to develop the H-bomb gave a fresh impetus to the American nuclear program. It was decided, in parallel with the new research on this weapon, to increase substantially the production of nuclear materials. Further installations were built, including in particular a number of large heavy water reactors for plutonium and tritium production by neutron irradiation of lithium, and two huge isotopic separation plants each costing around one billion "1950" dollars.

At that time, this gigantic industrial complex, which became operational in 1956, consumed 10 percent of the total national electricity production. From its start in 1951, the program weighed very heavily on the annual budget of the U.S. AEC, which rose sharply to reach a maximum of $4 billion in 1953. This was well beyond the rates of wartime expenditure: building work for the U.S. nuclear program at this time represented over 5 percent of the entire national construction effort.

The program involved, of course, increased uranium requirements. An extensive prospecting effort was called for that was speedily undertaken and successfully completed by the Anglo-Saxon countries. Extraction began of large quantities of uranium from gold-bearing sands in South Africa and these, together with new and important deposits discovered since 1953 in Canada and the United States, assured adequate and continuing production despite the approaching exhaustion of the first mines in the Belgian Congo and in Canada.

From 1950 onward, the British and American governments were both investing large sums of money in South Africa in order to create an industry for uranium production. By the end of the decade, this industry had an annual financial turnover equal to one-quarter that from gold production, witness to its importance for the local economy.

The U.S. AEC was also encouraging uranium prospecting on American home territory, offering generous guaranteed purchase prices to successful private prospectors. By 1954 certain areas in Colorado, Utah, and New Mexico were the scenes of a veritable "uranium rush," with new towns of several thousand inhabitants springing up, such as Moab in Utah where the main street was a parade of drugstores, public notaries offices for issuing legal prospecting permits and registering claims, and suppliers of mineral prospecting equipment and radiation counters.

Several prospectors made fabulous fortunes, which stimulated enthusiasm for the search. Indians from the reservations took part with considerble success thanks to their remarkable gifts of observation. Then the large private companies joined in, and in the end results were far better than had been hoped for.

However, not all the projects launched in the early 1950s in search of uranium were crowned with such success. A case in point was a Franco-American venture, during which first contacts were established between the atomic energy commissions of the two countries.

In view of the ever-increasing U.S. uranium requirements for the nuclear defense of the Atlantic Alliance, in 1952 Washington proposed to the French government a secret, joint prospecting project for uranium in Morocco, then believed to be a geologically promising area. If the project proved successful, production was to be shared in accordance with the capital invested by each party and their respective needs. These needs were naturally rather modest in the case of France, then seeking a few tens of tons, whereas U.S. requirements were measured in hundreds of tons. No restrictions on use were envisaged for the uranium from this joint enterprise, so that the French share could possibly have been used even for a military program.

The operation appeared to have begun largely as a result of a request to an American laboratory for an analysis of some ore samples, which were said to have come from a Moroccan mine where other metals were being extracted. The samples were rich in uranium. However, when this mine was later bought by French interests, it was found to contain no trace of uranium, which — to use the mildest terms — casts a doubt as to the nature of the whole affair. The project was in any case a failure, for prospecting in the rest of the territory yielded disappointing results, and the operation was called off shortly before Morocco became an independent kingdom in 1957.

The new prospecting projects in Canada and the United States, however, met with considerable success. New Canadian mines were rapidly opened and by 1956 their production made Canada, briefly, the Western World's leading uranium producer. At the end of the decade, the United States deprived it of this position, holding the leading position up to the present.

This highly productive race for uranium, the object of which was essentially military, was subsequently to influence the peaceful development of nuclear energy, particularly from the early 1960s onward when the American weapons program requirements began to drop. A plethora of uranium then appeared in the Western World, while the United States retained their virtual monopoly of uranium enrichment and the supply of uranium-235 for all research reactors and power plants using enriched uranium fuels in the Western World.

The military uses for increased production of enriched uranium were not confined to weapons, but also included supplies to the nuclear-powered section of the U.S. Navy, which had been thriving since the mid-1950s.

In 1955, in fact, the nuclear-powered submarine, which Joliot-Curie had foreseen as long ago as 1940, had become a reality. This application of uranium fission, no doubt the most revolutionary after the bomb, quickly became the essential complement of a great power's atomic armament, particularly in the form of a mobile and almost undetectable launching platform for missiles with nuclear warheads.

American and Soviet Tests

In these same early years of the 1950s, while the Soviet Union, unbeknown to the rest of the world, was also developing a hydrogen bomb, and as the Korean War was reaching stagnation, American efforts toward the production of their superbomb yielded rapid results.

In May 1951 a bomb boosted with a mixture of tritium and deuterium was successfully tested at Eniwetok. The explosion was the most powerful to date. For the first time on earth, man had achieved the fusion of atoms of light elements.

The results of this test comprised an important step on the road toward the superbomb, providing crucial data for establishing a new theory elaborated shortly before by Teller, the great proponent of H-bomb development, and by another eminent theoretician of Polish origin, Stanislaw M. Ulam. Suddenly, the status of the superbomb had advanced from a doubtful ''possible'' to an encouraging ''probable.'' Even Oppenheimer had to admit this new factor.

It was then decided to proceed as quickly as possible to a practical test of the thermonuclear system proposed by Teller and Ulam and if this were successful, to go directly for a military device. The two stages were successfully completed in November 1952 and March 1954.

In November 1952, one of the islands of the Eniwetok Atoll group disappeared from the map during a test explosion whose power and nature were kept secret for a long time. The power was equivalent to that of 10 million tons of conventional explosives (10 megatons in the specialist's jargon), 800 times more powerful than the Hiroshima bomb (13 kilotons). But it was a bulky device, using liquid deuterium as ''fuel'' that had to be maintained at an extremely low temperature. The complete device was housed in a specially designed contraption of considerable size.

Sixteen months later, on March 1, 1954, a true air-transportable bomb, this time using solid ''fuel'' (lithium-deuteride), was exploded at Bikini. Its power was officially revealed as 15 megatons, or more than five times the total power of all the Allied bombs dropped from the air on German territory during the whole of the last war. This comparison strikingly illustrates the ''quantum jump'' achieved in the science of destruction. The American superbomb had become an operational military reality.

This last explosion — for which the bizarre code name ''Bravo'' had been chosen — had left a crater nearly half a mile wide in the island, throwing many tons of radioactive coral and other debris into the stratosphere. Much of this fell back on the Rongelap Atoll some 90 miles away, over 200 inhabitants of which suffered health effects. The crew of a Japanese tunny-fishing boat inappropriately named *The Happy Dragon,* which was about the same distance from the explosion, suffered similarly; the radioactivity caused two deaths and severe health problems for the other sailors. There was an emotional reaction in Japan, reopening the ever-present scars from Hiroshima and Nagasaki. These effects on the innocent victims of ''Bravo'' shook public opinion throughout the world, and were ultimately to help initiate the process that, nine years later, was to lead to a treaty banning all but underground nuclear testing — a treaty made possible as a result of American and Soviet military equilibrium.

A vital step toward this equilibrium had been achieved a few months earlier when, on August 12, 1953, the first Soviet thermonuclear explosion had been detonated. The Americans learned of this dramatic development through their atmospheric radioactivity monitoring, which detected the resulting radioactive cloud. Analysis revealed not only that there had been fusion of light elements, but that these elements had been in the form of lithium deuteride, which as we have just seen was being considered by the Americans but had not been tested in either their 1951 or 1952 experiments. In addition, the Russian device must have been transportable by air, probably a boosted bomb, with a power of less than a megaton. The physicist Andrei Sakharov, who has since become one of the leading Soviet ''dissidents,'' appears to have played a vital role in this success. His age at the time was less than 30.

November 1955 saw the final stage of this rivalry between the Soviet Union and the United States when the Russians finally exploded a ''true'' superbomb, with a power of several megatons.

It is only today, on the basis of recently published information as to the nature and power of the first American thermonuclear tests, that it has become clear that the United States had in fact won the technical race with the Soviet Union by several years. But they had lost the psychological race, for the Russians had easily profited from the secrecy enshrouding the more advanced American experiments, which allowed them to claim they had themselves been first to explode a real weapon using the phenomenon of fusion.

The effect of all this in American political circles was dramatic. A striking demonstration had just been made, and clearly espionage was not the main factor involved in the Russian success: The Soviet scientists had demonstrated that they were fully capable of developing original technology without outside help.

Thus was the qualitative equilibrium restored between the two Great Powers — but at a higher level of terror, a 1000 times more destructive.

The Oppenheimer Affair

In the United States, the surprise and shock of the first Soviet H-bomb explosion was followed by a further drama, this time on the internal scene.

An inquiry was opened into the political integrity of the U.S. AEC's chief scientific adviser Oppenheimer who, as far back as the months preceding President Truman's 1950 decision, had advised against, if not impeded, the efforts toward the American superbomb.

This affair had its origins in the U.S. AEC in December 1953, six months after Strauss, confirmed partisan of the H-bomb, had become chairman. On several occasions already, he had had differences with Oppenheimer.

Strauss's first move was to have his scientific adviser informed that it had been decided to deprive him of his right of access to secret documents. Oppenheimer reacted by asking that the matter be referred for arbitration to a committee of inquiry; this was agreed, and the spring of 1954 saw several weeks of tedious and unpleasant hearings in an atmosphere not unlike that of Senator Joseph R. McCarthy's earlier "inquiries" into security and loyalty.

Nothing remained unexposed of the life and most intimate activities of this scientist who, after the war, had become one of the most valued advisers of the Democratic administration, who indeed was also an idol of the public, fascinated by his role in the development of the atomic bomb and by his remarkable intelligence. Oppenheimer was particularly attacked over his prewar associations with American organizations of the extreme left — knowledge that had long been public — and for having persuaded his scientific colleagues in the AEC's consultative committee to support his initial opposition to the proposed H-bomb program.

In a tense atmosphere resembling in many ways that of a full-scale legal trial, the prosecution attempted to prove that the eminent scientist could not offer all the security guarantees required by the country's national interests so that, under the law, he was precluded from access to the AEC. The prosecution had particular success in drawing attention to an isolated incident that had occurred over 10 years earlier, at a time when scientists generally had yet to appreciate the full implications and obligations of the strange world of secrecy.

What had happened was that in 1943, Oppenheimer had waited six months before informing General Leslie R. Groves of an incipient attempt at espionage, on behalf of the Soviet Union, then being contemplated by one of his university colleagues. Then, when he finally disclosed what he knew, he had given two different versions of the incident.

Following a majority decision by the committee of inquiry (two votes to one), which was confirmed by the U.S. AEC (four votes to one), Oppenheimer — despite his past services — was deprived of his position as a government adviser in 1954. He nevertheless remained at the head of the Institute for Advanced Study of Princeton University.

American scientific and university circles were for a long time deeply disturbed by this miserable affair. During the years that followed Oppenheimer made no attempt to have the government decision reversed. However, President John F. Kennedy in 1963, shortly before his assassination, recognized the injustice done and decided to award the scientist the government's highest distinction in the nuclear field, the Enrico Fermi Prize. The prize was presented to Oppenheimer personally by President Lyndon B. Johnson. However, Teller, who had always been in favor of the H-bomb project, had received the same award the previous year.

But even Strauss did not have a happy ending to his career. At the expiration of his presidential mandate as the head of the U.S. AEC, he was nominated by President Dwight D. Eisenhower as Secretary of Commerce; but the Senate, where he had made many enemies, refused to approve the nomination, an action without precedent for over a century in the ''advice and consent'' procedure for a ministerial appointment.

These personal dramas reflected human reactions in the United States to the revolutionary changes brought about by the new and terrifying developments in the world's capacity for destruction, including the second Soviet success. Strauss's reaction had been precisely that of an American leader obsessed by the Soviet threat, while Oppenheimer behaved as a citizen of the world, haunted by his previous responsibility and trying in vain to halt the race for more and more megatons, the inevitable consequence of this Cold War period.

The ''ordinary'' atomic weapon or A-bomb, from the point of view of destructive power, is equivalent to the heaviest air attacks during the Second World War; whereas the H-bomb is beyond human reckoning, being roughly comparable to the entire total of airborne bombing during the last war, and easily capable of annihilating a large city and its outskirts.

Both the American and Soviet thermonuclear explosions had once again changed the world balance of power; division into two opposing camps was now unavoidable. The test explosions had demonstrated the failure of the policy of secrecy, and the impossibility of a monopoly on know-how; they led to a change in direction for American policy, and to a degree of détente characteristic of the new balances in the Cold War. This détente was in fact the beginning of the end of secrecy, heralding a new openness in the civil applications of atomic power. In the military field, it led to negotiations to avoid further atmospheric nuclear tests.

Eisenhower's Proposals

The 1953 Soviet thermonuclear explosion was such a sufficient psychological shock to the American leaders, as to the general public, that it persuaded the former to lift the barricades of atomic isolationism that they had maintained since the end of the Second World War.

During the following months, President Eisenhower, whose election a year earlier had been largely won on the strength of his promise to end the Korean War, instructed his administration to prepare proposals that might at least reduce the pace of an apparently irreversible nuclear arms race, and that at best could lead to reopening the still deadlocked negotiations on atomic disarmament. Among the resulting suggestions to the president was the idea of an international collaborative project in which increasing quantities of fissile explosives should be withdrawn from military stocks and used for peaceful applications. American political initiative had resurfaced.

The idea was put to the British during the summit meeting in Bermuda at the beginning of December 1953. This meeting took place between Eisenhower, Churchill, and French Prime Minister Joseph Laniel to discuss a number of problems raised by the end of the Korean War in July, by the explosion of the Soviet H-bomb in August, by plans for German rearmament, and by the Soviet agreement to hold a 4-power meeting.

When it became known in Paris that Churchill was to be accompanied by Lord Cherwell, his scientific adviser and a member of his cabinet, it seemed appropriate also to include Francis Perrin, the scientist then in charge of the French atomic program, in his country's delegation. This proposal was dropped because the U.S. State Department, on inquiry, told the French that atomic issues were not scheduled for the talks. This proved to be inaccurate, for Eisenhower was in fact accompanied by U.S. AEC Chairman Strauss and thus the French were deliberately excluded when the first moves on abolishing the policy of secrecy were discussed and made in Bermuda by the Americans and British.

On December 8, 1953, immediately after his return from the Bermuda meeting, Eisenhower delivered a major address to the General Assembly of the United Nations. He began by referring to the balance of terror resulting from the existence of atomic bombs now 25 times more powerful than the Hiroshima bomb, with H-bombs several tens of times more powerful still, and he confirmed that the United States, even if subjected to a devastating surprise nuclear attack, would nonetheless still be able to retaliate with a lethal blow to their aggressor.

Faced with such a possibility, the destruction of all civilization and all human values, Eisenhower put forward a proposal, modest indeed albeit symbolically positive, for a first step toward nuclear détente. He proposed that the powers mainly concerned, the producers of uranium and other fissile materials, should progressively give up increasing quantities of these materials by withdrawing them from the stocks available for military uses. These materials should then be taken over by an international body whose task, under U.N. authority, would be to ensure their peaceful use in accordance with the general interest.

Eisenhower insisted, naturally, that his country would contribute to this "materials bank" only if the Soviet Union did the same, but he pointed out that the plan did not require (as had earlier disarmament proposals) acceptance of a worldwide system of control and inspection. Governments would not be asked to deprive themselves of fissile materials beyond what elementary prudence allowed. This notion, that every country has a legitimate right to security, was to reappear a quarter of a century later as one of the bases of French disarmament proposals put forward to the United Nations in 1978 by President Valéry Giscard d'Estaing.

For the first time since the war, a plan for nuclear disarmament was not dependent on the two preliminary conditions that until then had been insisted on by one side and refused by the other: the American requirement that the Soviet Union be opened to international inspection; and the Soviet insistence on the prohibition and prior destruction of all nuclear weapons.

The novel character of Eisenhower's proposal was warmly welcomed in the United Nations. A new era of openness was beginning in the history of atomic energy at the very time when technical progress was strengthening hopes for early nuclear solutions to the problems of electricity production and ship propulsion.

The Soviet government agreed to discuss the proposal directly with the United States through normal diplomatic channels. Their initial reaction, however, was reticent and they continued to insist on prior renunciation of the use of the hydrogen bomb and other weapons of mass destruction.

Later, toward the end of 1954, the Soviets proposed an expert meeting to examine technical possibilities for preventing diversions of fissile materials from civil to military uses, and to examine possible ways for rendering these materials unsuitable for military purposes without hindering their civil utilization.

Such a meeting, between experts from the main atomic powers, took place in September 1955 in Geneva, without, however, resolving the same problem that was to arise again during the International Nuclear Fuel Cycle Evaluation program proposed by President Jimmy Carter in 1977: namely, the impossibility (emphasized as early as 1946 by Lilienthal) of developing atomic energy for peaceful purposes without increasing the potential for production of materials that could be used for military purposes. Unfortunately, a true and sufficient "denaturization" of these materials is not possible.

Meanwhile, the new policy of openness had made rapid progress in the civil field. In August 1954, the McMahon Act was amended to allow the transfer of American fissile materials to friendly states subject to the conclusion of a governmental agreement, known as an "agreement for cooperation," between the United States and the country benefiting from this assistance. Each agreement was to include a commitment not to use the materials provided in any way for the production of a nuclear weapon or for any other military purpose.

The new law also allowed, within very strict limits, the transmission to allied powers of certain information indispensable for their defense. Such information was to be strictly confined to the tactical uses of nuclear weapons, and was to exclude any significant data on the design and production of the nuclear components or of any other important elements of these weapons.

The secrecy veil was to be further lifted, this time on a worldwide scale, at a vast and unprecedented scientific conference proposed by the United States, organized by the United Nations, and held in Geneva in the summer of 1955. Détente was under way, and American and Soviet nuclear scientists were meeting again for the first time in 15 years. This, as described in Part Two, was largely the end of atomic secrecy.

Thus did Eisenhower's proposal open an era of international exchanges on the peaceful applications of atomic energy. The slogan "Atoms for Peace," then believed fully justified, was coined and widely adopted. To many it was a true "breakthrough" for peace. Twenty years later, the same liberal policy was to be severely criticized, particularly in the United States, as having caused too great a dispersion of nuclear technology in "sensitive areas" from the viewpoint of the production of atomic explosives.

Nevertheless, this is the moment to emphasize the vital part played by Washington in the creation of two basic factors in international nuclear policy: an international control over the peaceful utilization of atomic energy and the organization whose task it was to carry out that control, the International Atomic Energy Agency (IAEA). These factors have made possible both the expansion of world atomic trade, and the development of a nonproliferation policy for nuclear weapons. Because of that dual influence, they are examined in each of the two parts of this book: briefly in this first part, and at greater length in Part Two.

International Control

A clause on peaceful utilization — a highly original notion in international exchanges — had been proposed by Eisenhower in his U.N. address on the use of fissile materials that might be handed over to a future "international nuclear bank" (the IAEA) for subsequent redistribution. However the American president had not mentioned the possibility of a surveillance system to ensure that the clause was respected. The idea of such a system had in fact been rejected in the Acheson-Lilienthal plan as almost certainly unacceptable to sovereign nations.

By 1956, however, when the first agreements for cooperation were being concluded by the United States, the American government decided to require countries receiving assistance to accept a clause granting right of access to relevant materials and installations either to American inspectors, or possibly to inspectors of another nationality, for the sole purpose of verifying that the aid was not being used for military ends.

This right of surveillance was at first accepted without complaint by nations concluding agreements with the United States and therefore liable to inspection. These nations had just passed through a 15-year period during which access to nuclear materials had been virtually impossible, while enriched uranium, for which the United States had a quasi-monopoly in the Western World, had until then been totally unavailable. To these countries it did not therefore seem particularly unreasonable that, if the American government agreed to transfer some enriched uranium, it should also ask the right to verify its peaceful use, especially when the transfer concerned quantities or qualities of military significance.

The insistence on and acceptance of this right of inspection, by inspectors who were not nationals of the country concerned, was nevertheless a revolutionary innovation in the history of relations and trade between nations. The innovation was justified by the exceptional nature of the dangers of nuclear armaments. It reflected a certain mistrust on behalf of the exporting government — here the United States — toward a solemn undertaking by the assisted country. From the opposite side, its acceptance by the latter was a renunciation of sovereignty, accepted because of the advantages of the assistance received in return.

In any case, the surveillance was restricted solely to the aid received and to its direct consequences. Neither the amended McMahon Act nor the agreements for cooperation required the assisted country to abandon any form of nuclear military program. Such a program was still possible provided it was carried out in installations and with materials totally distinct from those provided by the United States. To have made assistance conditional upon the renunciation of all military nuclear activity, or on the acceptance of safeguards on all the nuclear installations within the country concerned, was out of the question then, in contrast to the situation 20 years later.

The American policy was, of course, aimed at persuading an increasing number of countries to devote progressively more of their nuclear programs irreversibly to peaceful objectives, subject to verification by inspection. The policy was thus favorable to nondissemination of nuclear weapons by somewhat reducing, through verifiable restrictions on the programs, the military options available to the countries concerned. Activities subject to the peaceful utilization clause and to international inspection could not, in fact, be used for a military program other than in flagrant breach of given commitments.

The creation in 1956 of the first international organizations devoted to nuclear energy, in particular the IAEA, was to give Washington the possibility of institutionalizing the concept of control into a truly international safeguards system.

On the eve of the big U.N. scientific conference in Geneva in 1955, the Soviet government had in fact made it known that it would be prepared to take part in the future agency proposed by Eisenhower, to supply certain fissile materials to the new organization, and to accept as a basis for discussion the latest draft statutes proposed by the Americans.

Following this, in the autumn of 1955, the U.N. General Assembly asked the American government to organize a conference of the countries mainly interested in the new agency. The meeting was held in Washington in February 1956, and was marked by the conciliatory attitude of the Soviet Union; it produced a plan for an organization that was to possess very extensive powers for the control of nuclear activities and that would act as a broker rather than as a banker in matters related to the distribution of nuclear materials.

The new organization's draft statutes were finally submitted to a plenary conference of 81 countries, held in New York in October 1956, at U.N. Headquarters during the U.N. General Assembly.

The conference lasted for a month, and came close to failure over the issue of the extent of the new organization's powers of safeguards of nuclear activities. This time the very principle of international inspection was challenged by many countries, especially those of the Third World, joined toward the end of the meeting by the Soviet Union.

Those opposing the controls saw in them a form of neocolonialism; for the less-advanced countries needing the agency's services the most, or requiring external assistance, would be the most closely controlled. On the other hand, the advanced industrial states, especially those already equipped with nuclear weapons, would hardly need agency assistance at all due to their technological advance and, in the case of the most important nuclear powers, due to their having already appropriated needed uranium resources within their respective spheres of influence.

It also seemed particularly unjust to countries that did not possess indigenous sources of uranium that their future nuclear activities should be placed automatically under international safeguards, whereas countries possessing their own uranium resources could avoid inspection.

New discrimination in the atomic field, based on the extent to which the nuclear activities of the various countries were subjected to safeguards, was to be added to the existing discrimination linked with possession of nuclear weapons. Both led to the existence of two categories within which countries could fall: those having nuclear technology and materials and therefore in the best position to keep open a military option; and those largely dependent on external assistance and therefore likely to find most, if not all, of their nuclear activities placed under a guarantee of peaceful utilization and subjected to international inspection. Despite this, the conference ended with a compromise that enabled the IAEA and its system of safeguards to be set up.

The IAEA, with headquarters in Vienna, Austria, was inaugurated in 1958, but its safeguards system did not come fully into effect until 1963, at the end of five years of opposition from the Soviet Union.

Thus Eisenhower's proposal had paved the way toward an international system to guarantee the peaceful utilization of atomic energy on a worldwide scale. In the field of nuclear disarmament, however, for which it had initially been envisaged, the proposal proved less successful.

In fact, the amounts of fissile materials transferred to civil applications were for a very long time negligible compared with the even greater quantities produced and assigned to the American and Soviet armament programs. Neither was weapons proliferation limited to the United States and the Soviet Union. The United Kingdom, having exploded a first bomb in 1952, acquired a thermonuclear weapon five years later; while France in her turn launched a military nuclear effort and carried out a first test explosion early in 1960. Finally, in 1964 the People's Republic of China, seven years before being admitted to the United Nations in place of Taiwan as a permanent member of the Security Council, was to join the club of countries possessing an atomic arsenal.

4. France, Europe, and the Atlantic Alliance

The Early Days of the Commissariat à l'Energie Atomique

Of the five great powers having an atomic arsenal, France is the sole country not to have decided on a military program from the start of her work in the nuclear field. It was only 8 to 10 years later that the decision was made, the option having been kept open since 1945.

In their brief meeting with General Charles de Gaulle in Ottawa in July 1944, the three French members of the Anglo-Canadian project had drawn his attention not only to the military and political importance of the future atomic weapon, but also to the necessity for France to resume, as soon as possible, the research interrupted by the war.

In March 1945, Raoul Dautry, who in 1940 had given his ministerial support to the research program at the Collége de France, found himself once again at the head of a government ministry. Seizing the opportunity he immediately reminded General de Gaulle, the president of the Provisional government, of the importance of relaunching the atomic research effort without delay, and emphasized the part that Norway could play in supplying heavy water.

Two months later, Pierre Auger, aware of developments in the American nuclear enterprise, and Frédéric Joliot-Curie were able to convince the general of the need to set up an organization in France to deal with atomic energy. On October 18, 1945, at the moment when the state took over the production and distribution of gas and electricity, the Provisional government made an order placing all nuclear responsibilities in France, not only in the various fields of science and industry but also in national defense, in the hands of a national authority.

France had been, in 1936 under the then leftist Popular Front government, the first country in the world to set up a separate Ministry for Scientific Research. Nine years later, she was again the first country to establish a civil authority to take charge of the development of the applications of fission: the Commissariat à l'Energie Atomique (CEA).

Placed directly under the responsibility and control of the prime minister, the new organization was given an original form and statute unique in France. Although legally having a civil "personality," the CEA was to enjoy administrative and financial autonomy. Its statute was modeled on that of the nationalized Renault car manufacturing company.

As for the direction of the new body, the government, unable to choose between an administrator and a scientist, gave the administrative and financial responsibilities to a general manager who was a delegate of the government, and entrusted the scientific and technical direction to a high commissioner. The first holders of these posts were the two men who had been responsible for the country's atomic program in 1940: Dautry, who since the liberation hd been minister for reconstruction, and Joliot-Curie, then director of the Centre Nationale de la Recherche Scientifique. Each of them declared that this unusual dual leadership had been created at his own request, de Gaulle having designated for each a managerial partner who would relieve him of matters outside his own special competence.

The two leaders were assisted by a committee that included a representative from the national defense services. This committee, which was in fact the CEA's board of management, was subsequently enlarged on several occasions to include directors from the principal state bodies.

In this year, 1946, with France deprived of uranium resources by the Anglo-Saxon powers, the CEA was fortunate in having at its disposal some 10 tons of uranium, which came partly from the original research studies at the Collége de France (having been hidden in Morocco since the 1940 Armistice) and partly from a Belgian railway wagon, loaded with sodium uranate, which was found by chance at Le Havre after the end of the war. This stock was just sufficient for the first two French atomic piles. It was to take more than three years to produce the same tonnage from indigenous uranium resources in metropolitan France, resources that in 1946 had yet to be discovered.

The quantity of uranium available was too small for a graphite reactor, so the only possible choice was one using heavy water. Fortunately, this choice was reinforced by the available knowledge and experience of the four French scientists from the Canadian team* who were to take part in relaunching the French program.

Unfortunately, the Belgian company running the uranium mine in the Belgian Congo was now completely tied to the Anglo-American supply agency and could no longer furnish France with uranium as agreed in 1940. However, Norwegian industry was prepared to maintain its 1940 contract for the supply of heavy water. The Norwegians in fact gave this commitment priority and the first tons they could produce were delivered to the CEA and used in France's first three reactors.

These reactors, built for research purposes, were of very low power. For the first of them it was decided, following my suggestion and in order to advance as quickly as possible, to make no special arrangements for cooling and to use uranium oxide fuel rather than metal, which was more difficult to produce. The ZOE experimental pile [Zero power, Oxide fuel, moderated by Eau lourde (heavy water)] was thus built in 1946 under Kowarski's direction at an improvised site inside an old fortress on the outskirts of Paris. The pile went into operation in 1948.

Just a few weeks earlier, the indispensable condition for an independent French atomic development program had been fulfilled when the first uranium-rich ores were discovered near Limoges in the central region of the country.

The year 1949 saw the inauguration of the first large national atomic research center, on the "plateau" at Saclay not far from Versailles. In November of the same year, my team had isolated, by rather makeshift methods based on the extraction technique I had developed in Canada, our first few milligrams of French plutonium.

The early stages of the French atomic development program were thus devoted to work with no military significance, and we were still a long way from the point of divergence toward either weapons or controlled combustion. In June 1946, the French delegate to the United Nations, Alexandre Parodi, in an official declaration to this effect during the first meetings of the U.N. Atomic Energy Commission had said: "I am authorized to state that the objectives assigned by the French government to the research carried out by her scientists and technicians are entirely peaceful. It is our hope that all the nations of the world will as soon as possible take the same course; and with this end in view France will readily submit to those rules

*Pierre Auger, Bertrand Goldschmidt, Jules Guéron, and Lew Kowarski. Hans Halban had decided to remain in the United Kingdom after having been led to abandon the atomic energy field.

which shall be considered the best for ensuring worldwide control of atomic energy.''

The issue of a possible military orientation in the French atomic effort had not yet arisen when the CEA was shaken by a severe political crisis. With the intensification of the Cold War during 1948 and 1949, the Berlin blockade, and the signing of the treaty that led to the creation of the North Atlantic Treaty Organization (NATO), Frédéric Joliot-Curie, the high commissioner for atomic energy, was taking an increasing part in extreme left-wing political meetings. Joliot-Curie, bound by his communist affiliation, made speeches condemning this Atlantic Alliance, encouraging workers to resist any project for making nuclear weapons, and declaring that he, as a scientist, would never help in any way to prepare for a war against the Soviet Union. Joliot-Curie could not foresee that, 15 years later, the French military atomic program to which, despite himself, he had made his contribution, would by an unexpected paradox of history create severe friction within the Atlantic Alliance he was then opposing.

In 1949 a world congress of ''Fighters for Peace,'' a pacifist movement widely supported by the Communist party, was held in Paris, and Frédéric Joliot-Curie was designated as its chairman. In March 1950, this same oraganization issued the ''Stockholm Appeal'' in reply to the American announcement that they were starting research on the H-bomb. The appeal reiterated the Soviet thesis that all atomic weapons should be prohibited, and declared that the first government to use such a weapon would be guilty of a crime against humanity. This caused a considerable stir, and the appeal was said to have received the support of several million signatures in a few months. Truman's decision on the H-bomb had provoked the first manifestation of antinuclear propaganda on a worldwide scale, although this time it was concerned only with atomic weapons.

The Communist party seized on this, seeing in it a way of holding back American supremacy. In France, High Commissioner Joliot-Curie was inevitably pressed by the party to make a stand against his government's Atlantic policy, thus forcing the government to make a difficult choice. Either the high commissioner would retain his position, showing governmental weakness in dealing with an influential civil servant, or the commissioner would be dismissed before the end of his term of office in late 1950, thus making him a victim of his political ideals.

Joliot-Curie's fate was in fact sealed on April 28, 1950, and his dismissal was officially announced in the following terms: ''The prime minister has made it known to the government that he has, regretfully, been obliged to end M. Joliot's term of office. M. Georges Bidault has pointed out that, whatever the scientific merits of this scientist, his public statements and his unreserved acceptance of the resolutions voted by the Communist party's recent congress make it impossible for him to remain as high commissioner.''

Scientific colleagues and friends of this great French nuclear physicist were all deeply moved by this unhappy affair, which was to take Joliot-Curie permanently out of the atomic energy field in which he had played so vital a part for his country.

Dautry died during the summer of 1951. Pierre Guillaumat, a mining engineer and oil specialist, replaced him and along with Francis Perrin, one of the pioneer discoverers of the chain reaction, took over the leadership of the CEA during the 1950s, a crucial period in France for both the industrial and military applications of atomic energy.

The First French 5-Year Plan

The person politically responsible for this new orientation was Félix Gaillard, a young deputy who, at the age of 30, had been appointed secretary of state in the prime minister's office only a few days before Dautry's death. Two years before, I had shown him around the Châtillon Centre and the ZOE pile before introducing him to Joliot-Curie who had, no doubt, helped to reinforce his enthusiasm for atomic energy.

At his request, Gaillard was given ministerial responsibility for the nuclear program from 1951 to 1953 under four successive governments. In this role, he played a decisive part in the French development of nuclear power. Well aware that no important country could afford to stand aside from the atomic revolution, he decided that France ought to have a truly long-term plan, which among other things, should ensure the production of substantial quantities of concentrated fissile materials.

The progress of the CEA's research and mineral prospecting program determined, more or less automatically, the direction to be followed, namely, that of plutonium production in natural uranium reactors. To undertake in parallel the separation of uranium-235 in an enrichment plant was not possible; neither the uranium nor the necessary technology was available, and above all sufficient financial and industrial means could not have been found to develop simultaneously both methods of producing the two fissile substances.

From September 1951 onward, a series of meetings was held to decide the nature and possible scale of the future plutonium production, and the financial investment that would be required. Perrin was afraid that too large a production would inevitably result in the military becoming interested, leading to their involvement in the CEA management. On the other hand, the minister, encouraged by those in favor of a possible French atomic weapon, was much more ambitious.

The two attitudes came into conflict in early October. Gaillard wanted to secure public support through a radio broadcast announcing an annual production of not less than 50 kilograms of plutonium. Perrin made known his distaste for political bluff, in view of the difficulties he could foresee in rapidly building enough reactors of the required power to produce the plutonium. Finally, Gaillard was able to convince the high commissioner who then, with the help of Guillaumat, prepared the technical outlines of the first 5-year French nuclear plan.

The plan, which was to cost 40 billion francs (about $100 million in 1952 dollars) involved building two graphite reactors and the corresponding plant for extracting plutonium.

The aim of the program was clearly stated to be plutonium production for subsequent use in reactors for electric power stations and engines for propulsion. There was no mention of possible military utilization. Nevertheless, this aspect of the matter was recognized, and undoubtedly figured prominently in the minds of most of those who had proposed or who were otherwise responsible for the plan. An explanatory memorandum on the whole project ended with these words: ''It is our duty, today, to ensure that in 10 years' time France will still be an important country in the modern world.''

Parliament understood, and in July 1952 the plan was officially adopted following a debate in which the CEA's objectives were for the first time subjected to a serious public discussion. Criticism came from two sides: right-wing supporters who wanted any of Joliot-Curie's previous colleagues who shared his political opinions to be dismissed from the CEA, as well as from the Communists who were critical because they saw a military orientation in the plan.

The Communist party in fact submitted an amendment, designed to make the government guarantee that plutonium produced under the plan would never be used for making bombs, either in France or elsewhere. The amendment was rejected by a large majority.

In presenting the plan, Gaillard stressed the comparatively modest investment called for, which he said proved that construction of a bomb was not contemplated. He also maintained that France's unilateral declaration of a peaceful program, given six years earlier in the United Nations, still applied. However, he admitted that this position adopted in 1946 had no contractual value, and he saw no reason of principle why France should renounce the right and the possibility of producing nuclear weapons when these weapons were being produced by others on both sides of the Iron Curtain.

The Socialists also voted against the Communist amendment. One of their star speakers, Jules Moch, explained that although his party was completely in favor of eliminating atomic weapons, it nevertheless opposed measures that could entail unilateral disarmament, and it would not vote for any resolution that was not in some way a step forward toward peace and security.

This debate on France's atomic program, which was to prove typical of others to come, left open all options and already hinted at future developments.

The adoption of this first 5-year plan marked a turning point in the CEA's history. Until 1952 it had been an essentially scientific organization with scientists playing a dominant role. From 1952 onward, its program took on a full-scale industrial outlook, with the entire country contributing, at the moment when world circumstances were changing in favor of the lifting of atomic secrecy.

But the launching of this first 5-year plan was to raise international political complications along the road to French production of plutonium.

Early European Problems

In 1948 the three Allied high commissioners in Berlin had begun negotiations designed to allow the Federal Republic of Germany a restricted program of atomic activity, in place of the total prohibition initially imposed by the occupying powers. Experts from the CEA, consulted on several occasions in the course of these negotiations from 1948 to 1952, had suggested that future German plutonium production should be restricted to 500 grams per year, since any larger quantity would be militarily significant.

Meanwhile, in 1950, Franco-German reconciliation took a big step forward with the "Schuman Plan," conceived by Jean Monnet, one of those responsible for the French postwar economic recovery, and the creation in 1951 of the European Coal and Steel Community, which included France, Germany, Italy, Belgium, the Netherlands, and Luxembourg. At the same time, the intensification of the Cold War brought a growing emphasis to the issue of a possible German contribution to the defense of western Europe. It was for this reason that Monnet proposed the creation of a European army, within the framework of a European Defense Community (EDC) and including the same six partners.

In February 1952, at the beginning of negotiations for the EDC treaty, German Chancellor Konrad Adenauer persuaded the foreign affairs ministers of the allies not to impose discriminatory measures on Germany. Therefore, to maintain a restriction on future German nuclear activities, the limit of 500 grams of plutonium per year, originally proposed for Germany alone, was to be applied to all six partners.

A country wishing to exceed this production limit would have to obtain authorization from the EDC's directing body; and in principle, by a complex system of clauses in the treaty and additional protocols to it, all the partners other than Germany would automatically obtain such an authorization for civil purposes. Conformity with the rules was to be verified by inspectors from the member countries.

The texts were far from clear, and the lawyers could not agree among themselves as to the exact interpretation of the treaty itself or of its additional protocols. At the Quai d'Orsay, the secretary general, Ambassador Parodi, remarked with great truth: ''First we tie down Germany, then in the name of equal rights we bind ourselves similarly, and then we give ouselves headaches over how to unbind ourselves!''

The French Defense and Foreign Affairs services, responsible for the negotiations, never once consulted the CEA nor even informed it that all or part of the clauses proposed by its own experts, with a view to maintaining Germany's military nuclear abstinence, might also be applied to France's own atomic program.

Application of the treaty would have virtually precluded France from undertaking a military atomic program, for which the agreement of all of France's partners would have been required. And French civil nuclear activities, if not restricted, would have at least been subjected to controls in what at this time was a most secret area, the production and extraction of plutonium, which was at the center of the recently launched 5-year plan. The proposed controls, in which German inspectors among others would have taken part, would have provided Germany with access to technical data on plutonium.

The governmental services concerned were now thrown into a series of frantic discussions and innumerable legal consultations. Finally, the French government decided to ask its future partners to modify the treaty at least sufficiently to safeguard freedom in the civil atomic field.

But the problem of atomic armaments remained unsolved, and Gaullist Senator Michel Debré, during a parliamentary debate in July 1954, pointed out that '' . . . to the extent that France might wish to acquire, in the military field, the same relative advance she is seeking in Europe in the civil field, by this treaty she would be placed in the tragic position of automatically having also to authorize Germany to use atomic energy for military purposes.''

Modifications to the atomic clauses, together with many other amendments to various aspects of the EDC treaty, soon became a set of essential preconditions to ratification of the treaty by the French parliament. These modifications and amendments were grouped together in a ''Protocol of Application,'' proposed by France and, on August 24, 1954, rejected by her partners in the Brussels Conference. The following week the French parliament refused to ratify the treaty, and the joint defense project was dead.

One month later, the United States, Canada, the United Kingdom, and the Six — the proposed member countries of the EDC treaty — met in London and agreed to end the occupation of Germany, to allow the country to rearm, and to join NATO.

German Renunciation

Shortly afterward, on October 23, 1954, a series of agreements (the Paris Pacts) was concluded in the French capital. In particular, these agreements founded the Western European Union (WEU), which included the Six and Great Britain, by an extension of the 1948 Brussels Treaty grouping the same countries except Germany and Italy. Chancellor Adenauer, his country having regained freedom in the civil atomic field, had been persuaded to pledge officially and solemnly that Germany would forego the production of any nuclear weapon within its territory.

This pledge is contained in a letter written by Adenauer and annexed to the WEU treaty. Its definition of atomic weapons is stringent, covering any substance either produced or essential for such weapons. Plutonium, and uranium enriched in its 235 isotope to more than 2.1 percent, come within the definition. Their production was thus forbidden in Germany, although in the same letter there is a rather contradictory phrase that stipulates, among other things, that any products used only for civil purposes should be excluded from this definition.

Pierre Mendès France, the leader of the French government who had secured this renunciation from Adenauer, thought at the time that such an atomic barrier would be very unlikely to survive for more than about 15 years. This proved to be the case with regard to production of the ''forbidden'' materials for, in 1970, the Federal Republic of Germany brought a small plutonium-extraction plant into operation, and also joined in a project with the United Kingdom and the Netherlands to produce enriched uranium. Germany had recently signed the Non-Proliferation Treaty and therefore considered the pledge made by Adenauer to be outdated. In fact, none of the other signatories to the WEU treaty complained, least of all the French, who by then considered other aspects equally defunct.

Among these defunct aspects was a provision that the ''continental'' European countries (as opposed to the United Kingdom), which unlike Germany and Italy had retained full freedom of nuclear military action, could only use such freedom under WEU Council responsibility. This council was to decide the size of, and control the use of, any stocks of atomic weapons held on the European continent by these countries beyond the experimental stage. Insular Britain, now making its first atomic bombs following the successful test in 1952, was completely exempt from this measure. They could nevertheless have taken part, without risk of being affected, in a control system that, besides its use to verify the military atomic neutralization of Germany, would also provide a check on any weapons stocks held in Europe by any eventual producer.

The clause relating to determination of the size of weapon stocks held was less inconvenient for the French government than at first appeared. For it would have been quite possible, if necessary, to store atomic weapons in nonmetropolitan territorial departments such as those in Algeria, which would of course have been outside the continental control.

The unjustifiable discrimination between France and the United Kingdom explains why, in practice, neither the clause in the treaty nor the WEU inspection program was applied. The Euratom control system (see p. 136) subsequently adopted was entirely adequate for verification of the military nuclear abstinence of the European countries that had formerly belonged to the Axis.

Meanwhile France had retained her freedom of action, even in the military field. The French government was soon to examine this particular aspect.

The Mendès France Decision

Premier Pierre Mendès France, who was also his country's foreign affairs minister, on taking office in 1954 found himself faced with the problem of French independence over nuclear armaments; first of all during the final negotiations on the EDC, and then during the elaboration of the Paris Pacts. Shortly afterward, in November 1954, he decided to address the U.N. General Assembly in what proved to be a fruitless attempt to persuade the Americans and the Russians to stop their aerial bomb testing as a first step toward nuclear disarmament. In this attempt, Mendès France was following up a similar plea, some months earlier, by Indian Prime Minister Jawaharlal Nehru, after the first H-bomb tests and the irradiation of the Japanese fishermen.

On his return to Paris, deeply aware of the failure of his attempt, Mendès France called a meeting in his office at the Foreign Affairs Ministry on December 26, 1954. It was a meeting where every minister concerned, as well as other competent persons both favorable and hostile to French atomic armament, were asked to give their opinions. The CEA was represented by Guillaumat and Perrin, accompanied by myself.

Apart from calling on all those present to speak, whatever views they held, the prime minister said little during the three hours this meeting lasted. But at the end his conclusions were clear and precise. He had become very conscious of the difference in international influence, even in negotiations on disarmament, between the nuclear weapons states and the others; also of France's advantage in this matter over Germany as a result of the latter renouncing production of the weapon. He had therefore decided to launch, in secret, a research program for the development of a prototype nuclear bomb and an atomic submarine. He called on the minister for national defense, with the help of the finance minister, Edgar Faure, to prepare for parliament a draft decision to this effect. Faure, sitting through the meeting at the prime minister's side and facing all the other participants, about 40 in all, had remained silent.

The fall of the government in February 1955 blocked the implementation of this decision before it could begin. Later, Mendès France, without actually going back on his choice, somewhat restricted its scope. He had, he said, decided not to make an atomic bomb, which in any case would require at least four or five years if only to produce the necessary plutonium; but he had wanted to take the necessary steps to prepare for such an eventuality, in case the deadlock persisted in the nuclear disarmament negotiations between the Great Powers. In short, he had wanted to keep the nuclear option open in all directions, including the military one.

The next government, led by Faure, after some hesitation was persuaded by Gaston Palewski, the minister responsible for atomic energy, to authorize the transfer of substantial military funds to the CEA and to build a nuclear submarine. Two important overseas developments had preceded this: the British decision to manufacture the H-bomb and the successful entry into service of the first American atomic-powered submarine.

Euratom and Nuclear Abstinence

The EDC treaty discussions had obliged the French government to face the military issues of atomic energy. The relaunching of negotiations on Europe's unification, from 1955 onward, obliged the government to define its policy in this vital area.

On June 1, 1955, foreign ministers of the Six European countries met for the first time since the collapse of the joint defense treaty negotiations. The meeting, held in Messina, Italy, was to make a new start toward European economic integration. The Benelux Ministers (from Belgium, the Netherlands, and Luxembourg) submitted a memorandum to this effect, proposing several areas for integration: transport, postal and telecommunication services, energy in both conventional and nuclear forms, and finally, a common market.

Before the end of the year, only nuclear energy and the common market were still in the running, being apparently the only areas where agreement seemed possible. Negotiations continued in Brussels, Belgium, through an intergovernmental conference chaired by the Belgian Minister Paul-Henri Spaak. Somewhat reluctantly, the British took part in these discussions, in the role of rather hostile observers, until the end of the year.

All this took place against a background of veritable euphoria over atomic energy, which had been developing on a worldwide scale since the beginning of 1955. The U.N. scientific conference in Geneva had been an undoubted success and the lifting of secrecy was having a far-reaching impact. And so as the year ended, the three nuclear weapons powers, members of the most exclusive club in the world, were preparing to move forward under a new banner, that of the industrial atom for peace.

The main protagonists of the European negotiations believed atomic integration to be at once the most important and the easiest of the problems they had to resolve. They thought (incorrectly in this case) that because they were dealing with a technology still in its infancy, they would not be faced with particularly entrenched and irreconcilable national positions. Unfortunately, they failed to take into account the essentially political nature of the atomic adventure, whose military aspects were already ruling the world, and whose industrial aspects were to play an inevitable and vital part in world affairs over the coming decades.

However, all this rapidly became apparent during the first discussions in Brussels, in the course of which it was proposed that the future European supranational atomic agency be called "Euratom," short for the European Atomic Energy Community.

Great Britain, junior member of the Atomic Club, immediately made known that it could not take part in an organization wielding such powers. On the other hand, the United States greatly favored the idea of European integration within the Six, particularly since this alignment might put an end to the apparent French leanings toward nuclear armament. The fact that the nuclear integration of Europe could become an instrument of nonproliferation made it all the more attractive to the Americans.

In early 1956, Monnet, the Frenchman who inspired the earlier effort as well as this renewed effort toward European cooperation, organized the Action Committee for the United States of Europe, which was open to and included political and trade union supporters of integration from all parties in the Six countries. Monnet was personally convinced that the unification of Europe was of greater importance to France than her independent possession of nuclear weapons. He aired this view, at length, in several talks at his home to which I was invited.

In his opinion, one of the preconditions for any European nuclear organization was an "equality of rights," the criterion that had so complicated the EDC treaty negotiations and had finally contributed to their failure.

Although the conference in Messina had not dealt with the issue of nuclear weapons, there were clearly two possible ways open: either nuclear armament of all the European countries as a joint project, or European renunciation of nuclear weapons. The first road, being incompatible with the German pledge in the recent Paris Pacts, would have lead immediately back to the EDC problems. This left only the solution of joint renunciation, which Monnet's committee proposed and then adopted. Under this solution, the European Community, with sole ownership of all fissile materials in its member countries, was to develop atomic energy exclusively for peaceful purposes and was to be subjected to "watertight" safeguards effected by European inspectors. This would be an application, to the Europe of the Six, of the Acheson-Lilienthal plan proposed by the Americans nine years earlier in the United Nations.

A few days later, in his investiture speech on January 31, 1956, Socialist Prime Minister Guy Mollet, leader of the first government to be elected following the dissolution of the National Assembly by Faure, declared himself in favor of this proposal. "We must," he said, "make a preliminary choice. Are we to create a European industry to permit the production of atomic bombs, a production which would be virtually impossible at national level in each country concerned? My answer, clearly, is No!" In his concluding remarks he said that, among other objectives, the European organization should aim at establishing a control system that would rigorously guarantee the peaceful nature of all European nuclear activities. ''Whoever owns the fuel,'' he said, ''will be in a position to make atomic bombs. Consequently my government should call on Euratom to assume exclusive ownership of all nuclear fuels, and to maintain that ownership throughout any transformations the fuels might undergo.''

The combination of these conditions — exclusive peaceful utilization and Euratom ownership of all nuclear fuels — would have meant unilateral renunciation of military applications of atomic energy by any member countries that still had a choice in the matter.

A few days before the investiture debate of the Mollet government, General de Gaulle made his first visit to a French atomic center, that at Saclay. Before leaving the center he called together the leaders of the CEA and warned us of the probable dangers resulting from any loss of national independence due to the current European negotiations.

During the succeeding months, Prime Minister Mollet found himself faced with increasing opposition to any unilateral French renunciation both within parliament and from his own defense minister, Maurice Bourgès-Maunoury.

Even among the protagonists of European integration and those who had signed the resolution of the Action Committee for the United States of Europe, many thought the renunciation proposal went too far and that a compromise should be sought. A 5-year moratorium on the production and testing of nuclear bombs was then proposed by Spaak in Brussels in early 1956.

Under these conditions, the French government instructed its negotiators to support any plan that would allow the members of the community, except Germany, to retain freedom of nuclear action in the military field. (It seems to have been forgotten that Italy, through its peace treaty, had in fact already lost this freedom in 1947.) Two of the Benelux partners, Belgium and Luxembourg, were as a matter of principle opposed to their unilateral renunciations.

During these negotiations, the United States carried out an important action behind the scenes. Their dream was to see the creation of a zone of nuclear abstinence on the geographical scale of the Six, but they were aware they had to contend with a French attitude in favor of atomic armament. As a result, the U.S. State Department was cautious in its approach to this crucial question, avoiding the head-on attack that the nonproliferation policy of the 1970s was to call for on a worldwide scale. The Americans were for the moment counting on internal differences in France, combined with pressures from her European partners, to delay any final decisions: this tactic was made clear through an official memorandum that spelled out Washington's attitude toward Euratom.

Regarding military uses, the United States considered it inevitable that the Germans would raise the question of renunciation of the use of atomic energy for military purposes within the WEU in order to solicit their freedom from this discrimination. For this reason, the U.S. government was in favor of the moratorium proposed by Spaak; but since the French government was divided on this issue, the official attitude of the United States for the time being was to leave the Six countries to decide the matter among themselves.

However, the French differences were rapidly disappearing and the government, having initially considered giving up military atomic research, now authorized this both in the CEA and in the national defense laboratories.

The Road to Military Independence

In June 1956, the French Senate decided, by a very large majority, to set up a military division within the CEA. Two weeks later a wide-ranging debate on Euratom was held in the National Assembly, in which several protagonists of French atomic armament took part.

At the end of this debate, Prime Minister Mollet gave an unequivocal assurance that the country's freedom in the military atomic field was not and would not be restricted in any way. He did, however, make one concession, inspired by the Spaak proposal for a moratorium, by stating that France would not explode an atomic device before January 1961, the date when the parliament then in office would come to an end. But he insisted that no measures taken by Euratom, nor any agreement concluded by that organization, could restrict French freedom of action over military armament. Nevertheless, he foresaw preliminary consultations with France's partners, rather than simply informing them, before any final decision to manufacture nuclear weapons was made.

This was the first time the French government had admitted, within the National Assembly, that military research in the nuclear field was under way. The proposal that no nuclear explosion would be carried out before 1961 at the earliest was a rather academic restriction, for it was improbable that any French bomb could be produced before the end of the decade. This is how the clauses in the draft Euratom treaty, which could have resulted in French renunciation of military nuclear activities, came to be abandoned.

The treaty was finalized between May 1956 and the end of the year. It included the provision for a supply agency to ensure equal access by all community members to uranium ores and fissile materials, regardless of intended use. Euratom was given the right of ownership — more legal than commercial — of all plutonium and enriched uranium within the community beyond the requirements of national defense.

The "watertight" safeguards envisaged in the original Monnet proposal at the beginning of the year were included in the treaty, their rules of application being modeled on those of the recently adopted statutes of the International Atomic Energy Agency. However, the objective was different: It was not to ensure peaceful utilization, but to verify the use of all nuclear materials, whatever that might be. In the matter of weapons manufacture, the control did not extend (and did not need to extend) beyond the point of entry into the military factory concerned.

The system was thus a control of "conformity," verifying that nuclear materials were indeed used for the civil or military purposes declared by the user countries. This astute if somewhat contrived arrangement ensured that the control, while in no way discriminatory — very important from the political viewpoint — enabled the entire German and Italian nuclear programs, obligatorily restricted to civil activities, to be kept under surveillance so that military abstinence could be verified.

The system also meant that Germany, indirectly through Euratom, could henceforth take part in the civil atomic race. For this reason, the Soviet Union, already systematically opposed to European integration, unlike the United States, was hostile toward the new organization from the start. As for France, curiously enough the negotiations had obliged the government to renounce its indecision over a military nuclear program and declare itself more clearly in favor of independent civil development. However, even before the Euratom treaty was signed, in Rome in the spring of 1957, extremely serious international events had caused the French government to take further measures toward atomic armament.

Following the nationalization of the Suez Canal by Egyptian President Gamal A. Nasser in late July 1956, on October 30, Israel launched a lightning attack on Egypt. This was followed by a Franco-British landing at Port Said on November 5, the day after Soviet tanks had crushed a 10-day revolt in Hungary.

Confronted with Russian Premier Nikolai A. Bulganin's condemnation of the Franco-British action, with a barely veiled Soviet threat to use nuclear missiles to end the fight and reestablish peace in the Middle East, President Dwight D. Eisenhower on November 6 (the day of the presidential elections that returned him to office for a second term) officially warned British Prime Minister Anthony Eden to stop military operations in Suez so as to avoid giving the Kremlin any pretext for intervention. Under pressure from the Labour opposition, and faced with Nehru's threat to withdraw India from the Commonwealth, the British government gave way, and shortly afterward the French were forced to do the same.

One of the greatest postwar dramas had, in the course of only a few days, inflicted some of the deepest wounds conceivable between friends and allies.

French impotence in the situation had been clearly demonstrated before the whole world. Abandoned by Britain, held back by NATO, thwarted by the United States, and threatened by the Soviet Union, France found herself suddenly and terribly alone, at the very moment when the war in Algeria was entering its most difficult phase.

To many people, an autonomous French defense system based on national nuclear armaments then seemed the only answer to the failure and humiliation they had suffered. Although it was true that Britain, despite its own atomic weapons, had been the first to give way to the other two more powerful members of the Atomic Club, it was equally clear that British links with the United States had made it virtually unthinkable for Britain to contest the American veto.

The Mollet government, which with the Israelis had prepared the Suez operation in the utmost secrecy, felt the affront it had just suffered. Its previous hostility toward atomic weapons, the result of a passionate desire to see a United States of Europe created, was transformed overnight into a determined and positive interest in national nuclear armament.

On November 30, 1956, exactly 10 months after the investiture speech in which Mollet had declared himself in favor of a Euratom that would have entailed unilateral renunciation of atomic weapons, his armed forces minister and his secretary of state for atomic energy signed a new protocol, a fundamental document defining the objectives of a new atomic program for national defense. Under this agreement, the CEA was to carry out preparatory research into atomic explosions and, should the government then decide to proceed further, preliminary research leading to the production of prototypes and the staging of tests. The CEA, thus made responsible for providing the plutonium required for these projects, was also given the task of carrying out the necessary research for the construction of an isotopic separation plant to produce highly enriched uranium.

It was a long way from the EDC's definition of 500 grams of plutonium per year as the upper limit for safety. The combined determination of the successive governments of the Fourth French Republic had, despite their internal divisions, overcome the obstacle of the "European dream."

The Sahara Test

Following the fall of the Mollet government in May 1957, the two subsequent prime ministers, Bourgès-Maunoury, then Gaillard, were both in favor of national atomic armament. In April 1958, six years after the adoption of his 5-year plan, Gaillard decided on the initial steps needed so that a first series of nuclear test explosions could be carried out in early 1960 at an experimental site that had been under construction in the Sahara since mid-1957. In the absence of any progress with negotiations for nuclear disarmament, he considered himself no longer bound by the promise made two years earlier by his predecessor, Mollet, fixing January 1, 1961, as the earliest date on which French nuclear tests could take place. Nor was there a question of France consulting her partners in the European Community on this matter.

In June 1958, General de Gaulle was returned to power, first as prime minister and then, seven months later, as head of state. He immediately gave his full backing to the atomic weapons program, as well as to the anticipated date for the first experimental explosion, early in 1960. Under no circumstances was the general prepared to change the previous decisions. He confirmed this shortly before a unilateral moratorium on nuclear weapons testing became effective in October 1958. This moratorium, which was agreed to by the three existing nuclear weapons states, was intended to ease the path of the first American-Soviet negotiations on the banning of further nuclear explosions, negotiations that are dealt with in greater detail later.

It was under these conditions that neither the United States nor the Soviet Union showed much opposition to a campaign in the United Nations, which began in the autumn of 1959, to stop the French tests scheduled in the Sahara early the following year.

In fact, the U.N. General Assembly, under pressure from its African and Asian members, adopted a resolution opposing the forthcoming Sahara experiments and asked France not to proceed with them. The particularly strong U.N. opposition was principally due to the intense anticolonial emotions of the Afro-Asian countries, to whom the atomic bomb was still a white man's monopoly that had been used only against nonwhites.

France's position was defended by her representative in the disarmament negotiations, Jules Moch, who was not personally in favor of French nuclear armament. He explained that, in the absence of a general agreement on nuclear disarmament applying equally to all countries, France intended to go ahead with her test program, thus demonstrating France's determined resistance to any form of discrimination against herself or her interests. As to whether or not France should equip herself with nuclear weapons, the question concerned only the French and was not a matter for debate in an international organization.

Only a few years later, arguments such as this in the United Nations would have been considered quite shocking, for by then the limitation of the number of nuclear weapons states (the term nonproliferation was not yet in use) had become a major objective.

The General Assembly's vote on the Sahara tests was in fact no less than a manifestation in favor of nonproliferation even though the official reason given was the banning of aerial nuclear experiments and the corresponding radioactive fallout over a continent that so far had been free of such tests; and this at a time, it was stressed, when the United States and the Soviet Union were themselves both abstaining from testing.

The first French atomic bomb was exploded on February 13, 1960, at Reggane, nearly 1000 miles south of Algiers, which had barely recovered from the serious political troubles during the last week of January. It was three times as powerful as each of the first American and British bombs and, like those, it used plutonium.

Had the local army at that time succeeded, as had been threatened, in taking over Algeria, it could have led to the first instance of confrontation with an insurrectionist movement possessing a nuclear explosive device, which, however, was not yet in a form that could be dropped from an aircraft.

The second French explosion took place some six weeks later, on April 1, during a visit to France by Nikita Khrushchev, chairman of the Soviet Council of Ministers, while he was staying at the presidential residence in Rambouillet. The Soviet leader, after congratulating General de Gaulle on the French achievement, told him that the Russians had also been very proud of their first explosions, but less happy subsequently when the cost of producing a real atomic arsenal was fully appreciated.

In all, four different atmospheric tests were carried out at Reggane, between February 1960 and August 1961. The safety precautions taken limited the irradiation received by the sparse local population to barely detectable doses. Protests from neighboring African countries, some going as far as the temporary blocking of French financial holdings or, in the case of Nigeria, breaking off diplomatic relations, were essentially based on political motivations.

Following agreements giving Algeria independence, concluded at Evian, France, in 1962, the new Algerian government permitted the French atomic weapon tests to continue until 1966; and about a dozen underground experiments were carried out in the Hoggar region.

American Atomic Defense Agreements

France, in carrying forward a military nuclear program, had not been helped by either of her two North Atlantic Treaty allies — the United States and the United Kingdom — who had preceded her along this same road. However, a degree of softening of American policy in this field had developed over the past few years. In 1957, two incidents in particular had led to a change of policy in Washington: the British H-bomb explosion in May at Christmas Island in the Pacific and, above all, on October 4 the dramatic launching by the Soviet Union of the world's first artificial satellite, the *Sputnik I*. This considerable achievement of Russian technology had been preceded in August by their announcement that they had successfully tested an intercontinental ballistic missile. The surprise and psychological shock to the American administration was considerable.

One of the results of these Soviet achievements in armaments and modern technologies was an American effort toward reinforcement of defense relations with its allies, in particular the United Kingdom. So, in accordance with what had now become standard procedure, the British prime minister was soon on his way to Washington hoping to obtain from the American president a more truly genuine collaboration in the military nuclear field.

This time, Prime Minister Harold Macmillan held a new trump card in his British hand: the technical characteristics of his country's successful H-bomb test. Only plutonium had been used in this test, for as yet the United Kingdom had no supply of the preferred explosive for detonating such a bomb, namely pure uranium-235. The use of plutonium was believed to be both difficult and delicate, yet analysis by the Americans of the radioactive cloud resulting from the explosion had shown how well their ex-partners had mastered the technology of such a nuclear weapon.

For their part, the British were, once again, determined to be treated no longer as poor relations. They were also wary of a possible international treaty to end nuclear tests, which would clearly have impeded the further development of their various atomic weapons. Such a test ban was favored by the Americans, from whom the British requested access to both the technical know-how and the fissile materials they needed for weapons making. Their persistence in requesting this assistance, which they had now been pursuing for over 12 years, was this time backed by their own solid technical results. Their efforts finally bore fruit, and Eisenhower undertook to have the McMahon Act modified, particularly to allow renewal of Anglo-American military atomic relations.

Meanwhile, in December 1957 in Paris, the heads of the NATO

governments met for the first time since the advent of *Sputnik I*. The United States, wishing to make a gesture toward their other allies besides the United Kingdom, offered to provide NATO with a stock of tactical atomic weapons, on the condition that the nuclear warheads would remain under U.S. control. They also offered, for those allies prepared to accept launching bases on their territory, medium-range ballistic missiles, though again with the nuclear warheads remaining under American control. Finally, President Eisenhower announced an unequivocal offer by the American government to supply nuclear submarines to those allies who were interested or alternatively to supply the data and materials necessary for the construction and operation of such submarines.

The United States had in fact quickly recognized the revolutionary strategic and military potential of the nuclear submarine and by 1958 they had under construction or on the drawing board a fleet of some 50 vessels. These vessels comprised both attack units and missile-launching "platforms" capable, without surfacing, of launching Polaris missiles equipped with nuclear warheads and having a range of over 1500 miles. Such submarines, carrying up to 16 missiles, could remain submerged for months and, hidden off the coast of a possible enemy, would represent a redoubtable weapon for dissuasion or reprisal.

Apart from the United Kingdom, France was the only one of the Allies to inform the American government of her immediate interest in the nuclear submarine offer. This interest was all the greater at the time because France's own national project to build a submarine engine using natural uranium and heavy water had just been abandoned as a failure. The atmosphere with the Americans was relaxed, and the French even believed momentarily that they might receive transatlantic help with their atomic weapons production program. Thus, in early 1958, the chairman of the U.S. AEC, Strauss, one of the few American leaders who did not at the time believe that France's possession of the atomic bomb would make it more difficult to maintain Germany's renunciation, secretly authorized a French mission to make a detailed visit to the U.S. experimental proving grounds in Nevada. Following this secret mission, which bore the French code name "Aurore," the CEA was able to order from the United States some of the delicate and costly instrumentation needed to make the best use of the results of future French nuclear weapon tests. The American Congress, however, was unwilling to modify its atomic law in favor of allies other than the United Kingdom. In an attempt to change this attitude, President Eisenhower publicly declared in April 1958 that countries other than Britain, and in particular France, should be allowed to share in nuclear secrets.

The McMahon Act was amended in June by a virtually unanimous vote, but even then the door was really open only for the British. For although this second modification of the act provided for the transfer to certain countries of the nonnuclear components of atomic weapons, together with some secret information and materials relating to nuclear submarines, and although provision was also made for the transfer of secret information and materials for the production of the weapons themselves, this latter transfer was specifically restricted by the condition that the beneficiary country must have already achieved "substantial progress" in the field of atomic armaments.

At this time, this restriction was clearly intended to exclude all countries other than the United Kingdom for, during the discussions in the Joint Committee for Atomic Energy (JCAE), it was emphasized that the proposed amendments should in no way encourage a fourth nation to become a nuclear weapons power. "Substantial progress," in fact, was not accepted as accomplished until a nation had its own testing site, had carried out an important number of atomic weapons tests, and had the capability of manufacturing various types of weapons. At the time, only the United Kingdom fulfilled these conditions.

In 1958 and 1959, the British signed a series of agreements that secured for themselves once again, after a lapse of 13 years, the "special relationship" of a nuclear ally with access to American military secrets and materials. In addition, they obtained an engine for the first British nuclear submarine, a prototype for a series subsequently built by British industry under American license. One of the agreements provided for a supply of American uranium-235 for British weapons in exchange for plutonium produced in British reactors. The renewed alliance applied only in the defense field; there was no provision for collaboration over nuclear power plants, where competition between the two allies continued.

France, excluded from this new alliance, pursued her own efforts toward producing her first nuclear weapons. In September 1958, General de Gaulle sent a letter to the American and British heads of government, spelling out his thoughts on the organization of the Western World and what he deemed was France's too restricted role therein. At that time it was very clear that the NATO decision-making body was more and more completely under Anglo-American domination in its every action. The nations of the Atlantic Alliance therefore tended to be in two categories: those who were merely consulted and the two members of the Atomic Club who made the decisions. Incidentally, the situation was no different in the Warsaw Pact group of Eastern countries, all being merely consulted — or simply informed — by the Soviet Union.

The two members of the Anglo-American directorate, with the approval of the other partners in the alliance, seemed in no hurry to open their doors, and in even less hurry to make their atomic data available to a third partner. This was soon to be demonstrated in the evolution of French relations with the United States.

During the summer of 1958, following further American encouragement confirming Eisenhower's 1957 offer of a nuclear submarine, France informed the United States that she would like to take up the proposal and conclude an appropriate agreement. In February 1959, a French mission went to Washington for this purpose. It met with only partial success.

The JCAE, hostile to any such deal, was against the supply of a submarine engine, of technical data related to such an engine, or of any data on the submarine itself. Admiral Hyman G. Rickover, father of the nuclear submarine, had convinced the members of the JCAE that the Russians were far behind in this field, and that it was necessary for the Americans to preserve their advantage and avoid any risk of leakage as a result of communicating information to France. Rickover did not, however, oppose the sale to France of the uranium-235 necessary for a land-based prototype of a submarine engine, for he was convinced that French technicians would be unable to master so difficult and delicate a development. Hence, the barriers to information transfer remained firmly down, but those on nuclear materials were raised slightly.

The Americans thus showed their French ''allies'' that they were not real partners, and that this was so not only at the top levels of the Alliance, as proposed by General de Gaulle, but even within the framework of a specific offer officially made by President Eisenhower.

The enforced necessity of reinventing so much and of carrying out again so many research projects and programs that had already been completed long ago by France's principal ally, who now refused to communicate the results, were to lead to a greatly reinforced French scientific and industrial potential. This encouraged the nation further along the road to independence vis-á-vis the United States. This independence was hardly the objective sought by Washington. It should be emphasized that this American doctrine of refusal to communicate data and provide their allies with arms that for the most part had long been in the possession of their Soviet rival, could only weaken the cohesion among all the NATO members. Inevitably, the French attitude toward NATO progressively began to change.

The first step toward withdrawal was in fact taken a month after the failure of the French mission concerning the acquisition of U.S. nuclear submarines. In March 1959, the French government notified the NATO Council of its decision to retain its naval forces in the Mediterranean under national command in the event of war (until then, they would have been placed under NATO's command.)

Nevertheless, in early May 1959, the agreement for supplying enriched uranium for the French land-based prototype submarine engine was signed and subsequently approved by Congress. In the end, Washington provided only half the agreed upon amount of uranium-235, but even so such a supply saved France several years in the development of her engine.

During the same session of Congress, the first agreements had been approved to proceed with Eisenhower's proposal of December 1957 for the deployment of American nuclear weapons within the framework of the NATO Alliance. Following this, and with the exception of France, Denmark, and Norway, who either set conditions unacceptable to the United States or simply refused to have nuclear weapons stocks within their territory, all the NATO countries were supplied with American tactical atomic weapons, bombs, or missiles, which they held under a ''double lock'' control system (see below). Until July 1966, when France withdrew from the NATO's Military Committee, French forces in Germany were also equipped with these weapons, in accordance with a 1961 Franco-American agreement.

The nuclear warheads allocated to NATO were to be stocked in the various countries concerned under American supervision, to be made available to the combined forces if a conflict arose. Each country was responsible for its own ''delivery vehicles'' (aircraft or missiles) but the nuclear warheads could not be mounted on their vehicles without express orders from the Americans; in addition, the weapons could only be made ready for use following reception of a special radio signal that had to be authorized by the U.S. president himself. Consequently, these weapons were not under the legal responsibility of the countries where they were held, but under that of the United States. This important juridical position was to feature prominently in future negotiations on nonproliferation.

The same complicated ''double lock'' control system was to apply to strategic weapons held in Great Britain, Italy, and Turkey, the only countries that in 1959 had agreed to provide launching sites for U.S. medium-range Thor and Jupiter missiles.

Of course, the major part of the American strategic nuclear armament remained under the sole responsibility and control of the United States, maintained in perpetual readiness 24 hours per day whether in the form of bomber aircraft constantly flying around the earth, or Polaris submarines — for which a base was provided in Britain, and later in Spain — or of the most recently developed intercontinental ballistic missiles.

The French Deterrent Force

The "double lock" control system illustrates the complexity of the NATO military integration. This concept was to be rejected by General de Gaulle during a speech at the École Militaire (Military Academy) on November 3, 1959, when he declared that "of course, if the need arose, the French defense system would be joined with those of other countries. This is in the nature of things. But it is indispensable that we have a defense which is our own, so that France can defend herself, by herself, and in her own way."

The speech called on France to acquire a "force de frappe (strike force) . . . a great defense effort over the years to come . . . based on atomic armament which, whether we make it ourselves or whether we buy it, must belong to us."

And so in July 1960, six months after the Reggane explosion, the government of Michel Debré proposed a draft law providing for a vast military program of modernization in the armed forces. In particular, his proposal called for the creation of a so-called "strike force" or "deterrent force," comprised of both nuclear weapons and their delivery vehicles — supersonic aircraft, missiles, and nuclear submarines.

Opponents criticized the project for being too costly for the country and because they believed that in the end it would result in an inadequate strike force compared with that which the other nuclear powers could deploy. Nevertheless, the law instituting a long-term plan was adopted in December 1960, following some heated debates. France was to follow in the footsteps of her predecessors to produce an H-bomb and various types of other more effective and less cumbersome atomic devices.

The Debré government had resolutely secured the passage of the French atomic armament program from the qualitative to the quantitative stage, and this was to become one of the master achievements of Gaullist policies. But it was no less clear that the qualitative stage owed its realization to the opposition during its period of office in the Fourth Republic. But the members of the opposition were now careful not to recall this fact, for their hostility to the deterrent force had become a leitmotiv of their campaign against de Gaulle and his government.

The military program was based on plutonium produced in the Marcoule piles or, should it prove necessary, on that formed in the first nuclear electric power plants. In addition, it used uranium-235, separated in an enrichment plant that had been envisaged in principle for the past four years. The final decision to build this enrichment plant at Pierrelatte, also in the Rhône Valley, was made by the government in 1960 despite tempting murmurs from the Americans who, opposed to the project, had on several occasions since 1958 hinted at the possibility of supplying France with uranium-235 for her armament program.

This expansion of French nuclear military activities inevitably led to problems over the Euratom security control. As has been seen, this was based on verification of the civil nuclear activities of member states, among other things to monitor, without any apparent discrimination, the complete renunciation of nuclear weapons by the Federal Republic of Germany and by Italy.

Euratom's managing body, the Commission, initially attempted a strict application of the treaty, wanting to inspect all the nuclear installations in France with the sole exception of those actually making weapons. In fact, the treaty excluded from safeguards only "materials destined for defense requirements while undergoing special fabrication processes or which, subsequent to such processing, are held ready or stored within a military establishment according to an operational plan."

Faced with clearcut opposition from the French government, in 1961 Euratom's Commission agreed not to enforce a strict application of the treaty and to forego safeguards on any materials allocated to the "defense cycle," in particular, exempting from control the Marcoule and the future Pierrelatte enrichment plants, which were to produce the two atomic explosives, plutonium and uranium-235. This exemption demonstrates the limitations of the fiction of "conformity control," for clearly international safeguards can scarcely make sense in connection with materials that a government has officially designated as reserved for military purposes.

The Cuban Crisis

Two years later, at the end of a remarkable series of international developments, the French government was presented with an unsought offer from Washington concerning France's military nuclear program. The offer came about as a result of a new and troubled chapter in the story of Anglo-American relations, which followed in the footsteps of a world crisis.

The crisis began on October 15, 1962, after an American U-2 high-altitude spy flight over Cuba had discovered a partially built launching ramp for medium-range missiles, similar to those installed by the Americans in Britain, Italy, and Turkey over the past four years.

From the moment of Fidel Castro's successful bid for power in 1959,

Cuba had been the most westerly point of Soviet advance in the Cold War, as well as — since the abortive Bay of Pigs American landings in April 1961 — the Achilles' heel of President John F. Kennedy's foreign policy. For all Americans, Cuba was a particularly sensitive subject.

Three days after the U-2 discovery, further extensive reconnaissance flights revealed a large number of launching ramps under construction, together with adjoining silos for stocking nuclear missiles. In addition, a fleet of Soviet freighters, detected in the Atlantic and clearly heading for Cuba, included several ships capable of transporting such missiles. It was undeniable proof that the Russians were breaking their often-repeated assurances that they would only supply defensive weapons to their Central American ally.

The American fear of a surprise thermonuclear attack had suddenly acquired a reality far more tangible than the threat of Soviet intercontinental ballistic missiles, the size and status of which were still uncertain.

President Kennedy was convinced that removal of the missiles and dismantling of the launching ramps could be achieved peacefully only if this could be done without too great a loss of face by the Soviet Union. He therefore announced publicly, on October 22, the establishment of "quarantine" measures in the form of a maritime blockade limited initially to offensive weapons bound for Cuba. America's principal allies had been warned of this action in advance: former Secretary of State Dean G. Acheson went to France to inform the president of the French Republic of Washington's decision. De Gaulle gave his unreserved approval of this action, which incidentally he saw as a justification for his doctrine of independence for each country inside the Atlantic Alliance.

Despite the denials of the Soviet delegate in the United Nations — denials that were disproved by American photographic evidence — the organization's Secretary General U Thant sent identical messages to the U.S. and Soviet presidents on October 24, asking for a voluntary suspension both of the weapons shipments and of the quarantine.

Khrushchev, who had underestimated the capacity for resistance of the American leaders, had to accept the U Thant proposal. An exchange of letters and messages, some secret and some public, took place between the two presidents between October 25 and 28. In one of these messages, the Soviet leader emphasized that the weapons being installed in Cuba were in the hands of Soviet officers, so that their accidental use against the United States was impossible. In the end, Khrushchev gave in completely, and agreed to restrict all future Soviet military assistance to Cuba to defensive weapons. He was obliged to order the dismantling and return to Russia of the "offensive" armament, and to accept U.N. verification of this operation. For their part, the Americans renewed their pledge not to invade Cuba.

The Soviets had not even secured the removal, which they had requested at one stage, of similar American launching bases in Turkey. In fact, these bases were already obsolete (following the entry into service of the American missile-launching submarines) and they had no great strategic importance, even when the Cuban crisis was beginning. They were dismantled shortly after the crisis ended, as were the bases in Britain and Italy.

Thus ended the most dangerous crisis the world had lived through since the end of World War II. Armed conflict between the two Great Powers had been avoided. The danger had been caused by nuclear weapons, and had been circumvented thanks to a fear of using them. The balance of dissuasion was no longer merely an idea in the minds of theoretical nuclear strategists — it had become an undeniable political reality.

The Kennedy Proposals

Two months later there was a further nuclear crisis, this time between Western allies and fortunately implying less lethal possibilities.

In the late 1950s, the British had begun a modernization program for their atomic armament involving acquisition of missiles with nuclear warheads that could be launched from aircraft far from their targets. In the framework of their 1958 Anglo-American nuclear agreement, the British had decided to buy an advanced rocket from the United States, the Skybolt rocket, which was then being developed for American aviation.

In December 1962, the Americans informed the United Kingdom that, in view of the forthcoming entry into service of long-range ground-to-ground strategic missiles, their air force was no longer interested in Skybolt and was considering termination of its development. This was a hard and unacceptable blow for the British, who had abandoned their own rocket development and were completely dependent on the Americans for the missiles essential to their nuclear strike power.

The issue was raised at the Anglo-American summit meeting in Nassau on December 18. British Prime Minister Macmillan had spent the two previous days with President de Gaulle at Rambouillet, where they had examined problems of Great Britain's entry into the Common Market, now under discussion for over a year, and of its possible membership in Euratom. The broader issue of an overall common European policy was also broached; the possible participation of Britain's strike force in a joint European defense system was an important aspect of this proposal.

At Nassau, Kennedy confirmed to Macmillan his decision to abandon Skybolt, although the United Kingdom would be welcome, of course, to continue the research and production program on its own, but in view of the very substantial cost of the operation, this was not possible for Britain.

Thus, once again, the Anglo-American ''special relationship'' in the nuclear field had proved actually harmful for the British, ending in impairment of their military atomic independence for which Clement R. Attlee had fought 15 years earlier. Suddenly, the British found they had nuclear weapons, but no transport vehicles — without which the weapons were useless.

Macmillan therefore had no choice but to accept a Kennedy offer to supply the British defense services with American Polaris missiles, to which the British nuclear warheads could be fitted. However, it was specified that these missiles, intended to be launched from submerged submarines, were to be used only in British units placed at the disposal of a multinational NATO force. The Americans had been hoping to set up this force since 1961, and had undertaken to make a contribution to it equal to that of its European members. It was, nevertheless, understood that the bulk of the American nuclear devices would, of course, remain outside this hypothetical international force and thus remain under the direct orders of the U.S. president.

The United Kingdom was, however, to be allowed to use these missiles independently in exceptional circumstances if the British government had decided that ''overriding national interests'' were at stake. But the agreement effectively put an end to British long-term hopes of having an autonomous national nuclear strike force.

Following this, Kennedy and Macmillan decided to propose to de Gaulle the immediate creation of the NATO nuclear force. With the proposal came the offer of Polaris missiles for France, on similar terms to those accepted by the British.

But such an offer had none of the same significance for France, since it would be several years before she would possess either the submarines from which to launch the missiles, or the nuclear warheads to arm them. This was where the British, the only ones to have benefited since 1959 from special American assistance, were privileged in relation to both nuclear submarines and weapons.

Three weeks later, on January 14, 1963, President de Gaulle rejected the American offer during the same world-shaking press conference at which he announced his decision to break off negotiations on the admission of the United Kingdom to the Common Market and to Euratom. In complete contrast to the British, who had only just decided to join the American camp in matters of nuclear armaments, de Gaulle declared his determination that France should have her own deterrent force. He considered that ''to put our own fighting resources into a multilateral force under foreign command would be to go against all the principles of our defense and our policy.'' He further declared that the theoretical possibility of ''taking back into our own hands, in extreme circumstances, our elements in the multinational force'' would be practically unrealizable ''in the unprecedented moments of an atomic apocalypse.''

The rift in military atomic relations between France, the United States, and the United Kingdom had become deep. It was to become still deeper some months later, following the spectacular American-Soviet reconciliation that followed the Cuban crisis of the preceding autumn.

This reconciliation made possible the conclusion of the first nuclear renunciation treaty, prohibiting all but underground explosions. The agreement, signed in Moscow on August 5, 1963, represented the outcome of five years of animated negotiations, the story of which is recounted later (see p. 165).

The United States had been trying, right until the last moment, to persuade France to take part in the new treaty. To do this, however, would have affected France more than any other nation, for she had just reached the stage of producing powerful bombs that could then be tested only in the atmosphere. The other Great Powers had, of course, already passed this stage without being restricted in any way.

The clear lessons learned from the constantly repeated difficulties in Anglo-American nuclear collaboration were already enough to prevent France from considering any last-minute offer, however tempting, made in an effort to secure her adherence to the new treaty, or at least to persuade her to keep open the option by not immediately and finally refusing such adherence.

The American offer had come in a personal letter from Kennedy to de Gaulle, sent 10 days before the official signing of the Nuclear Test Ban Treaty in Moscow by the three Great Powers, the Soviet Union, the United States, and the United Kingdom. The American president, recognizing the handicap that the proposed renunciation would represent for France, suggested to the French head of state — while emphasizing the political and technical difficulties — that they might explore alternative arrangements that would make unnecessary any French tests that would come under the new prohibition.

In his answer, President de Gaulle, while expressing his full understanding of the U.S. and Soviet decision, made possible by the stage their own nuclear tests and armaments had already reached, pointed out that France — having started in this direction later and alone — could not now discontinue the necessary experiments before she had equipped herself with a smaller but nonetheless comparable armament. Referring to the American offer, which as he pointed out was somewhat vague, he declared (no doubt with the Nassau negotiations in mind) that he failed to see how the offer of U.S. assistance could be accepted by France without the imposition of restrictions on the French right to the free use of her weapons, restrictions that would be incompatible with her sovereignty.

It was under these circumstances that France refused to become a partner in the Nuclear Test Ban Treaty. They also rejected the Americans' proposal to join in exploratory talks to study possible political and technical solutions that might have made unnecessary any future French tests under the forbidden conditions.

The Multilateral Nuclear Force

Throughout this same year of 1963, discussions were taking place within NATO concerning the proposed multilateral nuclear force. The French government did not participate.

The idea of a joint nuclear force operating on behalf of the NATO countries originated as a result of the Cold War, one of the earlier consequences of which had been the rearmament of West Germany. Certain sectors in the U.S. State Department, believing that individual German nuclear abstinence would be more difficult to maintain following France's acquisition of her own atomic armament, were therefore advocating some form of shared ownership and responsibility for Germany. The underlying philosophy was, in fact, a compromise between maintaining total nuclear prohibition for Germany and allowing it full freedom in this field.

The inspiration for the joint nuclear force was similar to that behind American advocacy of international management for the sensitive parts of the nuclear fuel cycle: Here also, compromise between freedom and prohibition had in the past led to proposals (in the Acheson-Lilienthal plan) for international management of the "dangerous" stages. This concept has more recently appeared as support for multinational plants for fuel enrichment and reprocessing. However, in the case of the suggested NATO multilateral force, the internationalization was to be only partial, for Washington would reserve the right of veto over use of the force's nuclear weapons.

Throughout 1963, talks continued between the governments of the United States, the United Kingdom, and the other main NATO allies interested in the plan (Germany, Belgium, Italy, Greece, and Turkey). It quickly became apparent, however, that the United States had no intention whatever of attaching nuclear submarines to the multinational force, but rather, at this stage, of equipping internationally manned surface units with Polaris missiles. The vulnerability of such surface vessels, the difficulty of creating the complex control system needed to enable missiles still subject to an American veto to be used, the lack of British enthusiasm, inevitable language problems in the command chain for such ''ships of Babel,'' and the French abstention were all severe handicaps for the project. Against these handicaps was to be set the main objective: namely, to offer some measure of satisfaction to Germany by allowing it a ''finger on the trigger'' and so, it was hoped, reducing its possible inclination to acquire military nuclear autonomy. The fear of German nuclear armament, which was a real obsession with the Soviet Union, had also become an important factor in the policies of the Americans. The French consistently opposed the projected joint nuclear force, believing it likely to awaken a military nuclear appetite among all concerned, but the Germans supported it and were prepared to contribute 40 percent of the cost.

In the autumn of 1964, the issue was further complicated by the election in Britain of a Labour government hostile to any German nuclear armament, even in the attenuated form of a multilateral force. This British position coincided with that of the Soviet Union, also resolutely against the project.

This was the situation at the end of 1964, when the sudden and dramatic accession of China to the Atomic Club forced the Americans and the Russians to close ranks in order to concentrate on what was to them the overriding issue — the danger of an increase in the number of countries equipped with nuclear weapons. The multilateral nuclear force was to be the first victim of the new reconciliation, and in due course the project was abandoned.

From then on, the only beneficiaries from American offers in the nuclear weapons field were the British, although as a consequence they lost some of their freedom. France, on the other hand, was well on the way to having her own independent nuclear force, destined to become the second most important in the Western World, both militarily and politically.

5. The Chinese Bomb

When the Moscow Nuclear Test Ban Treaty was signed in August 1963, Nikita Khrushchev was much less considerate in his attitude toward his closest ally and rival, Mao Tse-tung, than John F. Kennedy had been toward Charles de Gaulle.

Faced with Chinese protests that the treaty reinforced the nuclear monopoly held by the United States, the Soviet Union, and the United Kingdom, the Soviet government bluntly replied that it could not resist the acquisition of atomic weapons by further capitalist countries and by West Germany while it was simultaneously supplying China with the same weapons or encouraging other socialist countries to acquire them. This was the Russians' argument justifying their determination to remain the only Communist member of the Atomic Club.

In actuality, the Russo-Chinese break, which was now revealed, went back several years. On October 15, 1957, the Soviet Union and the People's Republic of China had signed an agreement for technical military cooperation that included providing China with the design for an atomic bomb and the necessary data to build it. However, the Russians probably expected, just as the Americans had in NATO, to retain the right of decision on any possible use of the weapon by the Chinese, who themselves wanted full freedom to use it as they might see fit. Two years later China denounced the agreement, and once again determination to promote nonproliferation had clearly been the cause of a complete breakdown in relations between two allies.

The breakdown was openly confirmed in August 1960, when all the Soviet experts were suddenly recalled, causing a serious slowdown in Chinese industrial growth. Peking has never forgiven Moscow for this action: It is always described to foreign visitors as "treachery by the social imperialists who overnight tore up their contracts and recalled their technicians."

Four years later, by an ironic turn of fate, the fall of Khrushchev and the Chinese revenge occurred within hours of each other. On October 15, 1964, seven years to the day after the conclusion of the treaty signed by the Soviet leader with Mao and subsequently denounced, the People's Republic of China became the fifth member of the Atomic Club.

The Chinese intention to produce an atomic bomb as quickly as possible had been publicly declared in 1963, following the refusal of the People's Republic to sign the Nuclear Test Ban Treaty, and during a violent ideological debate then under way with the Soviet Union. The country that had given gunpowder to civilization, now faced with the prospect of isolation by the Western World and at the same time deprived of assistance from its former ally — which had become its rival for worldwide Communist leadership — was certainly not going to abstain from producing its own nuclear explosives.

The Chinese atomic bomb was exploded in the Sinkiang desert in northwestern China. Unlike the first test carried out by the other four nuclear weapons powers, it was based on uranium-235, a fact that added to the general surprise.

Virtually nothing is known about the Chinese research program, and the Soviet Union has never officially admitted giving direct assistance. However, in private conversations with Western colleagues, high-ranking Soviet scientists have said, "We gave them everything." It is uncontested that there have always been Chinese physicists of great merit; some of whom, trained in modern American laboratories, returned to their country of origin following the witch hunts of the early 1950s. But even this, together with the highest priority that was surely given to the work, does not fully explain the speed with which China achieved the various steps leading to the most sophisticated types of explosions, in particular, that of the H-bomb. The success was all the more remarkable since years of occupation and revolution had left the Chinese laboratories for advanced physics in a state of obsolescent decay, which largely persisted despite Soviet help during the early years following the founding of the People's Republic by Mao.

The Russians probably found themselves, some 5 to 10 years later, in the same position vis-à-vis the Chinese specialists as the Americans were in relation to scientists of the British team during the war. The Soviet need to catch up with the United States as quickly as possible must have overruled considerations of nonproliferation toward the new and close ally. The need to have the best physicists, in the greatest numbers possible, for research into the mechanisms of the bomb must have swayed the innate Russian mistrust of foreigners. So, it seems certain that the Soviets must have used Chinese help in achieving their atomic and thermonuclear bombs.

It is also a fact that, sometime around 1954, the Russians had started building a uranium enrichment plant in the Sinkiang area, the most distant region in the world from all the American bomber bases. The Russo-Chinese disenchantment occurred as the plant was nearing completion, so that the Chinese would have been able to finish the work on their own, thus becoming equipped to carry out the final concentration of the product.

In any case, it is certain that China discovered large quantities of uranium within its vast territories, and that the hundreds of specialists trained in Soviet universities must have been invaluable for their country's military atomic effort.

Another single Chinese bomb, similar to the first, was experimentally exploded in 1965, and this led to the belief that production was very limited. But the three tests that followed in 1966, by their nature and by the greatly increased power of the last one (15 times Hiroshima), surprisingly showed that China must have established considerable production of uranium-235, and moreover that it was well on the way to producing the H-bomb. In addition, one of the 1966 tests involved the use of a radio-controlled missile carrying a nuclear weapon, demonstrating not only that the Chinese could produce "miniaturized" transportable atomic bombs, but also that they could build missiles with a range of some 400 miles. Fifteen years later, they successfully tested an intercontinental missile that could reach every part of the Soviet Union.

The Chinese nuclear program had more surprises in store and, on June 17, 1967, China became a full member of the Atomic Club — a year before France — when it exploded a powerful thermonuclear device that had apparently been launched from an aircraft. China had succeeded in less than three years (i.e., more quickly than any of the first three nuclear weapons powers) in graduating from "conventional" nuclear weapons to the H-bomb.

This astonishingly rapid Chinese success, which was confirmed in late

1968 by a further powerful H-bomb test, constituted a world political event of the utmost importance; for it weighed the international balance of power in favor of a country that, due to its enormous population, its low level of industrialization, and its nationalism, had the least to fear from a nuclear conflict.

Act Three

RENUNCIATIONS
1963 to 1981

1. Limiting Nuclear Tests

A third of a century after the failure of the Acheson-Lilienthal plan for a nuclear weapons free future, there are thousands of atomic bombs unevenly distributed over the surface of the globe, owned and guarded by the five Great Powers to whom the U.N. Charter has entrusted the responsibility of keeping world peace.

Although these countries have had the wisdom not to commit any further nuclear sin, the destructive force of their total stocks of weapons is by now far beyond the power of human understanding. It is equivalent to nearly 20 billion tons of classical explosives, sufficient to destroy several times over all the main concentrations of human population throughout the world.

The two superpowers, the United States and the Soviet Union, unable to agree between themselves to slow the pace of the nuclear arms race, have nevertheless had some success in establishing an arms limitation procedure, which is applied, first and foremost, to other nations!

The first limitations, which implied renunciation of sovereignty, were connected with international nuclear exchanges in the civil field. For the countries concerned, the exchanges meant acceptance of two control

157

systems, the International Atomic Energy Agency (IAEA) safeguards system controlling peaceful utilization and the Euratom verification of conformity, both of which were inspired by Washington.

Soviet opposition, particularly in support of Third World protests, practically paralyzed the IAEA safeguards system during its early years. Throughout this period, the United States employed their own American inspectors to verify that U.S. international nuclear aid was not used for military purposes. They did, however, make one exception. For the six countries of the European Community, the United States accepted the Euratom security control, which they considered a multinational system. This was evidence both of the Americans' wish to encourage this European organization and of their lack of confidence that the community would soon become a single integrated political entity.

In 1963, circumstances suddenly took a turn in favor of U.S. policies. American industry concluded a sales contract for the first nuclear power station to be built in a Third World country, India, which accepted the principle of IAEA safeguards for the installation. Until then, India had resisted and refused the idea of international safeguards. At the same time, improved American-Soviet relations, following resolution of the Cuban crisis, resulted in a complete change of position by the Soviet Union, which was suddenly very favorably inclined to setting up international safeguards. This policy reversal opened the way to applying such safeguards to ensure the peaceful nature of world nuclear trade, and eventually to international agreements for the renunciation of atomic weapons.

In parallel, beginning in 1958, the two superpowers showed growing concern for avoiding damage to the biosphere due to their nuclear tests. In 1963, they concluded an agreement to this effect — the Nuclear Test Ban Treaty.

They now turned to the problem of limiting access to the new weapons by other countries, and in 1968 this resulted in the Non-Proliferation Treaty (NPT) under which nuclear abstinence is verified through the IAEA safeguards system. Finally, from 1969 onward, the Soviet Union and United States together attacked the problem of limiting the stocks of their most menacing devices — strategic nuclear weapons — leading in 1972 and 1979 to two Strategic Arms Limitation Treaties (SALT). Due to hostile reaction in the U.S. Senate, the second of these treaties was not ratified.

It will be seen that there are two distinct parts to the story of the explosive use of fission: the one just related, which described the accession to the Atomic Club of its five member countries, and a second part — the subject of the pages that now follow — that is devoted to nuclear limitation or renunciation agreements. However, these agreements have been developed against a background of constant growth in the arms stocks held by the weapons powers as well as a continued increase in the number of countries able, should they wish, to make such arms.

To tell the second part of this story, we must again go back in history.

Radioactive Fallout

In 1950, Albert Einstein, in an open letter to President Harry S. Truman, solemnly called on him not to pursue research on the H-bomb. The famous scientist, who ten years earlier had been the first to warn Roosevelt, now described the atomic menace: "Radioactive poisoning of the atmosphere, and hence annihilation of any life on earth, has been brought within the range of technical possibilities. The ghostlike character of this development lies in its apparently compulsory trend. Every step appears as the unavoidable consequence of the preceding one. In the end, there beckons more and more clearly general annihilation."

Four years after this spontaneous apocalyptic description, U.S. Secretary of State John Foster Dulles announced that the Americans were adopting a strategy of massive nuclear reprisals. This announcement was followed by reports about the Japanese fishermen who fell victims of fallout from the American H-bomb test at Bikini. Together these reports gave the general public — perhaps for the first time — a precise idea of the immense destructive potential of the new explosive, thus creating an ever-growing fear of the dangers of radioactive fallout from experimental explosions. Shortly afterward, Indian Prime Minister Jawaharlal Nehru proposed a suspension of nuclear arms tests, while the Great Powers of the Western World called for, and obtained, a resumption of disarmament negotiations.

The negotiations were restarted in 1954, in a disarmament subcommittee comprised of representatives from France, the Soviet Union, and the three Anglo-Saxon powers. The subcommittee met some 160 times over a 3-year period, but its work was for naught. The negotiations went through a series of advances and setbacks, hopes and disappointments, all faithfully reflecting the current fluctuations in the Cold War.

However, as the stocks of weapons and aircraft for strategic bombardment built up in the Soviet Union, a new type of equilibrium — the balance of terror — developed between the two great atomic powers; for they were reaching the stage where each had a sufficient quantity of atomic and hydrogen bombs to ensure mutual destruction in the event of conflict. For the first time, American cities were within the range of Soviet bomber aircraft, or of missiles launched from Soviet submarines. The era of reciprocal deterrence had arrived.

Paradoxically, possession of atomic weapons now became a factor for equilibrium and even peace in the world. The new situation was reflected in proposals made at the summit conference in Geneva, Switzerland, in July 1955.

Marshal Nikolai A. Bulganin, chairman of the Soviet ministerial council, proposed that the Great Powers should put a stop to all nuclear tests, as well as agree never to use nuclear weapons other than in self-defense, and even then only after a decision of the U.N. Security Council.

President Dwight D. Eisenhower, trying initially to create a climate of confidence, submitted a proposal for reciprocal photographic air observation over U.S. and Soviet territories, together with mutual exchanges of military plans. Bulganin did not appreciate these ideas any more than some photographs of Soviet territory, previously taken from American reconnaissance aircraft, which Eisenhower had chosen to show him on this occasion. Bulganin could not do other than reject the proposal, although this did not prevent the United States, some years later, from observing Soviet territory anyway — systematically and secretly by means of U-2 reconnaissance planes.

An additional factor at this time was the growing — particularly following the irradiation of the Japanese fishermen — protest movement, which was championed by many scientific and moral authorities that since 1954 had been opposed to the arms race and the dangers of nuclear testing. Einstein, the British philosopher Bertrand Russell, the French philanthropist Albert Schweitzer, and the American chemist Linus C. Pauling, all Nobel Prize winners, issued repeated appeals and petitions. They called for a halt to atomic explosions, insisting on the dangers of radioactive fallout to the human race, particularly the dangers of radiation-induced mutations in chromosomes, the bases of heredity.

In 1957, 18 of the greatest German nuclear physicists, with the backing of the Social Democratic party, signed a manifesto in which they refused to take part in any work on the production of atomic bombs (work that, in fact, their country had renounced in the 1954 Paris Pact).

In the United Kingdom, a minority group of trade unionists and members of the Labour party declared itself in favor of unilateral nuclear disarmament, in contrast to the ruling Conservative government that had no intention of stopping the tests — at least for the moment. In fact, Britain was just then preparing to explode its H-bomb, despite protests from both the Indian prime minister and his Japanese colleague, who publicly deplored the pursuit of ''these mad nuclear experiments.'' During the next few years, some thousands of British protestors, indifferent to the inclement English weather, organized many "peace marches" between London and the Atomic Weapons Research Establishment at Aldermaston.

It was already well known at that time that an atmospheric atomic explosion produces, besides extremely powerful instantaneous radiation, three different types of radioactive fallout. The most immediate comprises relatively heavy materials over an area extending some tens of miles from the explosion's center; then there is dust, initially projected some 6 to 10 miles into the troposphere; and finally there are clouds of very fine particles projected even higher into the stratosphere.

The absence of natural clouds at very high altitudes results in these last particles spreading around the globe and falling to earth slowly over a period of several years. On the other hand, the radioactive dusts in the troposphere are dispersed rather rapidly, in an opposite direction to the earth's rotation, and within a few weeks will have fallen out at various points largely dictated by rain or snowfalls.

For the first time in the history of technology, the consequences of a human action have had an easily detectable yet lasting effect over the whole world. Atmospheric atomic explosions inevitably increase the radioactivity in the air we breathe and in the food we eat.

The exact evaluation of the consequences for mankind of higher environmental radioactivity is a problem as important as it is difficult to resolve; it has been studied since 1958 by an expert U.N. committee. Their conclusions have shown that the additional radiation received as a result of experimental nuclear explosions adds only a very small percentage to the total radiation to which humanity is constantly subjected from natural sources. These sources are mainly radioactive elements in the earth's crust, such as uranium, thorium, and their radioactive "descendants," as well as cosmic rays from the sun and stars. One must also take into account radiation doses due to the medical uses of x rays and certain other artificial sources, such as color television.

Until now, the total radiation dose received by the populations of the northern hemisphere as a consequence of nuclear testing has been on the average equivalent to about two extra years of exposure to natural radiation. In the southern hemisphere, it has been some three times less, principally because most of the aerial tests have taken place over the northern half of the globe.

Public opinion, however, finds it difficult to accept these technical data that show how small the risk resulting from the tests is. People have a tendency to accept the most pessimistic statements of a few scientists who have chosen to renounce professional objectivity to become exaggerated propaganda agents; a distortion they justify on the grounds that there can never be too much of an effort made toward reducing the chances of an atomic war, and that anxiety over nuclear testing can perhaps help progress toward disarmament.

Unfortunately, in going to such lengths to increase public concern over this very small risk from radioactive fallout due to weapons testing (as has also been done more recently over another very small risk — that from civil nuclear power production) the end result is that attention has been diverted away from the real and cataclysmic horrors that nuclear war would inflict on humanity.

It was in the face of this terrible menace that in 1957 the French representative to the disarmament negotiations, Jules Moch, put forward a new idea. Because of the impossibility of verifying declarations concerning stocks of nuclear weapons, ever-growing quantities of which could be produced and secretly stored over the years, Moch proposed additional controls. While there was still time, he proposed that control measures be concentrated on the means for transporting and delivering the weapons — bomber aircraft and, particularly, ballistic missiles whose production was just beginning both in the East and in the West. Unhappily, this proposition went the same way as all the preceding ones, but the principle was to reappear in the American-Soviet negotiations, from 1969 onward, on the limitation of strategic armaments, the SALT talks.

In October 1957, the Polish foreign affairs minister, Adam Rapacki, submitted a proposal to the U.N. General Assembly for a "denuclearized zone" in Central Europe, in which the storing of nuclear weapons would have been completely prohibited. The zone was to comprise East and West Germany, Poland, and Czechoslovakia. The plan was set forth in more detail in December, at the time the United States announced their intention to supply the North Atlantic Treaty Organization (NATO) forces with tactical atomic weapons. The Soviet Union, trying to avoid further encirclement by the Western powers, was seeking the removal of peripheral military bases and therefore supported the Rapacki plan. But the Western powers rejected it.

Negotiating a Moratorium

A year later and with pressure mounting from a growing campaign to limit avoidable radiation, first negotiations began toward outlawing nuclear tests. On several previous occasions the Soviet Union had favored such a move.

The negotiations were preceded by a technical conference, held in Geneva in the summer of 1958, which brought together experts from the four principal Western powers (the United States, the United Kingdom, Canada, and France) and four Eastern powers (the Soviet Union, Poland, Rumania, and Czechoslovakia) to study methods for long-distance detection of atomic explosions.

The experts were soon in agreement. Detection and analysis of the radioactive fission products in the tropospheric cloud released by an aerial explosion could not only indicate the location and the date of the explosion but could also yield accurate knowledge concerning the type and power of the bomb tested. Only a virtually overlapping series of tests (such as the Soviet series at the end of 1961) could produce mixed clouds and thus make individual analysis practically impossible.

Other systems for the detection of atomic tests could depend on the acoustic shock waves resulting from aerial or submarine explosions; on the seismic oscillations occurring during explosions near ground level, underground, or underwater; or on the radio and optical wave emissions produced during explosions in the atmosphere. There was even the possibility of detecting a nocturnal test on a clear night by observing the moon, whose dark side could be briefly illuminated by the intense flash on earth.

By August 1958, the technical conference was able to specify the characteristics of a worldwide system for detecting all explosions except underground ones of low power (i.e., less than a quarter of the power of the Hiroshima bomb) that might be confused with seismic shocks.

With this specification in mind, the three nuclear powers now carried out quickly, during October, a series of very powerful experimental explosions, before applying themselves to the "real" Geneva negotiations on the test ban. The negotiations began with the Soviet Union announcing a unilateral moratorium, a move that was difficult for the United States (and, of course, Great Britain) not to follow also once their own test programs were completed. This American-Soviet moratorium was to last nearly three years, during the period of the French aerial tests over the Sahara.

The Geneva negotiations got off to a good start and during 1959 the principal clauses of a draft treaty were adopted, calling for renunciation not only of all further weapons testing but also of any help to other countries wishing to carry out such tests. Even the location of the future control organization was decided. It was to be in Vienna, Austria, and would employ more than 5,000 persons, each of its 180 surveillance and control stations requiring 30 technicians. The budget for the new organization was estimated at some $50 million per year.

With the relaunching of the Cold War in 1960, following the shooting down of the American U-2 spy plane over the Soviet Union, discussions on the test ban came to a standstill. Once again the immediate source of difficulty was the problem of inspection and control, the main disagreement being over the number of control posts to be set up in the Soviet Union, and the number of verifications necessary in cases where small seismic shocks could not be distinguished from weak underground atomic explosions.

The new American president, John F. Kennedy, following his election at the end of 1960, decided to continue the negotiations. He wanted them to succeed, despite growing hostility from the Pentagon and from part of Congress. To him, the signing of an agreement would be a token of Soviet good faith in seeking international détente.

However, when the discussions started again in the spring of 1961, it soon became clear that the Soviet Union was raising all sorts of new difficulties and no longer seemed interested in reaching an agreement. The Russians even attacked France about her first nuclear tests, claiming — in an accusation not without humor — that these tests were for the benefit of the United States and the United Kingdom and were therefore in violation of the test moratorium, which had been in effect since 1958.

France had, in fact, never concealed that she would not consider herself in any way bound by a test ban agreement concluded only between the three nuclear powers. She had not taken part in the political bargaining, which she could not accept as a genuine effort toward disarmament as long as these three powers continued to make atomic weapons and so increase the quantities they held in stock.

Just before the 309th meeting in the Geneva series of negotiations, the Soviet Union broke the truce. On September 1, 1961, at Novaya Zemyla Island in the Arctic, the Russians began a series of aerial tests — some of which were of very great power. In so doing, the Russians risked provoking opposition from most of the nonaligned countries, from Japan (where objections to all nuclear explosions were especially strong), and indeed from public opinion throughout the world. Two weeks later, the Americans also restarted their testing, but only underground so that the radioactivity remained in the soil and there was no fallout. At the end of November 1961, an attempt was made to relaunch the Geneva conversations, but once again this led rapidly to deadlock, for the Soviet Union remained opposed to any form of inspection.

Parallel to these events, disarmament discussions in the U.N. General Assembly were marked by a series of contradictory propositions and initiatives, giving the impression of a machine running with neither power nor purpose.

Disarmament negotiations nevertheless continued beyond the end of

1961 within the 18-member U.N. Disarmament Committee, which was comprised of five major Western powers, five from the East, and eight nonaligned countries. At the beginning of 1962, the French government, believing it indispensable to reach some form of preliminary agreement among the member countries of the Atomic Club and having therefore tried, but without success, to bring about the necessary ''prenegotiations'' on this important problem, refused to take further part in the meetings; the discussions continued with the French seat remaining empty until 1978.

The U.N. test ban conference adjourned indefinitely in January 1962, and in April the United States resumed aerial testing in the Pacific. It seemed that the labors of 350 Geneva test ban sessions were finally doomed to failure. The Cuban crisis of the following October changed this.

The new détente that followed this crisis soon brought about a resumption of talks between Kennedy and Soviet Premier Nikita Khrushchev. They began early in 1963, through diplomatic channels, and their first objective was to decide how many inspections per year should be permitted. The next step was the important decision to set up a direct communications link — the ''hot line'' — between the two heads of state.

Finally, in spring 1963, when the negotiators were still unable to resolve the problems of detecting low power underground tests and of carrying out related inspections, the Russians withdrew their demands for agreement on these points, and the negotiations culminated in a limited test ban treaty. This left the way clear for each country to carry out the tests needed to perfect their existing weapons.

The Moscow Treaty

A series of letters between Kennedy, Khrushchev, and Macmillan then led to tripartite discussions in Moscow, from which a treaty finally emerged that prohibited nuclear weapons tests in the earth's atmosphere, in space beyond the atmosphere, and underwater. Underground tests were not banned on the condition that no radioactive contamination resulted that could be detected beyond the frontiers of the country carrying out the tests. The parties to the treaty further undertook not to help or take part in any unauthorized test explosion arranged by or on behalf of a nonsignatory state.

This Nuclear Test Ban Treaty, also called the Moscow Treaty, was open to signature by all sovereign nations. Its duration was unlimited, but all signatories had the right to withdraw, on 3-months notice, if they considered that exceptional circumstances related to the treaty's objectives were endangering their supreme national interests. This last provision underlined the unilateral character of the engagement, which was similar to that conceded some months previously by the United States when they permitted Britain to use Polaris missiles independently of NATO in very exceptional circumstances.

On August 5, 1963, the eve of the 18th anniversary of the destruction of Hiroshima and following five years of negotiations, the foreign affairs ministers of the United States, the Soviet Union, and the United Kingdom signed the Moscow Treaty banning all but underground nuclear explosions.

Three months later, the U.N. General Assembly unanimously adopted a resolution on the nuclear neutralization of space, forbidding the placing in orbit of missile satellites carrying nuclear weapons — a provision that was also incorporated in the Treaty on the Peaceful Uses of Space, concluded at the end of 1966.

Since the end of the war, some 500 aerial nuclear tests had been carried out by the Great Powers: 300 by the United States, 180 by the Soviet Union, 25 by the United Kingdom, and 4 by France. The total energy released in these tests (650 megatons) was some hundred times that of all the explosives used during World War II, and one-twentieth of the combined total potential nuclear explosive power accumulated in the American and Soviet arsenals by 1981. Some 10 tons of "nonexploded" plutonium had been vaporized and dispersed in the atmosphere.

The Soviet explosions, although less numerous than the American ones, had double the cumulated power, due to the 1961 and 1962 series of Russian tests at Novaya Zemyla Island. The most terrifying Russian bomb was approximately 4,000 times more powerful than the larger of the two used against Japan.

The general public was evidently pleased with this first agreement, which not only represented an American-Soviet nuclear conciliation, but also offered a measure of protection for the world environment. In addition, some believed they could see signs of a relaxation in the atomic arms race.

The efforts of many great scientists, who for nearly 10 years had been campaigning against radioactive pollution in the atmosphere, thus began to bear fruit, for the halting of American and Soviet aerial tests was the start of a real and progressive reduction in the contamination of the biosphere. The very existence of the treaty, its precious precedent for future American-Soviet negotiations, its propaganda effect on world opinion, and the rapid adhesion of a great many countries made it a most significant instrument of détente.

When the treaty became effective on October 10, 1963, its success was assured by the fact that there were already over 100 signatory countries. The U.S. Senate ratified the treaty, 80 votes to 19.

France and China, of which one was continuing and the other just beginning to build up a nuclear arsenal, were the only two major powers who refused to sign the treaty, maintaining that it would inevitably reinforce the nuclear monopoly of the three principal atomic powers. The French government, despite President Kennedy's last-minute appeal, was determined to keep open all options for its future military program, although its nuclear testing at that time was being pursued underground in the Sahara and would therefore have been permitted under the treaty. But the decision had already been made, as part of the country's H-bomb development plans, to carry out aerial tests in the Pacific, such experiments being too powerful to conduct underground. France was not prepared to renounce these experiments which, because of her relatively modest program, could not have very much effect on the expected gradual reduction of atmospheric contamination following the end of the Russian and American tests.

The treaty was also to mark the first step toward an American-Soviet nonproliferation policy for atomic weapons, for the outlawing of aerial tests was regarded at that time as a way of blocking possible military programs. Underground experiments, because they were considered more expensive and also more difficult to analyze, were of much less value than aerial tests. Nevertheless, in 1963 the United States and the Soviet Union could not — or would not — agree on a complete halt to all forms of nuclear weapons experiments. Had they also been able to put a stop to underground testing, it is probable that many countries would have followed them, giving up from that moment any ambitions they may have had toward acquiring nuclear armaments.

In fact, there is no denying that a technically advanced country could produce an operable atomic weapon without the need for tests: None of the five members of the Atomic Club failed with their first experimental explosion. On the other hand, it is very difficult, without tests, to advance to the next stage and develop the range of sophisticated and powerful weapons needed to constitute a meaningful force either for dissuasion or for aggression.

From the legal viewpoint, the Moscow Treaty was not a discriminatory agreement, since all the signatory countries had undertaken the same obligations. Nevertheless, for a country that had not previously carried out aerial tests, to renounce the right to do so under the treaty implied a very real discrimination when compared to countries that had already profited from such tests in the past. These countries were, of course, the United States, the Soviet Union, and the United Kingdom, together with the two powers that hadn't signed the treaty, France and China, who remained free to carry out the tests other countries had voluntarily denied themselves.

In reality, therefore, the Moscow Treaty was neither a nuclear disarmament treaty nor a true instrument of nonproliferation. The hopes for disarmament that it awoke could not hide its limitations, emphasized during the week following its ratification when the United States became the first country to resume underground nuclear testing, and again highlighted 10 years later by the first Indian atomic explosion. This latter event fully demonstrated that, as long as the treaty could not be applied to underground tests, it could not prevent a new candidate for Club membership from trying out a first weapon, nor could it restrain existing Club members from improving their arsenals.

Nonetheless, the Moscow Treaty, as the first international renunciation agreement in the military nuclear field, was a considerable event in international atomic history. For the first time, and thanks to the specific support of the Americans and the Russians acting together, nations agreed voluntarily and unilaterally to accept restrictions on the pursuit of a nuclear armaments program, whether already under way or as yet only contemplated.

With the exception of a very few accidental escapes of fission products from insufficiently contained underground tests, both in the United States and the Soviet Union, the Moscow Treaty has been faithfully observed by its adherents. Nor has it provoked recriminations among its signatories, unlike the situation that arose with the most important nuclear renunciation agreement, the NPT.

Once concluded, the Moscow Treaty was used by Washington against the French government with a zeal that the latter thought both unnecessary and quite regrettable. The United States more than once invoked the clause forbidding the provision of assistance to the military program of a country likely to engage in test explosions other than those underground. In this context, Washington decided not only to prohibit the sale to France of American electronic equipment for the evaluation of results from the French nuclear weapon tests but also to block the transfer of sophisticated materials and apparatuses that could be used either for civil or for military purposes. This was the fate of a giant electronic computing machine, costing several million dollars, that had been ordered in 1964 by the Saclay civil nuclear research center essentially to analyze the results of experiments on the structure of matter. Such computers could be of great help in working out the extremely complex theories and calculations leading to the H-bomb, but these computers were by no means indispensable, for they had not yet been perfected at the time when such weapons were first being conceived in the United States and the Soviet Union.

Faced with the American embargo on equipment and materials, the French government decided to take the first steps toward a national computer industry. Then when Washington, anxious to improve the American balance of payments and to strengthen the dollar, finally realized that its measures were encouraging an independence that could result in loss of a fruitful market for U.S. equipment, it withdrew the embargo and authorized the sale of advanced computers to France, on the condition that they be used only for civil purposes.

The withdrawal of the embargo does not alter the fact that the U.S. efforts to prevent France from developing her atomic armaments were contrary to the spirit of the North Atlantic Treaty, which specifies that its adherents shall provide mutual assistance in maintaining and increasing their individual and collective capacity for resistance to armed attack.

In this sense, the U.S. interpretation of the Moscow Treaty could be considered barely compatible with American engagements toward France in the context of the Atlantic Alliance. Two years later, the French government was to decide to withdraw from the integrated command of the Alliance.

The Tests Continue

The Moscow Treaty had no effect on the rate at which the two superpowers pursued their efforts to improve their armaments. Both sides turned to underground testing, which was soon taking place at a rate one and one-half times greater than the aerial tests of the 1950s. From 1963 to the end of 1980, some 400 underground tests, with powers up to a megaton, had been carried out by the United States; and although there had been only about 300 Soviet explosions, most were of greater power. In addition, improvements in the cost effectiveness of these tests in relation to the analysis of results virtually eliminated the initial handicap created by abandoning aerial experiments. Thus the superpowers succeeded in giving real satisfaction to a world public opinion worried about atmospheric radioactive fallout without hindering the progress of their nuclear super-armament in any way.

In 1974 the Americans and Russians reached an agreement that from 1976 onward they would limit the power of their underground military explosions to 150 kilotons, 10 times the power of the Hiroshima bomb, such a limit being no great hindrance to the advancement of military test programs. The agreement, apart from a few rare Soviet breaches, has been relatively well respected.

By contrast, negotiations for a total ban on all nuclear testing, which have been under way since 1978 between the three initial nuclear powers, are nowhere near a conclusion. A 3-year moratorium has been envisaged, but even this unsatisfactory compromise solution has failed for two reasons: the insistence by the military establishment on following up the effects of aging on the efficiency of their weapons and, once again, the problem of local verification in case of possible confusion between test explosions and natural seismic shocks.

The largest American underground explosion had a power of five megatons. It took place at the end of 1971 in one of the Aleutian Islands between Alaska and Kamchatka in the Bering Sea. The ecologists made great efforts to have it banned, predicting a disastrous earthquake or a tidal wave. James R. Schlesinger, chairman of the U.S. AEC, went with his wife and children to the area at the time of the test to demonstrate his confidence that no such dangers existed. According to the specialists, the only and indeed minimal risk could have been that a "developing" earthquake might have been advanced by a few days. Had this happened, the tremor would have been less severe than if it had occurred naturally at a later date.

The other American tests were carried out in the Nevada desert, which has become a perfect demonstration ground for long-term storage of the resulting radioactive wastes. Indeed, regular checks have shown that the fission products from the explosions, together with the quantities of plutonium that have not undergone fission, remain fixed in the molten rocks and are not dispersed into the surrounding terrain. Curiously enough, opponents of the civil development of nuclear energy, for whom a favorite argument is that no solution has been found to the problem of final storage or disposal of wastes, have never seemed worried about this huge national polygon where for years the ground has been loaded with radioactive products.

Although Great Britain does not have access to the technical data obtained from American nuclear tests at Nevada, it is able to use the underground experimental area, which it has done some 10 times between 1963 and 1981 for the improvement of British Polaris equipment. Some of these explosions even took place while Labour governments were in power, much to the chagrin of the party's left wing, which has consistently opposed any form of nuclear armament.

China has for its part carried out some two dozen tests, mainly in the air, in the Sinkiang desert over the 16 years following the first Chinese explosion. The most powerful was of some three megatons. Other than at the United Nations, where aerial tests were regularly and virtually unanimously condemned by the General Assembly, these Chinese explosions excited little criticism. This was not the case with French tests. Initially carried out in the atmosphere at Reggane in 1960 and 1961, then underground in the Hoggar (another region of the Sahara) between 1961 and 1966, they were continued in the air over the Pacific in order to study thermonuclear charges too powerful, at this experimental stage of French development, to be exploded underground.

The French H-Bomb

At the beginning of the 1960s, the French Commissariat à l'Energie Atomique (CEA) was committed to the study, development, and realization of nuclear warheads for the country's deterrent forces. Bombs were being designed for delivery by Mirage IV bombers, for nuclear-missile-launching submarines, and for missile silos on the Albion Plateau in Provence, in southeastern France.

The work of the CEA in this field was financed by the military authorities, for which a "classic" atomic weapon was considered appropriate as a first step. Priority was therefore given to armament types already agreed upon for the air force, navy, and army. As a result, at the beginning of 1965, scientific work on the development of a thermonuclear weapon was at a standstill.

But the president of the French Republic, General Charles de Gaulle, had other ideas. For him, the H-bomb was of capital importance; but he wanted his country to master the technology by herself. When, as the Moscow Treaty was being concluded, he had refused President Kennedy's last-minute offer to study the political and technical conditions for possible American help with the French armament program, he had confided to his closest colleagues, "If they offered me the plans of the H-bomb, I would not take them at any price." Not only was he determined to avoid the dependence on Washington that would have been the price for a type of bomb that could not be effectively tested by an adherent to the Moscow Treaty, he was also particularly concerned that his own nation should completely master the technology, as she had already done in the case of nuclear-powered submarines.

General de Gaulle's impatience, which since 1966 had fallen particularly on the minister responsible, Alain Peyrefitte, and on the CEA authorities, reached its peak in June 1967 at the moment of the Sinkiang H-bomb explosion. "So, now the Chinese have done it too!" was his reaction.

Fortunately, Robert Hirsch, the CEA's general administrator, was determined to neglect no avenue toward the goal. He mobilized the best scientific brains in the CEA to carry out a systematic study of every solution previously envisaged — none of which by itself had seemed completely satisfactory. This study led to an association of three concepts; one of which, the most recently proposed, had been discarded, although in reality it was the key to the practical realization of the H-bomb. The association of these three concepts was to lead, in late summer 1967, to mastery of the thermonuclear system.

One year later, on August 24, 1968, the first French thermonuclear device was exploded at the Pacific test center. The device was in a cradle, suspended from a balloon at a height of 900 feet. By substantial use of lithium deuteride, it was possible to achieve a power well above two megatons.

This explosion represented a double satisfaction for the minister responsible for the CEA, Robert Galley, because the uranium-235 of the detonating charge had been produced in the Pierrelatte enrichment plant, which had been built under his direction. This explosion, which had been preceded by that of a boosted bomb and was followed by a thermonuclear plutonium explosion, confirmed French mastery of the range of technologies and of the phenomena concerned.

The first series of aerial tests took place in 1966, and General de Gaulle was present at one of them. With the exception of 1969, the tests continued each year until 1974 over two uninhabited atolls, Mururoa and Fangataufa, in the southern part of the Tuamotu Islands in the South Pacific. Among other things, these tests enabled a warhead with a power on the order of a megaton to be perfected for use with submarine-borne missiles.

Opposition to the French Tests

Altogether, some 40 tests were carried out in the atmosphere in the Pacific by the French. Their total power was approximately 10 megatons, whereas the previous American and Soviet tests had accumulated powers on the order of 150 and 450 megatons, respectively.

Despite the remoteness of the site from populated areas, and the limitation of radioactive fallout due to testing with the device supported from a balloon, these French tests raised protests in many countries of the South Pacific and several countries in Latin America, as well as in Australia and New Zealand. Nevertheless, it may be noted that in 1972 the U. N. Scientific Committee on the Effects of Atomic Radiation recognized that the additional contamination resulting from both the Chinese and French aerial explosions did not significantly modify the situation following the greater contamination from the earlier tests by the first nuclear powers.

At first, merely formal, the protests became more vehement with the passage of time. In 1971, a press campaign in Peru attributed the origin of an earthquake in Chile to the French tests; and although the situation calmed down in 1972 following discussions between responsible French and Latin American scientists (who agreed that no biological danger could result from the radioactive fallout), Peru nevertheless suspended diplomatic — though not cultural or commercial — relations with France from 1973 to 1975.

In 1972 the opposition moved to Australia and New Zealand. The situation was amplified by the approach of elections in both countries toward the end of the year. The two Labour parties, backed by powerful local trade unions and playing on the sensitivities of Pacific nationalism and anticolonialism, together with popular antipollution themes, succeeded in mobilizing public opinion. France was judged to have no business in the Australian hemisphere and was called upon to carry out her tests in her own area, for example, in Corsica. A campaign to boycott French products, which at times led to violence, was successfully organized.

The protestors' campaign ignored the fact that 15 years earlier the very coast of Australia had provided the site for the first British thermonuclear explosion. Nobody in Australia or New Zealand had protested then or during the very numerous and powerful American Pacific tests. By contrast, the offending French atolls were some 2,500 miles from the Australian and New Zealand coasts. In addition, the radioactive clouds from the French tests, moving from west to east, took approximately 12 days to circle the globe before passing over Australia, so that the biological effects of their fallout were insignificant, a fact that Australian experts readily accepted.

Nevertheless, the stimulated agitation paid dividends, for the two Labour parties won their elections in Australia and New Zealand, and so felt themselves entrusted with a popular mandate to take any actions necessary to halt nuclear testing in the Pacific. Following fruitless ministerial discussions in Paris in the spring of 1973, the New Zealand and Australian governments took the affair to the International Court of Justice at The Hague, Netherlands, accusing France of damaging their interests and rights and those of their dependencies. Having themselves agreed, through the Moscow Treaty, to forego nonunderground nuclear testing, New Zealand and Australia found it hard to accept that another country should carry out such tests, if not on their doorstep, at least in "their" ocean.

In this matter, France did not recognize the competence of the International Court of Justice, having decided in 1966 to exclude from that organization's jurisdiction "any differences over French activities related to national defense." In June 1973, the court asked France, in moderate terms, to exercise restraint, to abstain from nuclear testing leading to radioactive fallout on Australian and New Zealand territory, and to avoid any action that could further aggravate the dispute between these countries. These latter countries, however, were in no way restrained in their behavior: relations with France were suspended, the boycott of French products and of ships wearing the French flag became total, and among various vessels attempting to penetrate France's prohibited zone during the tests was a New Zealand frigate with a government minister on board — the minister for immigration!

Calm gradually returned, but for a long time there were deep scars in Franco-Australian nuclear relations, as well as in Australian public opinion, which remained particularly sensitive to military and even civil aspects of atomic energy.

The last series of aerial explosions in the Pacific took place in 1974. The French government, no doubt satisfied with the progress of her experimental program and also anxious to take into account world opinion, had been saying since 1972 that all possible measures were being taken so that only underground tests would be needed in the future. However, the government was not prepared to exclude completely the possibility of further aerial tests should these again prove necessary.

Meanwhile, the International Court at The Hague was in no hurry to pass judgment on such a delicate matter. It finally did so at the end of 1974, stressing that France had made several public declarations of her intention to stop aerial explosions once the 1974 series had been completed. On this basis, the court was able to state that "the Australian objective having been effectively attained, France having undertaken not to carry out further nuclear tests in the atmosphere in the southern Pacific, the subject of dispute has disappeared and there remains nothing on which the court has to give judgment."

Accordingly, starting in 1975, the first French underground nuclear tests were carried out at a depth of 1,800 feet in artificial cavities made in the coral formations inside the more resistant geological structures of an ancient volcano. France is the only country to have carried out such explosions in an atoll. The necessary techniques were quickly mastered and there were some pleasant surprises. All the radioactivity from such a test was completely trapped in the basalt melted by the explosion, and the shock presented no risk of starting an earthquake, only a slight tremor that would be detectable at ground level.* Above all, however, the cost of this method of underground testing proved significantly less than that of experiments in the atmosphere.

There was very rapid progress in the rate at which experiments could follow one another and in the reduction of their costs. Also, the quality of data obtained was continually getting better. Each successive experiment involved a modified explosive device, embodying new improvements resulting from sophisticated processing of information from the preceding tests. The data processing required "super-computers," which were also indispensable for further analysis of the results obtained and of the precise modes of operation of the nuclear devices involved.

In this way, the underground explosions, of which there were about 40 between 1975 and 1980, led rapidly to the "miniaturization" of weapons; to important improvements in their behavior, including their resistance to attack by "antimissiles"; and to new work toward the development of missiles with multiple warheads. In addition, a great deal was learned about the formation and structure of atolls.

Civil Underground Experiments

Under the Moscow Treaty, the door was wide open for every type of underground explosion. Both the Americans and the Russians made full use of this freedom, not only for their military programs but also in developing civil applications. The latter received much publicity in the United States, where they created considerable popular interest.

Indeed, the use of atomic explosions in the course of civil engineering work was an excellent way of justifying the pursuit of underground testing with military implications, but it became a source of difficulty in agreements on the renunciation of atomic weapons. The application was eventually abandoned, at least in the United States, where from 1974 onward it was considered unlikely to be used in practice and the Americans felt the less said about atomic explosions the better. Antinuclear campaigns and use of the argument by India to justify its first atomic explosion as having only peaceful objectives explains this reversal of American thinking.

This sequence of changing attitudes is a good example of the danger of premature emphasis on technical arguments in support of policies pursued for other reasons.

---

*On a few occasions, underwater rock movements following a test produced sea movements similar to a minor local tidal wave.

The idea of using the extraordinary power of atomic explosives for civil engineering is as old as the discovery of fission itself. As early as May 1939, Joliot-Curie and his team, in their basic patent on "the improvement of explosive charges," envisaged uses in mining and in public works. In 1949, some weeks after the explosion of the first Soviet bomb, the Russian delegate to the U.N. General Assembly announced, in a manner as political as it was premature, that atomic energy was being used in the Soviet Union to move mountains, irrigate deserts, and divert rivers.

Then in 1952, during discussions in the French parliament of the country's first atomic plan, one of the Communist deputies, Arthur Ramette, asserted that "in the Soviet Union, the release of atomic energy will enable the Turgay mountains to be moved so that the rivers Yenisey and Ob', which until now have discharged their waters into the frozen seas of the north, will henceforth irrigate and fertilize the sunbaked deserts of Uzbekistan, Kazakhstan, and Central Asia. Deserts will become cotton fields and rubber plantations, fertile soils for soft fruit cultivation, for orchards and for vineyards. So will a scientific miracle be put unreservedly at the service of the peoples of the Socialist countries."

This description of a science fiction miracle was greeted by applause from one side of the national assembly, by laughter from the other. Future American and Soviet projects, however optimistic they might be, could not be expected to reach such heights of imagination or even poetry! Nine years later, in fact, the Soviet Union was to denounce American work on the peaceful uses of nuclear explosions — very similar to its own work in this field — as military testing in disguise.

The arrival of the thermonuclear bomb brought about considerable changes in the problem of civil nuclear explosions. A completely new kind of power was suddenly available, at a lower unit price and also with less production of radioactive by-products. In short, a source of gigantic amounts of energy, concentrated in the most minute volume, became available at a moment's notice and without raising severe radioactivity problems.

A program to study these possibilities was launched by the U.S. AEC in 1957 under the name of "Plowshare," after the words of the prophet Isaiah, "They shall beat their swords into plowshares." The project was fathered by Edward Teller, the principal brain behind the American H-bomb, who also coined a slogan for this weapon, calling it a "clean" bomb. In truth, it was "clean" only to the extent that it produced less radioactive fallout, for a given power, than a conventional A-bomb.

A first series of underground tests was carried out in 1961 as part of the U.S. military program, the study having particular political importance in relation to U.S.-Soviet negotiations on banning all nuclear explosions. In case of such a ban, it would be vital to determine the level below which the detection equipment available in one country would be unable to distinguish an earth tremor from a weak nuclear explosion set off by the other country.

An underground explosion results in the instantaneous generation of temperatures of several tens of millions of degrees and pressures of several million atmospheres. The vaporization and fusion of the rocks nearest to the explosion are accompanied by a shock wave that fractures and crushes rock formations further away. The result is a cavity, higher than it is wide — in the form of a chimney, which can be over 300 feet high — that is filled with broken rocks and in which all the remnant radioactivity is trapped. The exact shape, nature, and dimensions of the cavity, which can be considerable, depend of course on the power of the explosion and the nature of the ground.

If the explosion is at a sufficient depth, it has no surface effect but produces a slight earth tremor. At a lesser depth, it is still contained but, due to subsidence of earth into the cavity, produces a form of crater in the shape of a funnel at ground level. In this way, in 1962 in an alluvial area of Nevada, the explosion of a charge some seven times more powerful than that of Hiroshima, at a depth of nearly 600 feet, produced an open crater 1,200 feet wide and 300 feet deep. However, there were no radioactive by-products at the surface since these remained trapped in the melted and vaporized materials near the explosion's center point. That was the largest hole ever made by man in a single operation: It had been achieved in a few seconds and had cost far less than an equivalent explosion with classical explosives would have.

If the explosion takes place only a little way below the surface, there is a risk that it will not be contained. The crater and the rocks thrown to the surface of the ground will be more or less saturated with radioactive products.

From this brief survey of the various effects in the formation of craters or cavities filled with crushed rock, a variety of possible applications for peaceful nuclear explosions may be envisaged. Two main categories have been considered: one concerning explosions at great depths, the other for surface excavation.

In the first category, the following applications have been proposed or already tried in practice: creation of underground storage capacity for liquid hydrocarbons; extinguishing fires in oil or gas wells; boosting of low-content hydrocarbon deposits by increases in temperatures and pressures and by the crushing of rocks; *in situ* cracking of heavy hydrocarbons contained in bituminous shales or sandstones; processing metallic minerals by pumping acids through a cavity containing crushed ores; recovery of geothermal energy by injecting water into an explosion "chimney" and extracting super-heated steam; and, finally, production of nuclear energy itself using the same system of water circulation in a large cavity, initially formed by a very powerful explosion and in which smaller bombs could be subsequently detonated at regular intervals.

Applications that have been considered in the second category, namely that of crater-forming explosions, include the formation of artificial ports, canals, hydropower reservoirs, and dams for changing river courses, as well as the removal of surface soil to permit opencast mining.

During the 10 years following the Nuclear Test Ban Treaty, the Americans carried out some hundred underground nuclear explosions for civil test purposes, at a total cost of approximately half a billion dollars. In some cases, several successive and connected firings were made in rapid sequence, as would have been necessary for the excavation of a canal such as that suggested as a "second Panama Canal" in Nicaragua.

These last tests were a rich source of information for the military program, for they yielded details of how atomic devices would behave under irradiation, and particularly in the face of shocks from nearby explosions. Such data were invaluable for the development of bombs that could resist intense radiations and shock waves from the explosion of nuclear antimissile defense missiles, which are launched toward the trajectory of an attacking device in order to destroy it.

The fact that the cost of drilling a hole deep into the ground to accommodate an explosive nuclear device at its bottom increases rapidly with the diameter of the hole, led to the choice of very small diameter devices. The United States built bombs that, although three times the power used at Hiroshima, were no more than some 6 inches in diameter, but between 6 and 10 feet long. Such a development was particularly useful for missile warheads, even for field artillery.

In the industrial sector, the main underground civil tests were designed to help in the exploitation of gas deposits. Several were successful but the last one, in 1973, which made use of three charges exploded at different depths, was a partial failure for the three "chimneys" did not join together as intended. This test sounded the death knell of the Plowshare project. From 1974 onward the antinuclear movement, with some help from India's atomic explosion, put it well and truly on ice.

The same considerations did not arise in the Soviet Union where an important program of tests was carried out, leading to real industrial experience. Many firings took place, some of which were publicly announced and reported on: they were concerned with oil, gas, mining, and hydropower applications. A particular success was the extinguishing of a serious fire in a gas well — a magnificent technical feat. Another important suggested Russian project was the Pechora-Kama canal, some 50 miles long and requiring between 300 and 400 crater-forming explosions. If realized, this canal would allow diversion of the waters of the Pechora River, now flowing into the Arctic Ocean, into the Volga River. Since the Volga flows into the Caspian Sea, this canal would compensate for the continual drop in the level of the world's largest inland expanse of water.

Such were the main aspects of the peaceful applications of nuclear explosions, which acquired unexpected political importance as a result of negotiations on agreements for the renunciation of atomic weapons.

2. Forgoing The Weapon

The First Renunciations

By 1965, 20 years had passed since the end of World War II, and 10 years since the end of nuclear secrecy. Considerable progress had been achieved in the field of controlled combustion.

Nuclear technology was still mainly the monopoly of North America and of Europe from the Atlantic to the Urals. Nevertheless, interest was dawning in Asia, particularly in China, Japan, India, Pakistan, and Israel.

Also, it seemed that production of electricity from nuclear energy, thanks to the construction of more powerful units, had become competitive not only in Europe — where energy was more costly than in the United States — but also throughout the North American continent.

Regarding international competition, the export market was running in favor of the Americans, builders of power plants using enriched uranium, for which they still held the monopoly of supply outside the Soviet sphere of influence.

Despite fears expressed 10 years earlier, the Russians had as yet made no attempt to penetrate either the Western or the Third World nuclear markets. Instead, they had restricted their activities to their own sphere of influence, taking great care — except in the unfortunate case of China — to avoid encouraging the creation of autonomous nuclear programs among their satellite states.

There was only one part of the fuel cycle, namely uranium enrichment, that was still covered by secrecy and remained the unique prerogative of the five Great Powers. On the other hand, some 15 nonnuclear weapons powers were already equipped or very soon would be — with installations (research reactors or their first nuclear power stations) capable of producing militarily significant quantities of plutonium. And all of these states possessed at least some measure of the delicate technology needed to extract this plutonium, though only a few had at their disposal the expensive and complex plants necessary for this process.

The accumulated technological and industrial nuclear experience was by this time considerable. Nuclear explosives were within the reach of many countries. Most of these countries, however, were to make, unilaterally and despite their having the technical and financial ability to develop atomic bombs, the historic decision to forgo these weapons — the most devastating in existence. These countries were to accept voluntarily, in the interest of preserving world peace, flagrant discrimination against themselves while the weapon was not only retained, but was still being perfected and manufactured, by five of the leading powers in the world.

This was surely one of the most remarkable of all the political developments that resulted from the explosive consequences of fission. Normally, it would have been expected that the number of new countries beginning nuclear explosives research and experiments would increase with time and in step with the security declassification of information and the general advance of national nuclear activities. What happened, in fact, was precisely the opposite.

In 1945, no one could have predicted the declining rate at which countries would be acquiring practical experience with nuclear weapons over the coming 30 years: three states in the first decade, two in the second, and in the third only one, namely India, who today, seven years after its 1974 test, still appears to have no nuclear armaments program.

It is certain that many industrial countries have been deterred from pursuing atomic armament both by the fear that this could render them priority targets in any nuclear conflagration and also because they might then be accused of increased responsibility for such a conflict.

Further, because any argument justifying the acquisition of nuclear weapons by one country could always be used subsequently by another, there was widespread concern that dissemination was likely to spread to the less powerful or less stable nations, some of which might be led by irresponsible governments or dictators (although Nazism has shown that irresponsibility is not necessarily confined to small or unstable nations!) Dissemination thus seemed necessarily synonymous with increased risk of nuclear conflict.

All of these arguments had been used to oppose the acquisition of nuclear arms by France: with France's renunciation it would be easier to persuade other countries, particularly the Federal Republic of Germany, to abstain also.

In the early 1960s, it was not generally believed — as became the case 10 years later — that any country exploding a first atomic bomb, or even launching an armaments program, was outraging international ethics and committing a sin against world peace.

Thus in 1962 Switzerland, even before the start of its civil nuclear program, had considered trying to obtain atomic weapons from another nation, despite the possible implications — in the highly unlikely circumstances of finding a willing supplier nation — of such an action with regard to the traditional Swiss neutrality. It was decided to hold a national referendum, which led to the rejection, by a two-thirds majority, of a proposition that would have forbidden the production, storage, or use of atomic weapons by Switzerland. At the time, this result was considered neither surprising nor scandalous.

Nevertheless, it is to be noted that public opinion throughout the world had a vague belief that each new country acquiring nuclear weapons could worsen the tension and instability in certain areas and could increase the risk of nuclear war breaking out by error or by accident. Public opinion had, in fact, been the unwitting target of propaganda by the nuclear weapons powers in support of their claim to be the nations who knew best how to avoid using their own weapons. Rightly or wrongly, the general public believed that the "balance of terror" between the Soviet Union and the United States was a protection against worldwide conflagration, and that the bomb was in the hands of internationally responsible powers — the Big Five of the U.N. Security Council.

The first country to set a world example by not pursuing a military atomic program was Canada, in the 1950s, in spite of the fact that it could easily have combined such an objective with its civil nuclear development program. However, Canada had no real need for such armament since it was already doubly protected by its proximity and alliance with the United States and by its links with Britain.

In 1965, Sweden found itself in the same position. Sweden's advanced technical knowledge and ability, the degree of independence of its civil nuclear program, and its indigenous uranium production made it a fully qualified candidate for the Atomic Club, which it could certainly have joined had it wished. But Sweden's political parties were divided over the issue of nuclear arms, a question debated several times without result in parliament. In the end, following the conclusion of the Nuclear Test Ban Treaty, and with the country playing an increasing part in the Non-Proliferation Treaty (NPT) negotiations, the opponents of nuclear weapons production won the day, and Sweden voluntarily and definitely joined the ranks of the nuclear abstainers.

These ranks also included three other most important countries: the Federal Republic of Germany, Japan, and Italy. All three could easily have launched military nuclear programs had it not been for legal obstacles established following the Second World War. In Germany's case, these obstacles were now in the form of pledges made by Chancellor Konrad Adenauer when his country became party to the treaty of Western European Union. Italy's renunciation resulted from clauses in its 1947 peace treaty, and for Japan the obstacles were embodied in its new postwar constitution, which forbade the possession of offensive weapons and the pursuit of any secret nuclear activities.

Another factor reinforced these three countries' renunciations: None of them had any significant indigenous uranium resources. All were therefore dependent on external supplies, which automatically entailed international control and verification of peaceful utilization, both for enriched uranium — for which the United States was the sole supplier — and for natural uranium from the Anglo-Saxon producers, South Africa, Australia, and Canada.

If the Second World War had been the cause of the original nuclear sin, its consequences for the three principal losers were a contribution toward making a repetition of the sin less likely. Precious breathing space during which a nonproliferation agreement could be established resulted from their enforced nuclear abstinence, together with the self-denial of Canada and Sweden. These five countries were precisely the five industrial world powers then most able, after the Big Five, to take part in the nuclear arms race.

India and Israel

In the mid-1960s, only two nonnuclear weapons powers had attracted attention and raised certain doubts as to the complete sincerity of their peaceful nuclear intentions. One was India, a champion of nuclear disarmament but equally of a national nuclear program as independent and free from control as possible. The other was Israel; suspicions arose as a result of the secrecy surrounding the construction of its largest reactor. These two countries, with one or more hostile neighbors at their frontiers, were the first after the lifting of secrecy to acquire natural uranium research reactors capable of producing militarily significant quantities of plutonium.

Indian Prime Minister Jawaharlal Nehru had long been in the vanguard of a campaign to halt further experimental explosions and had always been hostile toward the use of atomic weapons. Nevertheless, when in 1955 the chairman of the Indian Atomic Energy Commission, eminent physicist Homi J. Bhabha, suggested to the prime minister that India should make a public unilateral renunciation of the bomb, Nehru answered that they should discuss it again on the day when India was ready to produce one.

Ten years later, India had a plutonium-producing research reactor, and a reprocessing plant for extracting plutonium from irradiated uranium fuel. The latter had been built by Indians themselves, whereas for the reactor they had received very substantial assistance from Canada. Following the completion of the plants, Nehru and Bhabha, while proclaiming the purely peaceful intentions of their country, declared that India was now able to produce a nuclear weapon. In 1974, when both had been dead for more than eight years, their declaration was demonstrated to the world at large in a way that, as seen later, tried to show, however, continuing respect for their first claim of peaceful intentions.

For the Israeli reactor, France played a role comparable to that the Canadians played for the Indians. In September 1956, the Mollet government, in between the nationalization of the Suez Canal and the Franco-Anglo-Israeli military intervention, agreed that French industry, with the help of the Commissariat à l'Energie Atomique's technological know-how, should contribute substantially to the secret construction in Israel of a large natural uranium, heavy water research reactor. Israel, its existence threatened then as now, was anxious to reinforce its infrastructure by equipping itself with the most modern technologies. The envisaged reactor was to be similar to the one that, a year earlier, Canada had offered to install in India on extremely favorable financial terms and with the single condition that it would be used only for peaceful purposes.* This condition did not involve inspections, for the Indian-Canadian agreement was concluded before the first applications of international safeguards. The position a year later was no longer quite the same since international controls were just

---

*Canada had won the order in competition with a U.K. offer to supply a graphite-moderated natural uranium research reactor.

starting to be implemented. The French government, having only recently established closer military and political links with the Israelis, was not prepared to impose the safeguards on them. The Israelis would have been subject to the safeguards on the supply of natural uranium needed for their reactor. In the past, France had refused controls in such a case, but had recently accepted them with great reticence on her American supply of enriched uranium only because she was unable to do otherwise.

The Indian and Israeli research reactors, each in different circumstances, subsequently became examples of a certain laxity, whether intentional or accidental, toward what no one had yet called a strict nonproliferation policy. Both reactors were capable of an annual production of plutonium sufficient for the manufacture of one or two bombs, depending on the availability of an irradiated-fuel reprocessing plant to extract the plutonium.

In late 1960, when the press learned of the Israeli reactor and revealed that it was being built, there was a deeply emotional reaction in the Arab states, despite an assurance by Israeli Prime Minister David Ben-Gurion that the installation was to be used only for peaceful purposes.

The help provided by France for this project was essentially concerned with the construction and operational startup of the reactor. It had been limited to these activities on the explicit instructions of General Charles de Gaulle, who made the restriction clear at the time of Ben-Gurion's visit to Paris in mid-1960, after which de Gaulle nevertheless referred to Israel as "my friend, my ally." However, in his *Mémoires d'Espoir,* de Gaulle described one of the outcomes of this visit: "I put an end to a number of unwarranted collaborative practices which had developed at military level between Tel-Aviv and Paris since the Suez expedition, practices which had given the Israelis a permanent access to French military staff and services at all levels. In particular, this brought to an end our help with the early stages of construction, near Beersheba, of an installation to transform uranium into plutonium and from which, one day, atomic bombs might therefore be produced."

In fact, the general, three years earlier when he took over leadership of the government, had similarly blocked the beginnings of Franco-German-Italian cooperation in an area close to the military nuclear field. This cooperation had originated with one of the last governments of the Fourth Republic, with the idea of possible participation by the German and Italian defense services in the future French isotopic separation plant. The general never overcame his concern at the possibility of German nuclear armament, and he questioned me at length on this issue every time I had the honor of meeting him during the 1960s.

The Israeli reactor, at the country's Dimona research center, was scheduled to become operational in 1963. The United States, from whom the Israelis had never stopped seeking a formal military guarantee for the existence of their country, tried unsuccessfully to persuade Israel to submit its installation unilaterally to International Atomic Energy Agency (IAEA) safeguards.

The Israeli government had, in fact, constantly maintained that Israel would never be the first country to introduce atomic weapons into the Middle East. Nevertheless, by late 1974, a few months after the Indian explosion, the president of Israel, Ephraim Katzir, declared: "Should we have need of such weapons we could have them," thereby officially admitting that his country had the capacity to produce nuclear bombs, which no one doubted.

Israel did not seem to have any difficulty in obtaining nonrestricted uranium for this reactor, whether in limited quantities by extraction from its own low-uranium-content phosphates in the Negev, or by purchasing from South Africa, Argentina, or Belgium. As for India, there were sufficient uranium resources available within its own territories.

Apart from the two rather special cases of India and Israel, a review of the non-Communist world in the mid-1960s showed, beyond the five industrial powers that had either voluntarily or obligatorily accepted nuclear abstinence, a small number of other countries that had decided to launch, or had already launched, nuclear programs that were declared peaceful, but which in 15 to 20 years' time would give the countries the possibility of carrying out an initial explosion. These countries were: Argentina and Brazil in Latin America, both then contemplating the purchase of a first nuclear power station; Belgium, Spain, the Netherlands, and Switzerland in Europe; Pakistan in Asia; and finally, perhaps, the Republic of South Africa.

Thus at this stage, a negotiated worldwide renunciation of atomic weapons among all the nonnuclear weapons countries of the world would have meant a loss of national sovereignty, which in the short term would have affected only some 15 advanced countries. These countries, therefore, found themselves confronted by a coalition grouping the first three nuclear weapons powers together with all the smaller Third World nations for whom a nuclear program within the foreseeable future was out of the question. This last group of nations had no wish to see the gap that already separated them from the nonnuclear weapons industrialized countries (i.e., others beside the Big Five) get wider due to some of these countries also acquiring nuclear weapons.

Since the late 1950s, two approaches to this problem had been under consideration. One involved renunciation of the bomb on a worldwide scale by all nonnuclear weapons states (i.e., those not already in possession of their own nuclear armaments). This would not, however, prohibit the storage of nuclear weapons within their territories and under the jurisdiction of one or another of the nuclear powers.

The second approach to the problem was at a regional level, involving the creation of completely nuclear weapon-free zones. The creation of such a zone had, in fact, already been proposed by Poland at the United Nations in 1957, with a view to the nuclear neutralization of Central Europe. The Rapacki proposals had been rejected by the powers of the Atlantic Alliance because they would have prevented tactical atomic weapons from being situated in the Federal Republic of Germany, while allowing the Russians, who were already better equipped with conventional armaments, to maintain tactical atomic arms on nearby territory of their own.

The Antarctic Treaty of December 1959, which forebade any military activities in that area, was also the first agreement on "nuclear neutralization." Not only were atomic explosions prohibited in the area but also the storage of radioactive wastes.

Ireland also, as early as 1958, had proposed resolutions in the U.N. General Assembly aimed at preventing further proliferation of nuclear weapons. In 1961, Ireland received unanimous approbation for a draft text designed to establish an acceptable balance of mutual responsibilities and obligations between nuclear weapons states and nonnuclear weapons states, at the same time allowing any group of states to conclude regional treaties for the total prohibition of nuclear weapons in their respective territories — in other words, to establish "denuclearized zones."

Supported by many countries, and in particular, India and Sweden, Ireland continued to raise this matter during successive annual sessions of the U.N General Assembly. Resolutions were discussed and amended from year to year in attempts to obtain either the agreement of the United States or the Soviet Union. These attempts were resisted either by Moscow, if they left the Americans too much freedom to maintain weapons on the territories of their allies, or by Washington if they tended to restrict this freedom.

The American-Soviet détente following the Cuban affair, the conclusion of the Moscow Treaty on prohibition of further nuclear tests in the biosphere, the emotional repercussions of the Chinese explosion in 1964, and Soviet hostility to a proposed North Atlantic Treaty Organization (NATO) multilateral force provided as many reasons for a climate favorable to serious negotiations on the renunciation of the bomb by the nonnuclear weapons states.

The Tlatelolco Treaty

Since 1965 two main series of nuclear negotiations have engaged worldwide attention: one with the aim of concluding a world treaty of nonproliferation, the other for the creation of a denuclearized zone among the countries of Latin America, an area of the world which at that time was comprised of only nonnuclear weapons powers and in which none of the Big Five was holding stocks of atomic armaments.

This second series of negotiations resulted from the concern of five Latin American countries — Bolivia, Brazil, Chile, Ecuador, and Mexico — that they might have become involved in the nuclear maneuvering of the Cold War during the Cuban crisis of 1962. In April 1963, the presidents of these countries announced their intention to develop a multilateral agreement prohibiting anywhere within their zone the production, importation, storage, or testing of nuclear weapons or launching devices.

The following year, Mexico took the initiative and arranged a preliminary meeting, which led to the establishment of a Preparatory Commission under the chairmanship of Ambassador Alfonso Garcia Roblès. Under his competent leadership, the work of this commission led some two years later to the Treaty for the Prohibition of Nuclear Weapons in Latin America, known as the Tlatelolco Treaty, so named after the district of Mexico City where it was approved by 21 countries on February 12, 1967.

This Latin American denuclearization treaty was the world's first multinational agreement on the renunciation of atomic weapons. Like the NPT, which it preceded by nearly one and one-half years, its aim was to consolidate and perpetuate an existing situation, in this case a veritable and nondiscriminatory nuclear weapons free status throughout all the countries in the area concerned.

Each of these two treaties of nuclear weapon renunciation relied on inspections by the IAEA, to verify the peaceful nature of the nuclear activities of the participating nonnuclear weapons powers. But the Tlatelolco Treaty also provided for special inspections in any case of suspected noncompliance.

Another difference between the two agreements was that the regional treaty allowed the Latin American states to explode nuclear devices for peaceful purposes, even if these devices were similar to nuclear weapons, provided advance notice were given, preliminary details supplied, and the tests carried out under IAEA surveillance.

There were two protocols to the Tlatelolco Treaty. The first was intended to persuade the United States, France, the Netherlands, and

Britain — considered internationally responsible (*de facto* or *de jure*) for certain territories within the controlled zone — to underwrite the denuclearized status of these territories. The second protocol was designed to obtain from the five nuclear weapons powers a guarantee of respect for this denuclearized status and promise not to use or threaten to use nuclear weapons against any country party to the Tlatelolco Treaty.*

The treaty was signed by all but one of the states within the area concerned. The exception, Cuba, had from the start refused to participate in it on the grounds that one of the nuclear weapons powers, the United States, had an "aggressive" anti-Cuban policy, characterized by the American occupation of the Guantánamo naval base on Cuban soil.

It should be noted that the three Latin American countries with the most ambitious nuclear intentions (although none had ordered a nuclear power station at the time the Tlatelolco Treaty was concluded) proved to be the three most reluctant to submit to the treaty's conditions. They were Argentina, Brazil, and Chile.

In fact at the beginning of 1982, 14 years after the signing of the treaty, Argentina still had to ratify it, while Brazil and Chile were the only two countries to have invoked a clause allowing them, after ratification, to make implementation within their territories of both the treaty and its protocols subject to signature and ratification by all the other states concerned.

Negotiating the Non-Proliferation Treaty (NPT)

Negotiated between 1965 and 1968, this second agreement on the renunciation of nuclear weapons, the NPT, because of its universal nature and worldwide application, inevitably assumed far greater political importance than the Tlatelolco Treaty that preceded it.

Negotiations began in 1965, following the Chinese explosion of October 1964. In June 1965, a vote was taken in the U.N. Disarmament Commission calling on the 18-member Disarmament Committee (in which

*The first protocol raised certain problems for France. Already signed by the three other countries concerned, it received French approval in 1979 but at the beginning of 1982 still had to be ratified. The main difficulty was internal, due to the French overseas departments in Guadeloupe, French Guiana, and Martinique falling within the zone concerned, so that they would find themselves in a discriminatory situation compared with other French territories. This difficulty did not affect the second protocol. Already signed and ratified by the other four nuclear powers, it received Russian approval in 1978 when the Soviet government announced its intention to sign.

France refused to take her seat) to consider the matter of a treaty or convention on nonproliferation.

In fact, 1965 saw the first appearance of the term "nonproliferation." Used in its most general sense, it covered any increase in the numbers of atomic weapons in the hands of the nuclear powers, the geographical distribution of these weapons by those powers, and the manufacture or acquisition of such weapons by nonnuclear powers. Indian physicist Bhabha even suggested a distinction between "vertical proliferation," meaning an increase in the nuclear arms held by the five members of the Atomic Club, and "horizontal proliferation," meaning an increase in the number of countries possessing the new weapons.

At the beginning of the NPT negotiations, the following stands were taken by various groups of countries.

The least-advanced countries of the Third World were in favor of the nonproliferation objectives being pursued. They hoped to obtain, in return for their voluntary renunciation, promises that the nuclear powers would adopt specific measures for nuclear disarmament; would never threaten or attack them with nuclear devices; and would make available genuine help toward regaining lost time in developing peaceful applications of nuclear energy in the Third World countries.

Some of the more rapidly developing countries, such as Brazil and India (the latter had become more reticent in nonproliferation matters since the Chinese explosion and the death of Nehru) stressed the discriminatory aspect of the proposed surrender of sovereignty. They called for compensating sacrifices by the nuclear powers, such as promises not to manufacture new weapons and to convert part of the existing stockpiles and make them available for peaceful uses by the developing countries. These conditions had little chance of being accepted by the two superpowers and this effectively reduced the likelihood of accession by some countries to any future treaty.

The industrialized countries, for whom the renunciation of nuclear weapons was of real significance even in the short term, likewise called for disarmament measures on the part of the nuclear powers (Sweden wanted, in addition, a complete test ban). But above all, most of these countries were afraid of being handicapped in world civil nuclear competition compared to the powers able to retain complete freedom of action, and therefore likely to benefit in the civil field from experience and knowledge gained in their military operations. These industrialized countries were also keen to benefit from the peaceful applications of nuclear explosions. Furthermore, they were afraid that international safeguards might offer opportunities for industrial espionage, and therefore wanted the civil installations of the nuclear powers to be subject to IAEA inspection, in the same way as their own installations would be. In addition, the nonnuclear member countries of Euratom, supported by the Commission of European Communities felt that the community's safeguards system should be considered equal to that of the IAEA from which they should therefore be exempted.

For the Federal Republic of Germany, Japan, and Italy, who were already obliged to accept discriminatory abstinence, any further restriction, such as the imposition of international inspection, seemed unacceptable unless accompanied by an assurance that they would enjoy in the civil nuclear field exactly the same freedoms and advantages as the nuclear powers.

As far as France was concerned, the future treaty raised virtually no problems. Like the other nuclear weapons powers, France was not keen to see any further extension of the Atomic Club. France had criticized the Moscow Nuclear Test Ban Treaty as discriminatory and that it was evidently not a treaty of disarmament but only an agreement on the nonarmament of unarmed countries. So France could not, without self-contradiction, join in the negotiation of, or accede to, any nonproliferation treaty of the same discriminatory nature, even though such an agreement would in no way hamper French activities and would indeed confirm recognition of the country's status as a nuclear power.

The Soviet Union, for its part, had everything to gain and nothing to lose from the proposed treaty. The Russians were already in complete control of, and able to restrict as they chose, the nuclear activities of the Warsaw Pact countries, where nuclear abstinence was already established. Also, some of the countries, for example, Bulgaria and Hungary (as well as Finland), were committed to it by their peace treaties. Alone among these eastern countries, Rumania was the only one to have an independent stand in the negotiations, frequently supporting the countries of the Third World in their efforts to secure a more satisfactory balance between the concessions agreed to on both sides.

The Soviet Union was willing to allow IAEA inspection within the satellite countries, which did not particularly please them, provided the Federal Republic of Germany would be similarly inspected. Also, the Soviets were greatly interested in the possibility of following, through inspectors drawn from the Communist states, nuclear developments in most of the countries of the Western and Third Worlds.

Aware of the perseverance and dedication of American diplomacy over the past 20 years in pursuit of nonproliferation policies, the Russians were well-placed to secure valuable concessions in return for their cooperation, the more so since Washington was in contradiction with its own principles over its proposed multilateral force. Moscow, obsessed by fear of nuclear rearmament in Germany, was especially anxious to see the idea for a NATO nuclear force abandoned.

Finally, for the United States, what counted most was the adoption and universal maintenance of strict nonproliferation regulations, the observance of which would be effectively monitored by the IAEA inspection system. If, in addition, the proposed multilateral force could be saved by showing that, in reality, it was the U.S. president alone whose finger was on the trigger, so much the better.

These positions of the two superpowers were reflected in two different drafts for the proposed treaty, which were submitted toward the end of 1965 by the United States to the 18-member U.N. Disarmament Committee, and by the Soviet Union to the U.N. General Assembly. The American draft, while prohibiting the transfer of weapons to the "national control" of a single nonnuclear weapons state, left the door open for the supply of weapons to a number of countries in a group. The Soviet text on the other hand prohibited all nonnuclear weapons countries from having any part in the possession, control, or use of such weapons. During the U.N. discussions, the Soviet Union insisted that the real purpose of the proposed Atlantic nuclear force was to provide Germany with weapons.

American-Soviet negotiations continued throughout the next year. Following secret talks between Secretary of State David Dean Rusk and Foreign Minister Andrei A. Gromyko, the negotiations at last led to a compromise resolution of the deadlock at the end of 1966. Under this compromise, Washington abandoned the proposed multinational force, a move that was a real victory for the Soviet Union. In return, the Russians accepted the status quo and no longer opposed the presence of American nuclear weapons within the territories of America's Atlantic allies, nor objected to these allies holding consultations as to when and how the weapons might be used. A NATO committee for nuclear planning was set up for this purpose.

This American-Soviet agreement showed, despite the tensions created by the war in Vietnam, the extent of the political reconciliation between the two powers since the Cuban crisis. Now, with the main obstacle to preparing the treaty overcome, the two Great Powers were ready to face together the various coalitions of nonnuclear powers. It took a further 15 months before a final text was reached that was acceptable to the 17 countries taking part, without France, in the 18-member U.N. Disarmament Committee.

The 18-member committee in Geneva, within which the formal negotiations had been conducted, was now able to transmit a comprehensive report to the United Nations in New York, which examined the matter in a special session of the General Assembly. This session ended on June 12, 1968, with the adoption of a skillfully worded resolution supported by some hundred votes with 20 abstentions (including Argentina, Brazil, France, India, and Spain). The resolution went no further than expressing the hope that there would be as many accessions to the treaty as possible. It did not commit all countries voting for it to automatic signature of the treaty. As a supplementary gesture, the United States, the United Kingdom, and the Soviet Union jointly declared in the Security Council the following week that they would give help to any nonnuclear weapons state party to the treaty in the event of attack or threat of attack with nuclear weapons.

At the conclusion of the U.N. special session, France — having once again called for a halt in the manufacture of weapons and for the destruction of existing stockpiles and having again stressed that the treaty was not a disarmament treaty — nevertheless declared that although she would not be a party to the treaty, she would act in the future in exactly the same way as the states who were signatories.

Thus, after more than three years of negotiations, the NPT finally saw the light of day on July 1, 1968. Immediately following signature by the three ''depository states,'' the United States, the Soviet Union, and the United Kingdom, it was opened for signature by all the countries of the world. The depository states had meanwhile succeeded, throughout the previous year, in delaying a conference of nonnuclear weapons countries with the intention of modifying the treaty so as to obtain important concessions from the nuclear powers both in the field of disarmament and in terms of assistance to less-developed countries.

This nonnuclear weapons state conference finally opened in Geneva in August 1968, one week after Soviet troops had occupied Czechoslovakia. It lasted a month. The four nuclear powers attended as observers without making any statements. The conference was characterized by argument, bitterness, disappointment, and resentment: It gave the participant countries a needed chance to vent their feelings, if not to rebel, against the Great Powers whose constant pressure on them had been maintained throughout the last stages of the NPT negotiations.

Some of the resolutions adopted at this latest Geneva meeting concerned disarmament by the Great Powers. Other resolutions related to assistance that those powers should provide in the fields of technical training, the supply of fissile materials, financial matters, and the peaceful uses of atomic explosions. Some also referred to the IAEA safeguards system and its application, and called for a greater participation by Third World countries in the agency's board of governors, which is responsible for the administration of the IAEA.

Features of the NPT

The NPT, unique in the political history of the world, is intended to halt the course of evolution in a crucial field by fixing once and for all the number of nuclear weapons powers. Moreover, it defines a nuclear power as one that had made and exploded a nuclear weapon or any other nuclear device before January 1, 1967, and therefore covers only the five nuclear powers of that time. It classes France as a member of the Atomic Club in which she had so much difficulty, if not in gaining entry, at least in gaining recognition.

The treaty forbids the signatory nuclear weapons powers from transferring nuclear weapons or other explosive nuclear devices, or control over them, to any nation whatsoever. It was this clause that put an end to the proposed NATO nuclear force. The treaty also prohibits the nuclear weapons countries from helping any nonnuclear weapons state (whether or not a party to the treaty) in the manufacture or acquisition of such weapons or devices, or in obtaining control over them.

This prohibition is in keeping with the political line that, with a few deviations, had been followed from the start by the members of the most exclusive Club in the world, and is in fact the Club's unwritten law. It does not imply any real sacrifice by those bound by its conditions, for it leaves the transfer of technology or of weapons completely unrestricted among the nuclear weapons powers themselves.

Conversely, the other parties of the treaty, i.e., the nonnuclear weapons powers, promised not to accept the transfer of or control over such weapons or devices from any source, and also not to manufacture them.

The treaty refers not only to nuclear weapons, which it does not define, but also to any other explosive nuclear device. The renunciation therefore covers all systems intended for a peaceful nuclear explosion. This was the first time that one of the peaceful applications of fission had been prohibited to a great majority of the countries of the world. Previous nuclear contracts containing a restrictive clause had only prohibited the use of explosions for military purposes and not the explosions themselves.

The Latin American denuclearization treaty discussed earlier gives its signatories the right to carry out peaceful explosions in certain circumstances. The NPT, however, prohibits all nuclear explosions because the Russians and the Americans have accepted since 1966 that it is technically impossible to distinguish between the inherent characteristics of the two types of explosions. This is why both types are prohibited, for there is only one way of making nuclear explosions, whether for peaceful or for military purposes.

This requirement to abstain from peaceful nuclear explosions was deeply resented by many nonnuclear weapons powers, and was strongly resisted by India and Brazil. It came at a time when the two superpowers were engaged in a veritable orgy of military underground tests and were devoting considerable sums of money to the peaceful applications of nuclear explosions, the potential advantages of which they were stressing as valuable publicity.

All of this resentment obliged the United States and the Soviet Union to accept incorporation into the NPT of a clause promising the nonnuclear weapons countries, at minimal cost and on a nondiscriminatory basis, a share in all the benefits that might derive from the peaceful applications of nuclear explosions.

In the autumn of 1968, the U.N. General Assembly recommended that the possibility be studied of setting up, within the IAEA, a service for carrying out future peaceful explosions under direct international control. From 1970 to 1975, an international committee and many international meetings examined the technical, legal, and safety aspects of the matter, but the proposed clause in the NPT relating to such explosions for the benefit of the nonnuclear weapons countries remained a dead letter. The discontinuation since 1973 of all civil test explosions in the United States, followed in 1974 by the underground nuclear explosion in India and then — in complete contrast to the past publicity — the cloak of silence that enveloped this activity, led to the first case of nuclear energy for civil purposes falling victim to the cause of nonproliferation, at least in the Western World.

As for the measures of nuclear disarmament so insistently called for during the NPT negotiations as fair compensation for the restrictions voluntarily accepted by the nonnuclear weapons states, the nuclear weapons powers were able to ensure that they did not become obligatory. All that the treaty called for from these powers was that they should continue negotiations in good faith to search for an effective means to end the arms race as soon as possible and for a general and complete disarmament treaty with strict and effective controls. It was also recalled that these powers were "determined," as they had been in the Moscow Treaty five years earlier, to continue their negotiations with a view to the total prohibition of nuclear weapons tests.

The nonnuclear weapons countries signatory to the NPT undertook to accept IAEA safeguards solely as a way of verifying that nuclear energy was not diverted from peaceful applications toward weapons or other explosive devices. The safeguards were to apply to all peaceful nuclear materials and activities under their control, whether in their territories or elsewhere.

Further, all countries party to the NPT undertook not to provide nuclear materials or equipment specially designed for the processing, utilization, or production of uranium-235 or plutonium to any nonnuclear weapons state (whether or not party to the treaty) unless such materials or equipment were subject to IAEA safeguards.

NPT Inspection and Safeguards

The clause on international inspection and control was one of the most difficult for the nonnuclear weapons countries to accept during the NPT negotiations. These countries had agreed for the sake of nonproliferation to deny themselves nuclear weapons and, in practice, the civil uses of nuclear explosives. They even had to accept pious assurances, rather than concrete commitments, to nuclear disarmament from the military nuclear powers. In addition, they were obliged to submit all their nuclear activities to IAEA inspection on a permanent basis — an inspection from which nuclear weapons powers were exempt, underlining the discriminatory character of the safeguards system.

These nonnuclear weapons countries had in the past become accustomed, under the "Atoms for Peace" policy, to accept safeguards only in exchange for assistance. This time they were accepting, forever, full-scope safeguards in exchange for ... a sacrifice! This made them all the more insistent in their demands for compensating exemptions in their civil nuclear development programs. The most advanced countries wanted complete freedom of action, with no stage in the industrial chain denied them, from uranium mining to energy production. The least advanced countries wanted guarantees of assistance from the Big Powers. All wanted a share in the eventual civil benefits resulting from military research carried out by the members of the Club.

It is for these reasons that Article IV of the NPT obliges all states party to the treaty to facilitate the transfer of knowledge, materials and equipment, and to contribute to international cooperation, in particular with respect to the Third World. It is further specified in this same article that "nothing in this treaty shall be interpreted as affecting the inalienable right of all parties to the treaty to engage in research, production, and use of nuclear energy for peaceful purposes without discrimination and in conformity with Articles I and II of this treaty.'' Articles I and II deal with renunciation.

Article IV was of primary importance to powers such as Germany, Japan, and Italy, who were afraid of finding themselves handicapped in international commercial competition vis-á-vis the victorious Allied powers of the last war, who remained free from all restriction. They were prepared to make the unilateral concession of renunciation only on one specific condition — that they would have complete freedom of action in the civil field. From this standpoint, the NPT can be summed up very simply as follows: All nuclear explosions are forbidden, but everything that is not forbidden is permitted including all stages of the nuclear fuel cycle, even those that could enable nuclear explosives to be produced.

Article IV was particularly important for Germany, for it rendered null and void the clauses that could have been taken as prohibiting the manufacture of plutonium and uranium enriched to more than 2.1 percent, considered equivalent to the weapons whose prohibition was accepted in Chancellor Adenauer's letter annexed to the treaty of Western European Union (WEU) (see p. 129).

The signature by the Federal Republic of Germany of a tripartite international agreement with the United Kingdom and the Netherlands for the production of enriched uranium by the ultracentrifuge method on the very day of the NPT's entry into force, and the startup in the same year (1970) of a German pilot plant for plutonium production, were no less demonstrations of the German interpretation of the treaty. None of the WEU countries expressed any objections.

An argument that was voiced, particularly by Japan and the Federal Republic of Germany, against the application of safeguards inspections to all nuclear activities in the nonnuclear weapons countries party to the NPT was that the inspections involved a risk of industrial espionage. As a gesture, and in response to this argument, the United States and the United Kingdom in 1967 decided, through solemn declarations by President Lyndon B. Johnson and Prime Minister J. Harold Wilson, to voluntarily submit their peaceful nuclear installations to IAEA inspections, thereby running the same risks of industrial espionage as the other powers subjected to safeguards procedures. The Soviet Union however consistently refused to follow suit, so reducing the scope of the gesture, which was mainly psychological since the countries concerned remained the sole judges of which installations would be exempt from safeguards on the grounds of national security.

Moreover, it was doubtful in the case of the remaining facilities that the IAEA would wish to spend a great deal of manpower and money on inspections with no practical value since they verified the fulfillment of nonexisting obligations. Such voluntary submissions to inspection could be likened to the action of a traveller who had the right to tell customs officials which of his suitcases should be examined. In practice, following agreements concluded in 1978 and 1980 with the United Kingdom and the United States, respectively, these safeguards were implemented. However, for reasons of economy, inspections in each country were limited to a number of advanced installations using new technologies or having a key role in international competition. But these agreements did not fix once and for all the limits of the military domain. This was shown in the United States in 1981 when the Pentagon considered making use of civil fuel to meet new military requirements for plutonium.

In another direction, the member countries of the European Economic Communities (EEC), encouraged by American recognition of the validity of Euratom safeguards, wanted these safeguards accepted as adequate for the NPT, so as to avoid having inspectors from non-Community countries on their territories.

However a principal reason for the Soviet Union's negotiation of the treaty had been to ensure verification by the IAEA of German nuclear abstention following the abandonment of the proposed multilateral NATO nuclear force. In fact, the Soviets, now zealous supporters of IAEA safeguards, became formal and categorical. They considered that the IAEA alone should be responsible for ensuring the implementation of the NPT by and within its signatory countries. They refused, as a matter of principle, to acknowledge any value in safeguards operated by a regional organization — and above all by Euratom, which in their eyes would give the Germans a measure of control over themselves. In the end, the only concession offered to the EEC was recognition that the Euratom safeguards existed and could be regarded as equivalent to the national monitoring systems of other countries, through the verification of which the IAEA safeguards system operates.

The NPT, like the Nuclear Test Ban Treaty, gives every signatory state the right of withdrawal after 3-months notice if that state decides that special circumstances related to the treaty's objectives have endangered its supreme national interests.

The NPT became effective on March 5, 1970, after ratification by 3 nuclear weapons powers, the United States, the Soviet Union, and the United Kingdom, followed by 40 nonnuclear weapons countries. Ratification in the United States had been one of the themes in the presidential electoral campaign of autumn 1968. Republican candidate Richard M. Nixon had successfully prevented the outgoing president and the democratic party from winning prestige through its approval in the Senate. This approval was eventually obtained in March 1969 by a very large majority.

Despite pressure from the Americans and the Russians, ratifications by the key countries, Germany, Italy, and Japan, were not obtained until 1975 and 1976. In each of these countries, lengthy parliamentary debates showed how much importance was attached to freedom of action in the civil area, there being considerable reticence over the imposition of international safeguards and fears that adhesion to the treaty would be a handicap in commercial competition. One of the last obstacles to these ratifications, without which the NPT would have lost all political value, had been removed when France adopted, in 1975, a nuclear export policy in conformity with the letter of the treaty and with its methods of implementation.

At the beginning of 1982, 13 years after the conclusion of the treaty, 114 countries had acceded, and those having nuclear installations were submitting these to the IAEA safeguards system, though not without certain difficulties in the case of the EEC countries, as is seen later.

Through their adhesion to the NPT, a number of advanced countries — a quarter of a century after Hiroshima — had thus unilaterally accepted a limitation of their national sovereignty in order to promote international stability and the world development of atomic energy.

In conformity with the terms of the treaty, a review conference is held every five years. The 1980 conference in Geneva ended with a severe setback, for it proved impossible to agree on a final recommendation. The nonnuclear weapons countries, led by Mexico, the Philippines, and Yugoslavia, accused the weapons powers of having made no progress toward nuclear disarmament and a total ban on underground testing. They accused the industrialized powers of reticence in making available the full range of fuel cycle technologies. In short, the advanced countries were accused of failure to respect their part of the sacrifice called for within the general balance of the treaty, causing the aggrieved nations to threaten possible withdrawal from the agreement.

Despite these difficulties, it remains true that the conclusion and implementation of the NPT was a major success for the United States and the Soviet Union. These two Great Powers had brought off a *tour de force* in getting accepted, if not imposing, this "nuclear Yalta" by conjuring up the specter of the threat to world peace should additional countries acquire nuclear weapons. The two superpowers had established the principle of permanent division of the current world into two types of nations by trying to perpetuate the already existing division. Such a development, on so unprecedented a scale, was contrary to the course of history and a first demonstration of this was not long in coming.

The Indian Explosion

On May 18, 1974, India exploded several pounds of plutonium in the Rajasthan desert, a province bordering Pakistan.

The country was one of some 10 of the more important nonnuclear weapons powers to have refused the obligations of the NPT and to have decided to keep open the military atomic option. Besides India the group also included Pakistan; the two rivals for nuclear leadership in South America, Brazil and Argentina; two countries surrounded by hostile neighbors, the Republic of South Africa and Israel; two influential Arab powers, Algeria and Saudi Arabia*; and finally, Spain.

The Indian explosion came as no surprise to the specialists. As early as the 1960s, Bhabha, first chairman of the country's Atomic Energy Commission (AEC), had made no secret of his intention to study the mechanisms of nuclear explosions. However, following his untimely death in 1965, his successor Vikram Sarabhai, a sincere and committed pacifist, must undoubtedly have convinced the prime minister, Mrs. Indira Gandhi, that conducting a nuclear test would be politically inopportune. But Sarabhai died in 1971, and the following year Nehru's daughter gave orders for a test explosion to be carried out. The task was given to Homi Sethna, a talented engineer who had been responsible for the construction of his country's plutonium plant, and who in 1972 was appointed to direct India's AEC. His success considerably reinforced Mrs. Gandhi's national prestige, for in 1974 her country was facing increasing economic, social, and political difficulties.

In carrying out the test, the Indians had broken no formal international obligation, for they had declared the explosion to be in pursuit of peaceful objectives. The test had been carried out underground in accordance with the Nuclear Test Ban Treaty, to which the Indian government was party.

The plutonium used had been produced from indigenous Indian uranium, which had been irradiated in the civil research reactor built in the late 1950s with technology and financial aid from Canada. Extraction of the plutonium had been carried out at the national fuel reprocessing plant, in operation since 1965, which had been constructed with no direct external aid. India had chosen this independence, rather than accepting a British offer in 1958 to build such a reprocessing plant.

The Canadian prime minister, Pierre Elliott Trudeau, had warned

---

*Egypt, a signatory of the NPT, having at first decided not to ratify it until after Israel's accession, nevertheless did become party to the treaty in 1981 although Israel had still not done so.

Mrs. Gandhi several times during the preceding years that he could not consider as peaceful an explosion declared as such because there was no way of distinguishing it from a military test. However, in these early years of the 1970s, the reverberations of the American ''Plowshare'' tests were still fresh in the public mind, and only the countries already party to the NPT had as yet accepted that the nuclear weapons powers had a monopoly on peaceful explosion tests.

Canada, most obviously implicated in facilitating the Indian operation, was the only country to react strongly. None of the other leading nuclear powers made any complaint. There were, however, protests from neighboring Pakistan, where it was claimed that an increase in radioactivity had been observed. One way of considering the affair was to look at it only from the technical viewpoint, recognizing the competence of the Indian physicists concerned. Politically, however, it portended heavy consequences, for after a 10-year interval the monopoly of the Big Five had been broken, inevitably making it more difficult to maintain the exclusivity of the Atomic Club.

The intentions of the NPT had indeed been challenged by a Third World power. In addition, the Indian explosion was the first to have been made possible through the grant of foreign aid to a civil program. As shown later, these developments constituted a serious blow to the policies of safeguarded assistance practiced over 20 years, and they created doubts as to the validity of the rules of the international nuclear game.

But the most immediate requirement was to prevent any repetition of India's bad example; the Canadians and Americans vigorously pursued this objective. In 1976, the former's government stopped all atomic collaboration with India. The Indian nuclear program, based on Canadian technology and aid, was considerably affected. At the same time, Washington was trying to persuade Delhi not to carry out a second test explosion, which would be even more difficult than the first to represent as peaceful. Mrs. Gandhi refused to commit herself then, and again after her return to power in 1980, but, following the 1977 elections in which she fell from power and into temporary disgrace, her successor Morarji R. Desai officially abandoned carrying out further tests. Thus India, besides being the first country of the Third World to penetrate the lobby of the Atomic Club, was also the first not to acquire operational nuclear weapons during the years that followed. Nor, it should be noted, did India do anything to help any other Third World country in the realization of military atomic aspirations. Thus India behaved as a responsible member of the Club while remaining, together with France and China, outside the NPT.

In the end, the American government, fully supported by Congress, with Moscow's agreement and with help from the press, which orchestrated and amplified the general shock and anxiety caused by the Indian explosion, took advantage of the event to relaunch an energetic effort toward concerted international action on nonproliferation.

This reaction, which was essentially to affect civil nuclear programs, is examined in Part Two of this book. This effort at nonproliferation was to reach its zenith in 1977 under President Jimmy Carter.

A Satellite View of South Africa

The next alarm did not come from Asia or another Third World country, but from the most advanced and controversial power within the continent of Africa.

In the summer of 1977, Moscow advised Washington that Russian spy satellites had detected what seemed to be preparations for an underground atomic explosion in the South African Kalahari Desert, not far from Namibia. It was not the first time the South African Republic had been accused of such activities. Fourteen years earlier during my visit there to negotiate France's major purchasing contract for uranium, the local press had misreported my presence as a quest for a new test center to replace the Sahara site, then due to be evacuated following Algerian independence. Britain's future Labour prime minister, Harold Wilson, had even asked a question in the House of Commons in London concerning this imagined purpose of my South African visit.

In 1977, however, the matter appeared to be more serious, as American satellites seemed to confirm the Soviet suspicions, without of course being able to provide absolute proof (only a visit to the site could have done this). The timing was awkward, as difficult negotiations were then taking place between the South African government, the United States, and other Western powers concerned with the problem of Namibian independence. The South African government, from the administrative capital in Pretoria, denied that the photographed installations had any nuclear purpose, but admitted they were military installations and therefore refused to allow foreign representatives to visit them.

The affair caused a great stir, particularly in Black Africa, but nothing further could be done despite Washington's pressing though fruitless attempts to persuade South Africa to adhere to the NPT and immediately place its pilot enrichment plant under IAEA safeguards — this plant being the only possible source of fissile material for the hypothetical underground test.

Two years later the State Department announced, a month after the event, that during the night of September 22, 1979, an observation satellite had detected an intensely luminous double flash somewhere within a vast area comprising the southern part of Africa, the Atlantic and Indian Oceans, and parts of Antarctica. The flash could be interpreted as characteristic of a low power nuclear explosion although other usual indications (such as a radioactive cloud) that follow an aerial test had not been detected.

Despite immediate denials from Pretoria that any nuclear test had been carried out, the U.N. General Assembly, which happened to have begun a session on September 18, just four days before the reported observation, unanimously called for an inquiry.

The matter was to occupy the American experts through the whole of 1980. Some of them had no doubt that a nuclear explosion had indeed taken place but others, because of the absence of confirming evidence such as consequent radioactive fallout, believed that the satellite information was erroneous. There was also the fact that the production of a low-powered device based on uranium-235, the only nuclear explosive that South Africa could produce at the time, would have required a technological ability in the nuclear weapons field that the country at that stage was very unlikely to have achieved.

If the suggested hypothesis of a nuclear explosion were correct, the world would have been faced, for the first time, with such an event for which the responsibility was neither claimed nor determinable. The explosion could have been carried out either by a nuclear weapons power wishing to incriminate one of the nonnuclear weapons powers, or by one of the latter (for example South Africa or another) wishing to remain anonymous. The days when a country's first nuclear test was the subject of a proud announcement were well and truly over!

In reflecting on this incident, it may be noted that for the first time in the history of civilization the start of a war by an anonymous and unidentifiable action is no longer a theme of political fiction but a real possibility. Unlikely as it may seem, it is not unrealistic to envisage a sudden attack on a country's vital centers by nuclear missiles launched from the ocean depths by submarines of unknown nationality or allegiance.

Pakistani Ambitions

Meanwhile, another possible cradle for proliferation had appeared, this time once again in Asia.

Since 1976 Washington had shown serious anxiety over a plan for the construction by French industry of an irradiated fuel reprocessing plant in Pakistan, despite the fact that the proposed installation would be subjected to IAEA safeguards. The French authorities were certainly aware of the risks involved and two years later, following new discussions with the Pakistani government as to the precise purpose of their plant — difficult to justify from an economic viewpoint — the shipment of "sensitive" equipment from France was suspended.

Although this embargo delayed completion of the plant by several years, it did not put the plutonium option beyond the Pakistanis' reach for they already had a pilot reprocessing unit, supplied a few years earlier by Belgian industry, that was capable of both improvement and reproduction at a larger scale.

Furthermore, an unexpected revelation was to show that Pakistan was also pursuing the uranium-235 option. In the spring of 1979, shortly after the execution of the former head of state, Zulfikar Ali Bhutto (who several years earlier had announced that if necessary the Pakistanis would "eat grass" in order to maintain nuclear equality with India), it suddenly became known that they had been secretly building a plant for the enrichment of uranium. This inevitably brought their government's intentions under much more serious suspicion.

The plant was being built near Islamabad, at a site under military control and protection, using plans a Pakistani engineer had misappropriated from the Netherlands. He had been employed by a firm working for the multinational Anglo-Dutch-German enterprise for enrichment by the ultracentrifuge process, a top secret process only recently developed industrially. A network of fictitious companies had been set up to purchase the necessary equipment in various Western countries, their orders being divided and dispersed to avoid arousing suspicion. Pakistan had, in fact, never officially undertaken not to build such a plant capable of producing highly enriched uranium free from any safeguards control.

The U.S. government, despite a decision of Congress to reduce considerably economic and military aid to Pakistan, appeared to be taken by surprise and were unable to stop the development of the operation. The Pakistanis did not deny the existence of the operation, but they continued to claim it was part of a purely peaceful nuclear program.

During the summer of 1979, as a result of rumors in the American press about a possible plan to sabotage the plant, Washington sent a formal denial via the U.S. ambassador in Islamabad. The official wording of this denial specified that the ambassador "conveyed to the government of Pakistan categorical assurance that the United States had no intention to use force or any extralegal means (such as a paramilitary intervention) in Pakistan, nor does the U.S. government have any intention to encourage anyone else to do so.''

This extraordinary communication, especially remarkable because it referred to the completely hypothetical possibility of destruction of a foreign nuclear installation, demonstrated on the part of the U.S. government a depth of wisdom that two years later was to be found entirely lacking in the Israeli government's attitude toward atomic developments in Iraq.

In October 1979, Pakistan dictator General Mohammad Zia ul-Haq, who had dissolved the political parties and cancelled the general election due to take place shortly, issued a declaration refusing to give up the right to carry out a nuclear explosion should this be necessary as part of his country's energy program. In this, he was repeating the barely justifiable formula used five years earlier by India. In turn, the Indian defense minister now made it known that Pakistan's accession to the atomic bomb would be one of the reasons that could oblige India to change its former decision not to equip itself with nuclear armaments.

The full circle of this story was completed in late December 1979. Following the Soviet invasion of Afghanistan, the United States, now wishing to reinforce their ally Pakistan, foresaw a renewal of the military and economic aid that had been discontinued to Pakistan "because of the centrifuge project."

The continuing struggle in Afghanistan, together with the disorganization of an Iran weakened by its state of permanent revolution and its long conflict with Iraq, gave Pakistan an increased geopolitical importance for American strategy in the region. Despite growing rumors that Islamabad was preparing a "peaceful" underground explosion and despite a declaration by Sigvard Eklund, the IAEA's director general, concerning difficulties encountered in implementing the safeguards around the Canadian supplied natural uranium, heavy water power reactor near Karachi, the administration of President Ronald W. Reagan was able to secure a vote in the Senate in 1981 waving the "Symington Amendment." This amendment, as is seen later, forbade economic or military aid to any nonnuclear weapons country engaged in building a nuclear fuel reprocessing or uranium enrichment plant. The important military and economic package adopted by Congress for Pakistan, to be spread over several years, could in fact offer a method of holding back these national developments, for the program would very probably be stopped immediately following a test explosion. Indeed in late 1981 the American Senate decided to suspend all foreign aid to any new country exploding a nuclear device.

It would be difficult to find a better demonstration, in one of the most sensitive regions of the world, of the serious threat to peace represented by nuclear proliferation. But it equally shows the difficulty of reconciling the need for sanctions to block a risk of proliferation with the imperatives of a military alliance or even of maintaining a regional balance of power.

Furthermore, as with South Africa, it was difficult to estimate the state of advancement of the project or the date by which the quantities of concentrated uranium-235 or plutonium required for an explosion would have been obtained, if such was indeed the unacknowledged objective of the venture.

In any case, it is certain that the publicity that built up around the South African and Pakistani projects gave to those two countries almost the same "prestige" as they would have acquired from a first nuclear explosion. Consequently, it is to be hoped that they may be dissuaded from actually carrying out such a test.

It is impossible to foresee whether or not these projects will lead to the production of a few untested devices, or to one or more explosions possibly followed by the development and manufacture of sophisticated nuclear armaments. But the countries concerned have clearly, and for a long time, been determined to acquire nuclear capability; at the very least equipping themselves with a source of nonrestricted highly enriched uranium or plutonium.

In the case of Pakistan, the most advanced nuclear-oriented country of the Islamic World, this effort to explode a nuclear device could have been intended to redress the 1971 humiliation of losing its Eastern province of Bangladesh while at the same time counterbalancing India's achievements, Iran's ambitious nuclear plans (before they were cut short by the revolution), and Israel's advance. In this last objective, the Pakistanis must surely have had the approbation, if not the material support, of the Arab world.

The Destruction of the Iraqi Reactor

Near the beginning of the 1980s, two other Islamic countries, Libya and Iraq — despite the fact that both were party to the NPT — caused international concern as to the objectives of their nuclear programs.

The Libyan leader, Colonel Muammar al-Qaddafi, great protector of international terrorism, had never concealed his intention of equipping his country with the same weapons as Israel. In the absence of installations capable of producing concentrated fissile materials, he was obtaining uranium from his southern neighbor, Niger. He was planning to transfer some of this uranium at a suitable moment, together with the substantial financial resources necessary, to Islamic countries better equipped than his own to achieve the dream that neither his money nor his oil had enabled him to realize unaided.

As for Iraq, in 1975 it had ordered a replica of one of the large French research reactors. This order followed the abandonment of an earlier plan through which Iraq would have acquired a natural uranium, graphite-moderated and gas-cooled power station similar to one sold by France to Spain ten years earlier. The proposed sale had been dropped for industrial reasons and also because such a reactor would have produced considerable quantities of plutonium and so entailed a proliferation risk, despite the fact that Iraq was a party to the NPT and had placed all of its nuclear activities under IAEA safeguards.

The supply of highly enriched weapons-grade uranium for the research reactor's first core (despite the fact that two cores would have been needed to achieve a single explosion) led to violent press campaigns in 1980 and 1981, no less in Israel and the United States than in France herself, who was accused of helping her principal oil supply country gain access to nuclear weapons. These campaigns, inspired by the Israeli government and supported by Jewish communities in America and France, fruitlessly sought to stop the construction of the reactor and to obtain an embargo on shipment of the fuel.

The nature of the reactor, the fact that it was subject to IAEA safeguards, and above all the permanent presence of French specialists within the installation, under the terms of a later disclosed 10-year Franco-Iraqi joint research contract, would have made its use for military purposes practically impossible. The one reservation was that the project would inevitably result in the formation of a corps of Iraqi specialists having the high quality and competence essential for a possible military program. Such a result, however, cannot be avoided whenever advanced technical training in the civil nuclear field is provided by one country to another.

By spring of 1981, construction of the Iraqi reactor, at the Tammuz nuclear center at El-Tuwaitha on the outskirts of Baghdad, was progressing reasonably on schedule. This despite the sabotage of an essential major component in France, the assassination in Paris of one of the Arab specialists concerned, and an air attack on the installation site in autumn 1980 at the start of the war with Iran.

But on June 7, 1981, a squadron of Israeli bombers, in an astonishingly well prepared and effectively executed surprise attack, completely destroyed the reactor and killed one of the French technicians on the project.

Thus, the Israelis accomplished the preventive action that the United States had ''no intention'' of undertaking two years earlier against Pakistan, or a third of a century earlier against the Soviet Union. An action that the Russians equally had not dared against China in the 1960s was taken by Israel against Iraq, one of the Arab countries that still maintained a state of war against Israel, whose existence it consistently refused to recognize and whose more recent proposals to create a nuclear weapon free zone in the Middle East had remained ignored.

For the first time, force had been used in pursuit of nonproliferation. An IAEA member-state, nonsignatory to the NPT, that had refused to place its main nuclear center under the agency's safeguards system was denying the effectiveness of such a system when it was applied to another country.

A severe blow had been dealt to the political foundations of international nuclear trade, which the opponents of nuclear energy were bound to exploit. In the atomic field, resorting to the law of the jungle could very well confirm fears that, sooner or later, the last barrier of nonproliferation would give way — that barrier which so far had effectively prevented the nuclear weapons powers from passing on those weapons to others, even their closest allies.

The attack on the Tammuz center, only a few weeks before the Israeli elections that were to return Israeli Prime Minister Menachem Begin's government to power, provoked a great deal of emotional reaction. The affair, condemned by President Ronald W. Reagan, was taken to the U.N. Security Council, which unanimously "strongly condemned the air attack while also urging Israel to open her own nuclear installations to international inspection."

Condemnation by the IAEA's general conference in September 1981 was no less unanimous. While Israel avoided the immediate suspension of its membership requested by the Arab countries, such a suspension was to be reconsidered a year later if the Dimona center was still not placed under safeguards as urged by the United Nations. For the first time, the menace of a political sanction had been used to force a country to open up its nuclear facilities to the agency's inspection system. On the other hand, the Saudi Arabian government offered to finance the reconstruction of the reactor, which France had agreed in principle to rebuild.

A final reflection on this affair. In the aftermath of his reactor's destruction, the Iraqi president, Saddam Hussein, called on "all peace and security loving nations to help the Arabs to build a nuclear bomb to counterbalance the existing Israeli ones." Sooner or later, an Islamic nuclear bomb will appear on the scene and it is likely that several Islamic countries will have contributed to its production. Then a vital question will be whether this new nuclear power will be wise enough to keep strict control of the use of the weapon — as has been the case so far with all the nations to have carried out an explosion — or whether, contrary to tradition, the weapon will become available to all the Islamic countries concerned with its development. In the latter case, the world will be faced with a new and dramatic form of nuclear proliferation.

Other Possible Escapees from the Nonproliferation Net

If it thus appears inevitable that one or the other of the Islamic nations will eventually acquire nuclear weapons capability, the same must also be true for the more advanced of the Latin American countries where, although IAEA safeguards have been accepted on all publicly known atomic installations, a military nuclear option has nonetheless been kept open. This is the case with Brazil (assisted by West Germany) and especially with Argentina, the most advanced Latin American country from the nuclear viewpoint. Argentina's Atomic Energy Commission is patiently pursuing, in most cases initially no further than the pilot stage, every aspect of plutonium production from natural uranium, heavy water reactors.

Spain might also have considered an independent road except that this could have hampered its hoped-for membership in the European Community. The nonnuclear weapons countries of the community would find it difficult to accept further discrimination beyond that of the French and U.K. positions, and might well put pressure on Spain to persuade it to accept the NPT, despite the placing of all Spain's nuclear installations, including a Franco-Spanish power station, under IAEA safeguards in 1981.

As early as 1966, the Spanish government had suffered the consequences of an unexpected form of proliferation. Following an air collision between an American nuclear bomber and its refueling tanker-plane, four hydrogen bombs had fallen either onto Spanish territory at Palomares or into the nearby Mediterranean sea, causing plutonium contamination and leading to an extraordinary and costly U.S. submarine hunt to locate one of the devices, which was found intact at a depth of nearly 2,500 feet!

Nonproliferation policies inevitably had a worldwide impact for, ever since the Chinese explosion and the initial negotiations leading to the NPT, no country having kept open the military nuclear option had dared to admit publicly the intention of eventually acquiring nuclear weapons. On the contrary, those countries whose achievements appeared to point in that direction were at pains to proclaim their nuclear pacifism. The five members of the Atomic Club, for their part, had never concealed their intentions and hopes during the preceding period.

Such a change of attitude could be claimed by the five nuclear weapons powers as a success in their efforts to condition and control world public opinion. In fact, they had succeeded to some extent in deflecting attention from the risk of a nuclear holocaust due to their own megatons of armaments, to the dangers of a country without nuclear weapons possibly carrying out, or even considering carrying out, a single explosion. Any such act or intention was being considered more and more as a serious departure from international ethics.

The attitudes of governments party to the NPT, and of international public opinion, had been profoundly modified. All were now ready to accept, to a certain degree, the notion that the simple explosion of a supercritical mass of plutonium or uranium-235 by a nonnuclear weapons power was a kind of international sin, considered more and more as a serious deviation from the rules of correct behavior in the international community.

The international community has been fully convinced, particularly since the Indian explosion in 1974, of the seriousness of the threat to world peace that would result from any increase in the number of countries equipped with nuclear weapons. Such proliferation would be liable to produce not only various effects of regional destabilization, but also risks of a general contagion.

Any such proliferation would also cause special anxiety over its long-term effects on the existing German and Japanese renunciations, which those countries would surely review in the event of a series of new accessions to the Atomic Club by countries industrially and politically less important.

In this connection, it should be noted that, so far, the risks of further proliferation have been confined to a wider access to the A-bomb, the principles of which are now well known. The threat to world peace and to the balance of world power would be increased considerably if access to the H-bomb and its much more complex principles were to become comparatively easy. Unfortunately, this now seems to be the case.

In 1978 after a patient and lengthy inquiry of American atomic scientists and establishments, and because of involuntary leaks, a nuclear disarmament activist, Howard Morland, managed to piece together many of the facts relating to the most secret principles of the H-bomb's trigger system. These facts were later confirmed in a report, declassified by an unpardonable error, that was available at the library of the Los Alamos National Laboratory. Furthermore, in 1979, accompanied by an extravaganza of publicity, these data were released to the press and so became generally known throughout the world. Considerable legal efforts by the American government to oppose the publication had failed before the principles of freedom of the press.

The seriousness of this controversial release of information can be judged by the fact that the French nuclear weapon designers felt that they would have shortened their search toward the H-bomb mechanism by several years if these data had been available to them in the 1960s. Thus, by an ironic turn of fate, the greatest sin to date in nuclear proliferation has been committed by the United States. This breach of security evidently caused no particular concern in the U.S. Congress, which has always been so ready to seize on the slightest failing of other countries in the area of nonproliferation, the vital importance of which has been generally recognized by all nations.

However, in the past, one country had adopted an opposite attitude toward proliferation. Early in the 1960s, China was promoting the view that an increase in the number of countries in possession of nuclear weapons would be desirable, arguing that this would contribute to world stability by reducing the influence of the two great ''imperialist'' powers.

In more recent years, the Chinese government seems to have abandoned this thesis. There is no evidence of its application either in the significant export of Chinese nuclear materials and technology to friendly countries or, even less, in the transfer of weapons. Nevertheless it seemed strange, during the first mission of French nuclear specialists to China in 1975, to hear the point of view favorable to proliferation expressed by the various authorities we met, one of whom was Hua Guofeng. At that time, Hua was no more than eighth deputy prime minister; he later became the Chinese leader until 1981.

During the 15 years following the accession of the great People's Republic to the Atomic Club, China has remained aloof from all international nuclear negotiations and proceedings, and even from the IAEA, although to make way for its admission the IAEA's board of governors expelled Taiwan in 1971. This situation must inevitably change, for ultimately China cannot avoid, at least on the military level, becoming a world atomic power. China will therefore inevitably play an increasing role in the balance of world nuclear forces, a balance in which only the United States and the Soviet Union have as yet sought negotiated agreement.

3. The Balance of Terror

Overarmament

This book so far has been essentially devoted to the acquisition of atomic weapons and to efforts directed against the spread of such acquisition. Little attention has been given to the weapons themselves. This final section of Part One is devoted to an intentionally brief examination of this matter.

We therefore now leave the world of nonproliferation, with its rules and prohibitions, its suspicions and discriminations ... a world in which the idea of destroying a civil nuclear installation because it might possibly be used for military purposes suddenly and starkly became a reality with the 1981 Israeli attack on the Iraqi Tammuz reactor.

Instead we now take a quick look at the world of understood and accepted proliferation; proliferation that is not hidden (in any case, spy satellites now make secrecy virtually impossible); proliferation in which megatons are proudly proclaimed in grand public displays of national nuclear potential.

The members of the Atomic Club, having had such great difficulty in gaining admission, have subsequently allowed themselves — within the limits of their individual industrial and financial resources — to indulge in an orgy of production of all the types of weapons that are forbidden to other countries.

These weapons are designed and produced with the maximum possible level of technical secrecy, but also with the fullest amount of publicity given to the results. Such publicity, necessary internally to persuade national authorities to allocate the considerable financial backing that is required, is also a fundamental part of the international game of military deterrence, provoking a form of "nuclear exhibitionism," which underlines again the discrimination between the two worlds of nuclear weapons powers and nonnuclear weapons powers.

In the early 1980s, a total of 40,000 to 50,000 nuclear warheads were dispersed in different parts of the world, most of them more powerful than those used against Japan. About one-third of them were strategic weapons, the rest tactical.

At a time when terrorism, particularly in its more generalized form of guerilla warfare, seems to be continually increasing, when a devastating criminal attack can be made with a few pounds of "plastic," the quantities of nuclear explosives in the arsenals of the two superpowers correspond to a total power about a million times greater than that of the Hiroshima bomb. It is an amount equivalent to more than four tons of conventional explosives for every human being in the world.* (The French arsenal, which represents no more than about one-half of one percent of this total, is nevertheless redoubtable and generally considered adequate for its purpose.)

Such enormous explosive force is so excessive that the fact that two-thirds of it is in the possession of the Soviet Union (due to the greater power of the Russian H-bombs) and that the United States has only one-third does not seriously affect their capacities for reciprocal dissuasion or mutual destruction.

This completely nightmarish estimate of a million times the power of Hiroshima is well beyond human comprehension even for the specialist in these abominable devices, and gives some idea of the headlong rush that the nuclear arms race has become in no more than a third of a century. The man in the street is not sufficiently aware of this; on the contrary, he finds a certain relief in the fact that, so far, the horror of the atomic bomb has prevented any direct conflict between the nations equipped with it, in particular, conflict between the world's two superpowers.

---

*The total amount of fissile materials in the American stockpile of weapons must be in the region of 500 tons of uranium-235 and 50 tons of plutonium. Total Soviet production is probably somewhat less.

Experience in the two world wars of the present century has shown how difficult it is to forecast the characteristics of such conflicts. It was the case in 1914 to 1918, for the static front in which a million men confronted each other in hundreds of miles of parallel trenches, and in 1940 to 1945 for the mobile tank and air warfare.

Nevertheless, it seems no wild forecast that a worldwide nuclear conflict would lead to the annihilation of civilization as we know it, though it wouldn't destroy the human species as Einstein had suggested in his 1950 letter to President Truman (see p. 159).*

No one today can visualize the technical, psychological, and political development of a conflict in which the main opponents could use hundreds of ballistic missiles each equipped with one or more nuclear warheads. With any one of these warheads alone able to destroy a town, or an industrial or military center, their simultaneous use would inevitably lead to the loss of tens of millions of of human lives. It is impossible to predict the behavior of a population faced with the threat of such a cataclysm, the scale of which would be very little affected by "passive defense" based on underground shelters.

No Sanctuary

For the first time, even the home territory of the United States can no longer be considered an unassailable sanctuary, and this as a direct result of the weapon that, not so long ago, had — in Truman's 1945 words — made them "the most powerful nation on earth and doubtless the most powerful nation of all time."

This new U.S. vulnerability represents a political discontinuity of capital importance. But it has been accompanied by a second political revolution, namely the opening of Soviet territory to constant observation by the United States — not as a result of the multinational surveillance installations foreseen in the Acheson–Lilienthal plan, but thanks to "spy satellites" enabling the two superpowers to know at all times what is happening in any part of the world and especially within each other's frontiers.

---

*All the experimental aerial explosions that were carried out up to 1963 involved about 5 percent of the potential power of the current world nuclear arsenals and led to an increase of some 2 percent in the radiation to which mankind is exposed. From this, it can be deduced that the additional biological effects for man if the world's entire current stock of weapons were exploded would be on an order similar to those caused by today's natural radioactivity.

The present monopoly of such satellite use by the United States and the Soviet Union further accentuates the gap between the superpowers and the other countries, which is why, in 1978 when France returned to the disarmament negotiations, she proposed in vain to equip the United Nations with such a system of worldwide surveillance.

Since the 1960s, this further application of technological developments in the use of space has brought about a second revolution in the military applications of fission. Following the use of missiles as nuclear warhead carriers, satellites have provided the means for eliminating surprise attacks on the military potential of one country by another. American-Soviet relations have completely changed as a result.

In the United States as in the Soviet Union, the conquest of space and the introduction of ballistic missiles into defense systems progressed hand in hand, mutually influencing each other. From the start, the Russians based their strategies on the use of very powerful charges whereas the Americans, using smaller charges, sought and attained greater precision of trajectories and points of impact.

During the 1950s, American strategy was based on dissuasion by means of massive reprisals and the use of bomber aircraft to carry nuclear bombs to enemy territory. The introduction of nuclear missiles into the American strategic force was then developed in three different fields, the so-called "triad" of giant bombers, nuclear-powered submarines, and land-based ballistic rockets. The Soviet Union, with a few years delay, followed the same course.

The American triad came into being simultaneously with a tactical armament comprised of fairly short-range missiles intended to protect the Western World's European frontiers against the superiority in conventional weapons of the Eastern Bloc.

Throughout the presidency of Eisenhower, the American lead seemed unassailable. It was still considerable in 1962 during the Cuban affair, and this was undoubtedly a contributory factor in the Soviet leaders' decision to back down. At that time, those Soviet leaders vowed they would never again be in such a position of inferiority and ever since the Soviet Union has consistently sought to catch up with, and if possible overtake, the United States in nuclear armaments.

Toward Nuclear Parity

During the 1960s, American-Soviet nuclear equality was progressively established. In the middle of that decade, the United States still had a three or four times numerical superiority over the Soviet Union in each of the three fields of the triad. At this point, the Americans, anxious to encourage and exploit détente and aware of the pointlessness of pursuing further overarmament, officially decided to limit their strategic attack capacity. Their revised plans were for 41 nuclear submarines, each equipped with 16 underwater-launched Polaris missiles with a range exceeding 3,000 miles, 1,000 Minuteman missiles with twice that range, and some 500 giant bomber aircraft.

At the same time, both superpowers announced decisions to slow down production of fissile materials for their nuclear weapons. This was not a step toward disarmament, but rather open evidence that both powers recognized their overarmament.

In fact, both superpowers had succeeded, thanks to accelerated programs of underground experiments, in improving the effectiveness of their nuclear explosives, enabling them to increase the effectiveness and reduce the size of their warheads. As technology improves, it is always possible to redeploy the plutonium and uranium-235 in existing explosive devices in order to make improved warheads and perhaps even in greater numbers.

During this same decade, advanced research in the United States on missiles and their nuclear warheads was being pursued in two main directions (which the Soviet Union was also following): the development of an antimissile defense system and the possibility of fitting several nuclear warheads to a single ballistic missile.

It had long seemed that an effective antiballistic missile (ABM) defense system could not be realized. The problem was to divert or destroy a rocket traveling at a speed of about four miles per second before it reentered the earth's atmosphere. The envisaged solution consisted of sending a missile, guided by an advanced radar system and having a powerful nuclear warhead, to intercept the attacking missile. The explosion of the "counter-missile," even some distance from its target, would be accompanied by the emission of intense and penetrating radiation which, being "outside" the earth's atmosphere, could explode the attacking missile from a distance of a few miles, or at least sufficiently damage its electronic systems to divert it and render it useless.

The second area in which research was being undertaken was in multiplying the effects of a single missile by equipping it with several nuclear warheads, designed either to spread out to cover a single target, or to seek independently different targets that could be up to some 50 miles apart. Missiles of this type were known as MIRVs (multiple independently targetable reentry vehicles). They had the additional advantage of making

ABM defense much more difficult; also, it was impossible to verify by satellite whether or not any given missile was fitted with a single or a multiple warhead. This possibility of equipping a single missile with several independently guided warheads represented a major step forward in the escalation of nuclear armament because it radically altered the capability of an attacking force.

The first airborne trials of prototype ABM missiles with MIRVs, none of which carried nuclear warheads — in accordance with the treaty prohibiting all but underground nuclear tests — took place in the United States in late 1968. Virtually simultaneously, the two superpowers decided to start discussions on the limitation of their strategic armaments. These were known as the SALT (Strategic Arms Limitation Treaty) negotiations.

SALT I

The negotiations leading to the first Strategic Arms Limitation Treaty (SALT I) were without a doubt among the most extraordinary in the history of diplomacy. Among the motives behind them, the least important was by no means an attempt to limit the increase of military budgets, a highly significant factor in the political stability of the two superpowers. Another factor was the maintenance of a degree of détente in a divided world.

The whole effort was rather like a critical game of poker in which each player revealed the essential minimum of his hand, as well as what he knew of that of his opponent. The objective was to limit or, if possible, reduce armaments while maintaining effective mutual dissuasion.

During the negotiations, each player tried to gain the advantage by pretending that his position was the weakest. But as soon as an agreement was concluded, he had to try to show his own national audience that, on the contrary, he had played a skillful game and finished better placed than his opponent.

The negotiations were concerned only with nuclear warheads and vehicles for their delivery; they did not involve the performance of the national production plants for fissile materials and the nuclear fuel cycle as a whole. But in any case, the tens of tons of plutonium and the hundreds of tons of uranium-235, which each of the two powers was producing for military purposes, were far more than sufficient for their overarmament. Their consequent indifference concerning production levels of fissile materials (provided only these remained more than adequate) emphasizes once again the contrast with the world of nonproliferation and the complete uselessness of controlling peaceful nuclear operations within the territories of nuclear weapons countries.

Finally, these negotiations, from which the allies of the participants were excluded, underlined the exclusively 2-power nature of the current world balance and the military advance of the two "rivals and accomplices" as compared with other countries.

The decision to start the negotiations was made on July 1, 1968, during the signing of the NPT. The talks were scheduled to begin three months later. However the Russian occupation of Czechoslovakia that summer delayed the first discussions for over a year. Sessions lasting several weeks, followed by intervals of several months, were held late in 1969 in the greatest secrecy, alternately in Helsinki, Finland, and Vienna, Austria. The American negotiating team was led by Ambassador Gerard Smith, who had already been involved in international nuclear matters in the early days of the Eisenhower "Atoms for Peace" proposals.

While the Americans were equipping their submarine rockets and intercontinental missiles with MIRVs, increasing to 3 and 10, respectively, the numbers of warheads carried by each, and while the Russians were hurrying forward toward the moment when they would be able to do likewise, the negotiations came to a virtual standstill. Little progress had been achieved, even by mid-1971. The Americans wanted an agreement covering both offensive weapons and defensive ABMs; the Russians preferred to limit themselves to the latter. But on May 26, 1972, in Moscow in the year when U.S. President Richard M. Nixon's first term came to an end, he was able to conclude with Soviet leader Leonid I. Brezhnev the agreements that became known as SALT I.

The agreements were comprised of a treaty of unlimited duration for the ABMs, which were to be restricted to two sites (the capital and a zone for missile deployment), each to be defended by 100 ABMs and, second, a 5-year agreement "freezing" the number of ground-to-ground rockets but not forbidding the multiplication of warheads.* Both the treaty and the agreement were rapidly approved — almost unanimously — by the U.S. Senate.

In terms of launching rockets, the balance was unfavorable to the Americans. But on the other hand, the Americans were in a better position in terms of warheads, thanks to the MIRV technique in which they be-

---

*SALT I allowed the Americans to increase the number of their rocket-launching submarines from 41 to 44, while the Russians could build up to 62 of them. The freeze restricted the United States to 1,054 intercontinental missiles and the Soviet Union to 1,618, several hundred of these latter equipped with much more powerful warheads (25 megatons) than the American ones.

lieved, mistakenly, that they were well ahead of the Russians.

But in any case, such comparisons are of no significance since each superpower has more than enough weapons to inflict ruinous devastation on the other, even if one should have the initiative of a surprise attack.

Without observation satellites, these negotiations could not have been undertaken, nor could they have produced any results. The satellites had enabled each party to demonstrate that it had extensive knowledge of the other's armaments, a precondition to any useful talks. Satellites would also make it possible to verify that the agreements were respected, without any need for ground inspections on the national territories of the parties. One of the conditions set out in the agreements was a promise by both parties not to attempt to misrepresent the weapons systems concerned to their respective information-gathering services and not to interfere with the proper operation of these services, which were described, modestly, as "national means for technical verification."

The paradox in these agreements, supposedly "disarmament" agreements, was that their principal purpose was to limit defensive rather than offensive weapons. The logic of dissuasion had thus led to considering as a disarmament measure the provision of guarantees that an opponent's offensive weapons should be able to produce the expected devastation. The arms race was to go on, but regulated by certain rules and conditions that would give it a qualitative rather than a quantitative character.

Nevertheless, SALT I was an undeniable diplomatic success and a confirmation of détente. In a world where, in Winston Churchill's words, "safety is the healthy child of terror, and survival the twin brother of annihilation," the balance of terror had received official recognition.

SALT II

Since the section of the 1972 agreements that limited the number of intercontinental missiles was due to expire in 1977, the two SALT players returned to the gaming table and took up the cards again in 1973. Nixon, now under the strain of the Watergate scandal, was hoping to rebuild his image by concluding a further Moscow treaty in 1974. He was not successful: All he could obtain was a reduction in the number of permitted ABM sites from two to one in each country.

President Gerald R. Ford then took over Nixon's task. He had retained Secretary of State Henry A. Kissinger, who had not been directly involved in the previous lengthy SALT I negotiations but had intervened in their conclusion. In December 1974, Kissinger was able to settle on a preliminary agreement with Brezhnev in the Soviet Union at Vladivostok that was expected to lead to the future SALT II.

The Vladivostok agreement, which remains in force until 1985, reestablished parity for the number of intercontinental carrier rockets allowed each country, a parity that had been unbalanced in favor of the Soviets by the SALT I freeze. The new agreement set a limit of 2,400 for the total number of intercontinental missiles, submarine missiles, and strategic bombers that each country could possess. It similarly restricted the number of ground and seaborne missiles capable of carrying MIRVs to 1,320 for each country. Once again, these restrictions did not imply any disarmament, but merely set limits to overarmament.

Negotiations for SALT II continued behind closed doors for over three years; President Jimmy Carter, elected in 1976, was determined that this should be the most brilliant diplomatic success marking his term of office before his 1980 campaign for reelection.

In 1978, wishing to appear as a champion of détente and nuclear disarmament, Carter abandoned a project for the production of an "enhanced radiation warhead" or so-called "neutron bomb." The neutron bomb is, in fact, an H-bomb especially designed for tactical use against concentrations of tanks, the crews of which it would immediately incapacitate. Soviet propaganda against this "amoral" weapon (as if other nuclear weapons were any less so) was widely echoed by public opinion. The actions of the media, together with the hesitations of the American president, influenced the Western governments who might otherwise have been well protected by this essentially defensive weapon against Eastern superiority in conventional weapons. Despite pressure from the North Atlantic Treaty Organization (NATO), none of these governments in the end agreed to allow this possible future weapon within their territories. The main victim of the affair was the credibility of American nuclear protection for Europe.

Three years later, President Reagan, wishing to be in a position of strength before any negotiations with the Russians and indifferent to the growing peace movements in Europe, decided in mid-1981, without officially consulting the NATO allies, to go ahead with the full production of neutron warheads: these weapons to be stockpiled in the United States, with any future deployment in Europe to take place only after full consultation with the allies. France has also pursued research on this weapon following a decision made in the late 1970s and confirmed in 1981 by the new Socialist government, although opposed by the Communist party, a member of the left-wing coalition in power.

The neutron bomb was not the only innovation to appear in nuclear armament while the SALT II negotiations were dragging on interminably. The Americans were perfecting their Cruise missile, an updated version of the German V-1 device that appeared toward the end of the Second World War. This new missile is capable of being guided over considerable distances to a particular target with great accuracy, avoiding radar and anti-aircraft defense systems by flying at very low altitudes.

As for the Russians, they were making good progress, reducing much of the gap between themselves and the Americans in the field of MIRVs and equally in the precision of their most powerful intercontinental rockets, which have become formidable weapons of attack against U.S. intercontinental missile storage silos.

This obliged the Americans to consider a new intercontinental missile, the MX, which is equipped with 10 nuclear warheads. It would be the land-based equivalent of the Poseidon missile (also equipped with 10 to 14 warheads) with which their older submarines were armed. Just two of these older submarines, thus equipped, one stationed in the Mediterranean and the other in the Pacific, would be sufficient to destroy most of the Soviet cities and more than 100,000 people within half an hour, according to President Carter in his State of the Union message at the beginning of 1980.

Finally, the United States had begun the construction of 13 Trident-class nuclear submarines of some 20,000 tons. Each of these underwater giants (the first of which was due to be commissioned in 1982 at a cost of over a billion dollars) was to be equipped with 24 missiles all carrying 8, or later 14 independent warheads with an overall range of nearly 5,000 miles. One of these submarines alone represents a redoubtable and invulnerable dissausive force with a striking power equivalent to some 200 to 300 powerful atomic bombs.

However, the Russians clearly have no intention of being overtaken in the race. In spring 1980, American satellites identified a construction site where the first unit of a series of nuclear submarines of some 30,000 tons was being built.

The Soviets have also introduced an entirely new concept into this spiral of destructive power in the form of mobile ground-launching bases, and they have perfected a medium-range rocket carried on a mobile platform and known as an SS20. Its mobility makes the device easier to protect from attack. Furthermore, such a weapon falls outside the scope of the SALT agreements since its range is insufficient to reach the United States from Soviet-controlled territory. However, the range (2,500 miles) coupled with great accuracy is quite enough to present an increased threat to all the vital centers of Europe. To this threat must be added that of the recently introduced Soviet "backfire" bomber aircraft, which can carry nuclear missiles.

It was in this developing atmosphere of mutual distrust and military escalation that, on June 18, 1979, the SALT II agreements were signed in Vienna by Brezhnev and Carter, two failing presidents, one physically aging and as yet without a designated successor, the other suffering politically under the weight of a marked slump in popularity.

Compared with the Vladivostok agreement, SALT II provided for minimal weapon reductions. The total number of missile vehicles was changed from 2,400 to 2,250, and that of missiles capable of carrying MIRVs (maximum of 10 warheads for new missiles) was reduced from 1,320 to 1,200, of which no more than 800 could be intercontinental missiles. Each party, while free to develop further types of missiles for launching from nuclear submarines (which were still considered virtually invulnerable), was restricted to a single new type of land-based intercontinental device.

A week before these agreements were signed, President Carter had announced the start of production of 200 new MX missiles — the Americans being at this time 200 short of their permitted "missile allowance." The new weapons were to be installed on mobile platforms and circulated in a maze of shelters spread out across the Utah and Nevada deserts so as to reduce their vulnerability, and the project was to cost $30 billion over 10 years. The SALT II agreements were thus celebrated by an important addition to the U.S. arsenal. Two years later, this multiple shelter scheme was still undecided, and President Reagan was considering scrapping it and replacing it by 100 MX missiles based in fixed strengthened silos, as well as by a new generation of intercontinental bombers.

All this complex terminology and accountancy of destruction went completely beyond the comprehension of the general public, and no doubt also of the members of the U.S. Senate, whose two-thirds approval was essential for the ratification of SALT II. It soon appeared that this ratification was uncertain in contrast to that of SALT I, which had been quickly approved by 88 votes to 2. But this time the fate of the new agreements was dependent on developments in Soviet relations, which had deteriorated following the Russians' activities in Asia and Africa, the activities of their Cuban agents within the latter continent, and also the continuing Soviet presence in Cuba. Further, the SALT II issue was inevitably linked with the preliminary maneuvering in preparation for President Carter's campaign for reelection. The year 1979 ended without the Senate reaching a decision.

But the death knell of the ratification was sounded in 1980 when the Russians occupied Afghanistan. The issue of SALT II now became an important part of the U.S. electoral campaign, with the Republican candidate and his future secretary of state, General Alexander M. Haig, opposed to it. In fact, once elected, President Reagan announced his intention to renegotiate the agreements, but he would not consider doing this until there was some improvement in Soviet-American relations. No U.S.-Soviet disarmament conversations were foreseen before the end of 1981 at the earliest.

But the SALT II agreements, even without ratification, were going to be broadly respected by the two parties, for their terms roughly corresponded to the already scheduled U.S. and Soviet programs. Those terms bore witness to the undeniable reinforcement of the Soviet nuclear strike force, which very naturally helped to create feelings of insecurity in Europe.

Nevertheless, the poker game had to be played out between the two "rivals and accomplices," who were inevitably obliged to include in their "horse trading" the medium-range devices that were protecting or threatening the countries of Europe.

The Euromissile Affair

The deterioration of U.S.-Soviet relations quickly began to affect the various aspects of dissuasive policies that directly concerned Europe. This was no novelty, for in the early 1960s, Europe had already been in the front line of the balance of terror, before the two superpowers had perfected their intercontinental missiles.

From 1959 onward, each of the two rivals was occupied in setting up a network of medium-range rockets — the so-called "Euromissiles" — capable of annihilating the other's main European bases.

The American network, comprised of some hundred Thor- and Jupiter-type rockets based in Great Britain, Italy, and Turkey, was dismantled in 1963. Its surface-launching bases were far more vulnerable than the mobile platforms of the growing force of submarines equipped with Polaris missiles. The Russians, on the other hand, kept their network of some 700 devices (SS4s and SS5s), which were powerful but not very accurate. In 1977 they decided to replace them progressively, at the rate of about one per week, with SS20 missiles on mobile launching ramps, which were very much more accurate; more than 200 had been installed by the end of 1981. The Russians called this operation a "modernization," but in reality it was a fundamental change, for the new missiles made it possible to carry out selective attacks against NATO military targets, thereby destroying the military potential of the alliance while leaving unharmed the cities and populations of Europe.

Confronted by these weapons, NATO had no nuclear potential in Europe other than operational field weapons directed toward East Germany, Poland, and Czechoslovakia but incapable of reaching the Soviet Union. They were thus insufficient to counterbalance the new Soviet weapons.

Therefore, the NATO Nuclear Planning Committee (in which France was not represented) decided in late 1979, six months after the signature of SALT II in Vienna, to install in Western Europe by 1983 a network of highly accurate missiles: this despite Belgian, Danish, and Dutch reservations, and in exchange for an American promise to open negotiations with the Soviets to ease the dangerous situation.

The new network was to be comprised of a hundred or so Pershing II rockets, with a range approaching 1,500 miles, that would be based in Germany, together with 470 subsonic Cruise missiles, having a slightly greater range, that would be based in Germany, Great Britain, Belgium, and the Netherlands. These Pershing II rockets could reach Soviet military targets in under 10 minutes, whereas the Russian intercontinental missiles could not reach the United States in less than half an hour.

This "lack of symmetry" was considered unacceptable by the Russians, who regarded their own "modernized" missiles as tactical battlefield weapons since they could not reach American territory; but they considered the devices planned by the United States and their allies as strategic weapons since they could be used to threaten the Soviet homeland. This, in a nutshell, is the hub of the problem of European survival. It means that, for the Russians, Europe must be constantly exposed to attack, but can never be allowed the capacity to retaliate.

The Russians tried hard to nip the American proposals in the bud as they had managed in 1978, to stop for some years the production of the neutron bomb, and once again they succeeded in mobilizing the support of a considerable proportion of the general public in the countries most concerned. In addition, in February 1981 Brezhnev proposed a moratorium that would have officially recognized the existing inequality by allowing the installed Russian SS20s to remain in position while forbidding deployment of the planned American missiles. Because of this evident inequality, the Atlantic allies rejected the proposed moratorium.

In the case of the neutron bomb, it was President Carter's hesitation rather than Soviet propaganda that led to its development being abandoned in the United States for three years. In the early 1980s, the influence of the Soviet Union, added to the obvious reluctance of the Reagan administration to commit itself without prior reflection to any form of nuclear arms limitation, succeeded in reawakening the antinuclear weapons movement to a degree not seen since the mid-1950s.

Confident that their generalized knowledge of the problems involved was a better background for propaganda than the detailed expertise of specialists long and intimately concerned with the subject, the ecological movements opposed to civil nuclear energy joined forces with religious organizations and pacifist, neutralist, and leftist groups and in this way there appeared a veritable political force, especially in the NATO countries of northern Europe.

The neutron bomb, the Euromissile problem, and in late 1981 some unfortunate declarations of some of the highest American officials increased the fear in the European public that western Europe would become the battlefield of a nuclear war between the two superpowers.

This fear created a wave of neutralism and stimulated proposals for a European denuclearized zone, and in Britain the left wing of the Labour Party and the Trades Union Congress demanded that the United Kingdom should disarm unilaterally, while the British Liberals, together with the Socialist parties in the Netherlands and Belgium, opposed the planned deployment of ground-launched Cruise missiles in 1983. Opposition to the American decision to proceed with the production of the neutron bomb was another element of this campaign; it even spread to France where the Communist party, though favorable to the French nuclear deterrent, opposed research on this defensive atomic weapon.

But the most impressive demonstrations took place in the autumn of 1981 in Brussels, London, Rome, and even Paris, as well as in West Germany in West Berlin and Bonn. Here a quarter of a million people with placards hostile to nuclear armaments and to the United States but silent toward the Russian missiles (some claimed ''better red than dead'') marched through the avenues in the greatest, largest popular protest since the foundation of the Federal Republic. This peace movement endangered the cohesion of the ruling Social Democratic party already weakened by economic difficulties; it was strongly opposed by Chancellor Helmut Schmidt but was approved by some of the ranking left-wing members of the party and condoned by its former leader, ex-Chancellor Willy Brandt.

As a result, by the end of 1981 the installation of the proposed new missiles, at least in Germany and the Netherlands, had become a virtual impossibility until disarmament negotiations with the Soviet Union, which Washington had promised to initiate before 1982, had been tried and seen to fail. Such negotiations would of course inevitably include discussion of French and British nuclear armaments.

In November 1981 President Reagan offered to cancel the deployment of the new U.S.-built missiles in western Europe as long as the Soviet Union would dismantle all of its SS20s. On the other hand, Premier Brezhnev said the Russians would considerably reduce their medium-range weapons in the course of genuine negotiations with the United States. Finally, on November 30, just two weeks before the imposition of martial law in Poland, these American-Soviet conversations started amidst great secrecy in Geneva. A long new period of negotiations had commenced.

Other Nuclear Arsenals

The Chinese nuclear armament seems relatively limited though well conceived. Its capacity for reprisals against the Soviet Union must as yet be modest, and its further development will depend on the speed with which a long-range (over 5,000 miles) intercontinental missile can be produced. A missile with this range would also be capable of reaching American territory. A first test firing of such a device was successfully achieved early in 1981.

The development in due course of a nuclear submarine would clearly be an important additional factor in a Chinese dissuasive force, which for the present is inevitably 15 to 20 years behind those of the two superpowers.

Great Britain, for its part, has been content to equip itself with an "economical" strategic dissuasive force, acquired by means of its special links with the Americans to whom in consequence there has been some loss of independence. Four British nuclear-powered submarines assigned to NATO have been equipped with Polaris rockets armed with three warheads, each one designed for an attack on a single large target.

In 1980 the British government announced an important modernization plan, to be completed by the 1990s, involving the acquisition of American Trident missiles fitted with eight MIRV nuclear warheads and having a range of some 5,000 miles. Four new nuclear submarines would be built rather than modifying the existing ones. In the interval during the implementation of such a projected program, which because of its very high cost of $10 billion to $15 billion was strongly opposed by the Labour and the new Liberal-Social Democratic parties, the British planned to increase to six the number of warheads carried by their Polaris missiles. By the end of 1981, a new difficulty appeared with an American proposal to phase out the production of this Trident missile with eight MIRVs that the British had chosen in favor of the more advanced 14-headed one. Such a decision could lead the United Kingdom toward an agonizing technical, economical, and political revision of its initial plan.

As for France, her policy of atomic armament, pursued since the beginning of the Fifth Republic, had the support of all the main political parties. In 1977 it even received the endorsement of the Communist party, and in 1981 of the newly elected Socialist President François Mitterrand.

Although indisputably a member of the Atlantic Alliance, France has remained determined to maintain her own independent national defense. In terms of nuclear armament, she has become the third power in the world. In achieving this position, she has benefited, both in the military and the civil field, from a remarkable continuity in the role and responsibilities of her atomic energy commission, the Commissariat à l'Energie Atomique (CEA).

French nuclear defense is based on the conventional triad mentioned earlier, the least vulnerable weapon of which is the nuclear submarine: four of these have been in service since 1976, a fifth became operational in 1980, a sixth is under construction, and a seventh confirmed by President Mitterrand should be ready by the early 1990s.*

The continuing effectiveness of this strategic force through modernization is assured thanks to an ambitious stand taken in late 1972 by President Georges Pompidou, on the advice of the CEA, involving the launching of research into a rocket equipped with multiple warheads capable of independent guidance toward several targets. Satisfactory tests carried out in the Pacific in 1979 guaranteed operational availability for these multiple warhead missiles by 1985. This will result in a considerable change in the number of warheads available to the principal arm of France's deterrent force, thus leading to a significant increase in credibility and influence.

The French strategic force also includes squadrons of Mirage IV nuclear fighter-bombers, which can be refueled in flight, and a ballistic missile unit established beneath the Albion Plateau in Haute Provence. To this list should be added a number of nuclear artillery regiments equipped with Pluto tactical weapons, similar to bombs, which can also be delivered by Mirage III and Jaguar strike aircraft and, as of 1981, by carrier-based Super Etendards. From a strategic point of view, by 1985 this armament will be qualitatively comparable with the American armament of the early 1970s. Also in the tactical field, the year 1985 will see the entry into service of a new high-powered airborne weapon, the "air-sol moyenne portée" (medium-range air-to-ground), which could be used to equip Mirage IV,

---

*The French Navy is also to be equipped with 5 nuclear strike submarines, the first of which was launched in 1979. At that time, the British had 10 such submarines, the Americans over 70, and the Soviets around 90.

Mirage 2000, and Super Etendard fighter-bombers.

Finally, France has also advanced in the study of the "enhanced radiation" tactical nuclear warhead, the neutron bomb, though, by the end of 1981, no decision had been made as to its production or operational use.

The interest the Russians have shown in wanting the French deterrent force to be included in any further discussions beyond SALT II on the limitation of nuclear arms is proof of how seriously the French force is taken. Further indications of this are the new and repeated proposals for the establishment of an independent European dissuasive force, although such a force remains inconceivable outside a politically united Europe, which remains a distant dream.

Any European force of this kind would have to be based on Franco-British, Franco-German, or tripartite armament involving these three countries. But Britain shows no signs of willingness to give up its privileged links with the United States, while the Soviet Union shows no signs of being prepared to see Germany freed from nuclear abstinence.

The future of civilization therefore rests today in the hands of the five nuclear powers in the world, and depends on their individual and combined wisdom. In this, their main duty will be to prevent the development of conditions leading to a generalized conflict. One of these conditions could be a serious energy shortage in a world suffering an acute economic crisis. The controlled combustion of uranium offers a means of avoiding such a situation.

Part Two

THE POWER

It is not uncommon for the successful exploitation of a technological invention or discovery to take at least as long as the scientific evolution leading up to the invention or discovery itself. Hence, some 40 years elapsed between the birth of aviation and the establishment of regular commercial air services across the Atlantic. And this despite the fact that, from the very beginning, the technical problems to be resolved were clearly defined and recognized. Each one of these problems involved considerable difficulties.

The same has been true for the peaceful use of atomic energy. Some 40 years were needed between the discovery of uranium fission in 1939 and the production, in many of the industrialized countries of the world, of an appreciable proportion of their electricity from nuclear power stations.

Moreover, because prototypes require years of research and considerable financial investment, the construction of nuclear power stations or advanced reactors — like the construction of the largest and fastest aircraft — has become the monopoly of the leading industrial countries.

New technologies also present new dangers. In nuclear power there is radioactivity, which, like gravity in aviation, could lead to serious accidents following an error of judgment. Then, as with aviation, there are vital questions concerning materials and their behavior under their respective operational conditions.

But here the parallel ends. For while man is familiar with gravity from birth, radioactivity — with its rays denoted by Greek letters and its extremely complex modern alchemy by which elements of variable lifetimes can change from one to another — is entirely alien to him. Furthermore, aviation fuel is not divided into two categories, one of which may be freely used while the other is restricted in its civil uses by international safeguards. Production of high octane aviation fuel and the refineries concerned are not cloaked in official secrecy, nor are they the monopoly of a very limited number of advanced countries. No treaty has been signed that divides the world into countries possessing military air power and others that have either voluntarily renounced it or are under constant international pressure to do so.

Military and civil aviation have advanced together, the one continually benefiting from the immense financial investment in the other, in peacetime no less than in war. In nuclear energy, however, the roads to the weapon and to controlled combustion are two distinct branches of progress, stemming from the same roots but becoming all the time more distant from one another. Certainly the controlled production of energy from nuclear sources was, in its infancy, derived from wartime technologies and installations that had been developed in order to produce nuclear weapons. Subsequently, however, investments devoted to further advances in nuclear armaments no longer yielded information that was of use for building or operating nuclear machines and power stations.

Nevertheless, the expansion of nuclear electricity production in many countries has inevitably led to the spread of technological knowledge and equipment that could facilitate the acquisition of nuclear weapons. In particular, the nonnuclear weapons states most concerned for their energy independence have sought to replicate, as far as possible within their own control, the full range of installations — from mines to enrichment or reprocessing plants for fissile materials — from which the nuclear weapons powers developed their own civil programs. This reaction by nonnuclear weapons states has frequently been provoked by arbitrary and excessive precautions taken by exporters of nuclear materials and equipment who have attempted, by means of regulations and security measures, to ensure that their supplies could not be diverted to military purposes.

Thus the military factor, having provided the nuclear weapons states with an initial basis for their civilian developments and having subsequently become progressively detached from the expansion of their nuclear electricity programs, now seems again to be having some influence on nuclear development in the nonnuclear weapons states. Thus markets for the basic materials, uranium and enriched uranium, are not only subject to economic variations, but also suffer from fluctuations and changes in the political environment. The problem of international control of their peaceful use has become a factor of major importance.

Last, the road to the production of nuclear power has taken two separate directions, though this time very close to each other: that of naval and, in particular, submarine propulsion and that of electricity production. These are the only two energy-producing applications of fission to have truly reached the industrial stage.

But all things nuclear are dominated by particular psychological problems, linked principally to the radiation generated in the core of an atomic power station and to the treatment of the resulting radioactive wastes.

It is a fact that radioactivity excites much greater fear than any other dangers found in nature or among dangers that man has never ceased to create in his search for technical progress. These mysterious rays — that we cannot touch, feel, or see, and that can cure malignant tumors if correctly used, but can also cause leukemia in the event of excessive exposure — acquired a fearful and worldwide notoriety following the tragedy of Hiroshima. In genetic terms, the possible effects of low exposure levels, an issue that is constantly raised, are difficult to evaluate with certainty. Therefore to public opinion, particularly sensitive to any issue related to cancer, child health, or sexual and genetic questions, radiation appears both mysterious and dangerous.

Time is an important factor in this fear of the radiation linked with atomic energy because some of the radioactivity generated in a power station will persist well beyond the operational lifetime of the plant. Many by-products of fission remain radioactive for such a long time that their management and final disposal must remain effective for centuries; extensive accidental contamination by these residual materials could have serious and lasting effects on the environment. Apart from this, certain biological effects of radiation in the human being might not become apparent until many years after irradiation.

We fatalistically accept accidents in the air, in which the death tolls rise proportionally with the sizes of the aircraft concerned, but that, generally speaking, only involve the passengers. By contrast the risk of a nuclear accident, which in fact could result in comparable losses of human lives — but which has never yet done so — excites public opinion to demand the impossible guarantee that what has never happened yet never will happen. This is the basis of the psychological barrier, the basis of the antinuclear movement.

For more than a third of a century, this complex of obstacles has prevented the fullest realization of the benefits of fission in a world thirsting for energy in all its forms. It is to this aspect of the atomic adventure that the second part of this book is devoted.

The first act, from 1945 to 1954, covers the period of hope, marked by the construction of the very first prototypes of nuclear engines and electricity-producing installations. It was a period dominated by policies of secrecy and of Anglo-Saxon control of uranium resources.

The second act, from 1954 to 1964, is characterized by the euphoria that followed the lifting of secrecy. The first industrial power stations became operational and international safeguards made their appearance.

The following act, from 1964 to 1974, covers the period of industrial boom and of international competition in the realization of large-scale nuclear electricity programs.

In the final act, not yet over, nuclear maturity is achieved both technically and economically, while at the same time there is growing political and psychological confusion. The somewhat irrational problems of nonproliferation and of the antinuclear controversy have greatly complicated the further development of atomic energy production at the very moment when this has become indispensable in view of the oncoming world energy crisis.

Act One

HOPE
1945 to 1954

The first steps toward the civil use of the energy released by uranium fission were taken in a world completely disorganized by five years of war.

The United States had attained unequaled technological and industrial power. The United Kingdom had avoided invasion but in doing so had become exhausted and was obliged to begin the process of dismembering the British Empire. The countries of Europe, with very few exceptions, had been subjected in varying degrees to the ravages of invasion, the sufferings of occupation, and further destruction in the process of liberation. Stalin's Soviet Union had survived only because of the extent of its empire, now to be rapidly strengthened by new acquisitions beyond its frontiers. Finally, the fate of the defeated countries, and especially of partitioned Germany, was a focus of discord between the erstwhile allies of east and west, now sliding progressively toward a cold war.

All of these countries wanted to regain at least some of their lost ground in the scientific, technical, and industrial fields. Very substantial contributions were made by the United States toward this recovery in western Europe in all areas except one, that of nuclear energy, because of its military content, despite the fact that its potential for peaceful applications seemed particularly promising.

The remarkable accumulation of scientific and industrial data gathered by the Anglo-Saxon allies in their atomic undertakings was to be withheld from the other states for a decade, even from those that had been in the vanguard of modern physics before the war. The required uranium supplies were on the whole also denied them.

For the first time in the history of civilization, one of the greatest discoveries ever made was to be developed under conditions of unprecedented discrimination and restriction. This was in complete contrast to the free exchange of knowledge, materials, and equipment that characterized the flowering of science and industry in the decades on either side of the early years of the 20th century.

The American advance became even more firmly established in the field of armaments with the appearance of a new area, that of submarine propulsion, the forerunner of electricity production. Nevertheless, by the end of the 10-year period of secrecy, the Russians and the British succeeded in regaining a large amount of their lost ground in both the military and civil areas, which at this stage were still closely interlinked.

Far behind, France was struggling to recapture her prewar position in this field. Most important of all was the availability of sufficient quantities of uranium, without which nothing could be achieved. Fortunately, France's home territory was found to be much richer in this respect than had been thought.

Early Predictions

The course of history was perhaps changed as a consequence of the first estimates made concerning nuclear combustion. As we have seen, in 1940 the Germans (like the French) made the mistake of thinking that controlled combustion would be easier to achieve than an atomic explosion. The outcome of World War II was quite possibly a result of this error.

Following the end of hostilities, General Leslie R. Groves, who had led the American nuclear enterprise during the war, sought advice from Arthur H. Compton, Enrico Fermi, Ernest O. Lawrence, and J. Robert Oppenheimer, four scientists who had taken part some months earlier in the decision to use the bomb. He now wished to consult them on the possible peaceful applications of fission. Their conclusions and advice provided the basis for the Acheson-Lilienthal Report. These were the first such assessments resulting from the achievement of the nuclear chain reaction, the existence of plutonium, and the possibility of enriching uranium by isotopic separation.

The scientists first of all stressed the difficulties of evaluating the consequences of a discovery that was still in its infancy and that until then had been essentially directed toward military ends. They saw two main areas of civil application: the controlled production of energy and the use of radiation and radioisotopes in science and industry. They even suggested that this latter would prove to be the predominant future role. To them, the increased benefits to be expected from the use of radioactive by-products of fission, not only in biological, medical, chemical, and physical research but in industrial applications also, would prove more important than energy aspects — a prediction that has hardly proved to be the case!

The publication of the Acheson-Lilienthal Report, together with the scientific and technical data made available in 1946 and 1947 by the Americans during the first U.N. negotiations on international control, enabled these problems and difficulties to be better defined and understood. The political problems of controlling the peaceful uses of nuclear power were considered by far the most serious. On the other hand, the difficulties of protecting workers and the public against radiation and of managing the radioactive wastes from the future nuclear industry, although they required specific technical solutions, seemed by no means insoluble.

As then envisaged, the future energy-generating plants were expected to be comprised of the following main units: a nuclear reactor (the term "reactor" was beginning to be used in preference to "pile") in which the controlled fission chain reaction would generate in the fuel both heat and radioactive by-products of the reaction; a liquid or gaseous coolant fluid to extract the heat and transfer it, via a heat exchanger, to water that would then evaporate to steam; and a classical system with the steam driving a turbine linked either to an alternator or to a driving shaft, according to whether the requirement was for electricity production or for propulsion.

During the war, the heat produced in the large American reactors was an embarrassingly inconvenient by-product of plutonium production. With an electricity-producing reactor, however, the problem is to transform the heat generated in the fissile material into electricity with the maximum possible efficiency. The reactor replaces the more conventional furnace, which is the heat source in a classical power station. But besides the high temperature generated by the operating reactor, intense radiation is produced, some of which persists and decays slowly only after the chain reaction has stopped.

Protection of the operators from this radiation and the need to avoid any radioactive contamination in accessible areas of the installation, or of course in the surrounding area, comprise the most specific and delicate constraints on the exploitation of this new form of energy. This is why the fuel elements, in which the chain reaction takes place, must be enclosed in a protective cladding. The cladding prevents radioelements produced by the fission from escaping into and contaminating the coolant fluid.

Nuclear fuel may consist of natural uranium, of uranium enriched by the addition of more uranium-235 or of plutonium, or even of almost pure uranium-235 or plutonium. The critical mass, and therefore the size of the reactor, becomes smaller as the "concentration" of the fissile material increases. However, there are lower limits to both the size and weight of a reactor, dictated by the considerable weight and volume of the indispensable protective shielding, and these limits rule out completely any application to automobile propulsion.

Reactors are operated entirely by remote control. Operators regulate energy production by means of control rods, made of neutron-absorbing materials, which by greater or lesser insertion into the reactor core regulate the intensity of the chain reaction. Some of these control rods are inserted automatically to stop the reaction in case of abnormal functioning.

Technical advances over the three decades since the last world conflict have called for virtually no changes in these basic principles for the harnessing of controlled fission to produce recoverable energy.

Predictions took note of two particular characteristics of nuclear combustion: it was "combustion" that did not require oxygen and that produced an extraordinarily high energy density per unit weight. Both of these characteristics were invaluable for the realization of submarine engines; the second was of great importance for power stations built in areas where the transport of currently used fuels was both difficult and expensive.

It was foreseen at the end of the 1940s that within 10 or 20 years nuclear energy could be employed for special circumstances, such as power stations constructed in polar regions or in the desert. The following decades would then see atomic power providing massive additions to conventional power sources and starting to become competitive with them while in no way replacing them. American calculations made at the time, despite technical uncertainties, showed that the price of the nuclear kilowatt-hour would undoubtedly be able to compete with that of electricity from coal or oil.

These predictions were to prove relatively correct as to the time scale, but they placed too much emphasis on the possibility of installing nuclear power stations in sparsely populated areas of the world where, although the cost of electricity was high, energy demand was also very limited. It had not been appreciated how rapidly the investment costs per installed kilowatt of capacity decreased with increasing size in any atomic power station, which meant that electricity produced in small plants serving low-consumption areas would be relatively expensive.

Ecological considerations had already arisen, though in the opposite sense to those confronting nuclear energy today. In France, classical hydroelectrical projects were being resisted by objectors wishing to avoid the environmental modifications that would result from building the necessary dams. One of the arguments advanced in opposition to the dams was that nuclear power stations were only a short time away. In a debate in 1947 over a projected dam, which was to involve the flooding of the village of Tignes (in the Alps in southeastern France), Frédéric Joliot-Curie had to explain in a radio interview that atomic energy was as yet far from ready to replace conventional sources. In a similar way 30 years later, the distant and unsure promises of solar energy were, in their turn, to contribute to the antinuclear debate.

Another difficulty in evaluating the future possibilities of nuclear power concerned the economically available uranium supplies in the Western World. These were underestimated at that time. It should be recalled that uranium played no part in the prewar world economy, being no more than a residue from the production of radium; about 1,000 tons per year were being extracted, two-thirds in the Belgian Congo and one-third in Canada. Nevertheless, had all the contained uranium-238 then been completely transformed into plutonium and undergone fission in reactors, this annual production would already have roughly equaled the energy content of the total world coal production in 1946.

The utilization of fission seemed possible thanks to the extraordinary characteristics of uranium and plutonium, transcending the wildest dreams of the alchemists of the Middle Ages. As has already been mentioned (see p. 87), the nuclear combustion of plutonium can, under certain circumstances, be accompanied by its regeneration in the very machines in which it is "burnt." How easily this operation might be achieved and the time required to implement a large-scale breeder reactor program were soon regarded as particularly important for civil atomic development.

This fuel-breeding process can only take place in reactors in which the so-called "fast" neutrons are not slowed down by a moderator or cooling fluid, since the ratio between production and consumption of plutonium is more favorable with fast than with slow neutrons. Breeding must therefore take place in reactors that do not use a moderator and that use as coolant fluid an element that has very little slowing-down effect on the neutrons, such as molten sodium metal.

Solution of the various problems related to breeders was of capital importance because, if success were achieved, the quantity of energy

available from the earth's resources of uranium (and possibly thorium)*
would increase very substantially and could satisfy the needs of world
consumption for several centuries as opposed to a few decades.

These were the uncertainties and main problems to be resolved in the
years 1946 and 1947. Had the use of atomic energy in peacetime not been
subjected to political restrictions at that time, many industrialized countries
would have begun to tackle these problems through coordinated or joint
research activities. But in fact the scientific and technological knowledge
acquired during the war was kept secret by the three Anglo-Saxon allies, the
United States, the United Kingdom, and Canada, who also shared the bulk
of the Western World's uranium production. As previously explained,
following the adoption by the U.S. Congress of the McMahon Act
establishing the country's nuclear isolationism, even the British and the
Canadians were denied the benefits of American progress.

The restrictions resulting from this situation practically forced all
countries with access to uranium, other than the United States, to pursue (or
begin from zero in some cases) individual atomic efforts between 1945 and
1955. When the secrecy was lifted in 1955 and both uranium and even
enriched uranium became generally available, the most advanced countries
were already so deeply engaged in the technical channels imposed by
former circumstances that it was too late to change. This was the main cause
underlying the origins of the different principal types or "families" of
nuclear power reactors, which later were to find themselves in commercial competition.

Meanwhile those countries having no uranium were reduced to virtual
impotence in the nuclear field. Paradoxically, this was the fate of Belgium,
who found itself unable to obtain even the smallest fraction of the uranium
produced by its African colony in the Congo, which was bound by a 10-year
contract to supply all its output to the Anglo-American Combined Development Trust. In return for this, Belgium had been promised privileged access

---

*Among the more complex elements with which nature has endowed our earth, uranium is
not the only one capable of producing energy by breeding; another heavy element, thorium,
may also be used for this purpose. Thorium, like uranium-238, is not fissile but, when
subjected to neutron bombardment, gives birth by transmutation to the isotope uranium-
233, an isotope that does not occur naturally, and is both an explosive and a nuclear fuel,
similar to plutonium and uranium-235. But because thorium is not fissile, it must first be
transmuted into uranium-233 in a reactor burning either uranium-235 or plutonium. It
would then become possible by degrees to achieve breeding by burning the uranium-233 in
the presence of more thorium. However, it has not yet been necessary to use thorium
because relatively large supplies of uranium are available.

to the benefits of the future commercial and industrial uses of atomic energy. The promise was never more than a lost illusion!

Research and Prototypes in the United States

Despite the difficulties of reorienting their atomic effort during the first years of peace, the Americans, on the strength of their head start, quickly took the lead in research and development of controlled nuclear combustion. Their program at this stage, directed by the U.S. Atomic Energy Commission (U.S. AEC), was entirely financed from the federal budget.

The limited uranium resources available, on which in any case defense requirements had first call, clearly meant there could be no large-scale civil U.S. production of energy or electricity from nuclear sources unless it became possible to make maximum use of the uranium in fast neutron breeder reactors.

But the production of recoverable energy appeared to offer attractive possibilities in areas other than those of civil application. Both the air force and the navy were interested: The former believed nuclear propulsion would be essential for bombers designed to carry atomic weapons up to 10,000 miles without refueling and the latter had realized that a nuclear-powered engine offered the only way to make submarines autonomous and entirely independent of an atmospheric oxygen supply.

So the air force and the navy, in collaboration with the U.S. AEC and industry, each launched its own project. One of these was a total failure — the nuclear-powered bomber research program, with its substantial expenditure of money, was abandoned after 15 years. However, the submarine project was a brilliant success, thanks to which the United States was to start along the road to civil nuclear power.

In 1949 Navy Commander Hyman G. Rickover's efforts, which he had pursued tenaciously for two years despite lack of support from his superiors, finally bore fruit. It was decided to set up a naval division within the U.S. AEC to design and build a submarine engine. The big American-type electricity-producing nuclear power plants, as well as American leadership in this field, stemmed largely from the encouraging results of this decision.

At about the same time, in early 1949, a period of uncertainty over which types of experimental reactors to build and where to construct them came to an end when a large testing site of some 800 square miles was acquired at Arco, in the Idaho desert. Here priority was to be given to a fast neutron reactor with which research on breeding could begin. A land-based prototype submarine engine and a materials-testing reactor were also to be built, the latter for testing materials in the conditions expected so as to identify the most suitable for use in future nuclear power stations.

In 1951, nine years after the startup of the Fermi reactor, the fast neutron reactor at Arco produced the first electricity to be generated from nuclear power. The energy produced, extracted by means of molten sodium, was conveyed via an intermediate heat exchanger to a steam generator driving a 100-kilowatt turbine — the electricity obtained was just enough to supply the reactor building's lighting system.

Submarine Propulsion

In March 1953 the start of tests at Arco on a prototype nuclear submarine engine confirmed Rickover's belief that his project to build such a vessel could succeed.

Working simultaneously for the navy and the U.S. AEC, Rickover, who during the war had been responsible for the development of naval electrical equipment, was expert in achieving and maintaining delicate liaisons between research and industry — between scientists ever seeking perfection and constructors beset with delays and mounting costs. He knew how to make each understand what the other wanted, and so to obtain from each what he himself required. His perseverance in obtaining the materials he needed was equaled only by his insistence on their quality and by his firm conviction that other countries would find it extremely difficult to do the same.

The problems confronting Rickover were doubly complicated. Not only did he have to produce, by fission, some tens of thousands of kilowatts of mechanical power, he also had to do this within the strict limitations of size and weight imposed by installation in the hull of a submarine.

His reactor used fuel elements of highly enriched uranium, clad in zirconium, a new metal particularly resistant to the high temperatures in the cooling water under pressure. From this basic design, characterized by remarkable operational safety, came not only the reactors for nuclear submarines but also the principal "family" of American nuclear power stations (water-moderated and water-cooled reactors).

It was here that the race for atomic power stations was won by the United States: a world "first" that, however, was hardly recognized internationally, for the land-based prototype submarine engine was a military project and did not produce electricity.

The world's first civil nuclear power station was in fact built in the

Soviet Union. Inaugurated in June 1954, it had a generating capacity of 5,000 kilowatts or five electrical megawatts [5 MW(e)].* This was about the same as that of the American submarine motor that preceded it, though the Russian achievement — exempt from the limitations of volume and weight that Rickover had to observe — was more easily accomplished.

Work had been started on the first American nuclear submarine, the *Nautilus*, in 1952 without waiting for completion of the prototype engine or its operating information. The ship was launched in 1954, and completed her first underwater cruise in the spring of 1955.

Shortly after completing her first trials, the *Nautilus* made a voyage equivalent to two and one-half times around the world, at speeds that would have been inconceivable for a conventional submarine and with the consumption of no more than a few kilograms of the invaluable uranium-235. In August 1958, during a 1,000-mile mission under the polar ice cap, the *Nautilus* reached the North Pole. The speed and operating depth of such a submarine make it practically invulnerable, with a substantial chance of survival in case of nuclear conflict.

Rickover had asked each of the two largest American electricity-producing companies, Westinghouse and General Electric, to develop a different prototype engine. Both were to use highly enriched uranium, with ordinary water as the moderator, but the first was cooled by water under pressure (known as pressurized water), the second by molten sodium. Both proved successful, as did the two first nuclear submarines, each powered by a replica of one of the types. Rickover decided to use only the pressurized water type for all further submarines, but Westinghouse and General Electric shared the building of the engines. It was, therefore, with a wealth of acquired experience that these two firms entered the market for civil nuclear power stations.

In 1953, President Dwight D. Eisenhower refused the navy the funds necessary to finance research by the U.S. AEC on a nuclear-powered aircraft carrier engine. However, with support from the most influential members of the Joint Committee for Atomic Energy, and despite opposition from those members hostile to starting anything that might lead to electricity nationalization, Rickover obtained authorization to transform his shelved naval project into one for the first civil power station. He asked Westinghouse to build the plant, and chose a private utility (Duquesne Light

---

*Throughout the rest of this book, the expression MW(e) is used to indicate power in electrical megawatts. In 1979, 1 MW(e) of electric power was sufficient for the average industrial and domestic requirements of 1,000 people in France or 400 people in the United States.

Company) to finance its conventional part by buying the steam produced by the reactor at a price high enough to subsidize its operation. On December 2, 1957, 15 years to the day after the inauguration of Fermi's atomic pile, this power station using enriched uranium and pressurized water was brought into service at Shippingport in Pennsylvania. It had a power of 60 MW(e).

These years of the 1950s were the same years that saw the abandonment of the nuclear-propelled aircraft project on which over a billion dollars had been spent. This project had been partly promoted by inter-service rivalry and by Congress, and had been encouraged by repeated (though groundless) rumors that the Russians were nearing success in this field. It was even thought, in 1958, that the American project might be able to provide a psychological counter to the Russian success in launching the first man-made satellite, the *Sputnik*.

In reality, not one of the three key problems of nuclear-powered aviation has ever been resolved: that of the behavior under irradiation of materials in the propulsion reactor at its high operating temperatures; that of the great weight of the radiation shielding necessary to protect the crew; and above all, that of the danger to populations from the inevitable ground dispersion of intense radioactivity in the case of an accident with an aircraft whose only special asset would have been its great operational range without refueling.

British Work Restarts

Western Europe, which had been the cradle of prewar atomic research, found itself severely handicapped in its efforts to restart the work when hostilities ended. Germany was now under occupation and all nuclear research was forbidden. Prewar persecutions had led to the dispersal of many research teams from there and elsewhere, notably the galaxy of brilliant Italian physicists that Fermi had assembled in Rome in the 1930s, nearly all of whom took refuge in the United States.

In the United Kingdom there was both knowledge and access to fissile materials, as well as (even more significant) a large team of technicians trained in American and Canadian laboratories and factories. However the United Kingdom could not help the other European countries because of its adhesion to the policies of secrecy, accepted in 1943 at the Quebec Conference and reconfirmed in 1945.

Despite being several years behind the United States, from whom they were once more isolated, the British embarked on the same series of projects as their American precursors. Under the direction of Sir John D. Cockcroft, a large research establishment was set up at Harwell, near Oxford, at the site of a no longer used airfield where the hangars were adapted to house the first two research reactors using graphite and natural uranium. The first of these, of very low power, was brought into service during the summer of 1947.

The next step was the construction at Windscale, in Cumberland in the north of England, of two large plutonium-producing reactors, similar to those at Hanford but cooled by air rather than water. (A reactor of this type had already been envisaged at the Oak Ridge facility during the war and was subsequently built at the Brookhaven research center in New York.) A processing plant for the extraction of plutonium from irradiated uranium was to be added later.

At the same time, construction began, at Capenhurst in the Manchester area, on an isotopic separation plant that would produce uranium-235 by gaseous diffusion of uranium hexafluoride; that is, a plant of the same type as the principal one in America. The process used was based on British wartime studies, undertaken to a large extent independently of American research.

The Windscale reactors, which were completed in 1950 and 1951, respectively, not only produced the plutonium for the first British bombs, but also served as test installations for gas cooling of uranium fuel in an atomic power plant.

In 1953 and 1955, following a decision to increase military production of plutonium, it was decided to build, at Calder Hall in Cumberland, several dual-purpose reactors that would produce both military plutonium and electricity. Each of these reactors, cooled by carbon dioxide under pressure, was to produce about 40 MW(e). The first of the series, completed in 1956, was considered to be the first industrial nuclear power plant in the world. The reactors, of which there were eight, were the prototypes of the so-called "magnox" type, from the name of the magnesium alloy developed for the cladding of the uranium metal fuel rods. Here is yet another example of how civil applications of atomic energy have been developed initially from military projects.

Whereas in the United States the construction of the first prototypes was always entrusted to private industry under contract, in Britain a specialized industrial branch of the governmental atomic organization was responsible for the design, construction, and operation of the big industrial nuclear plants. The work was under the competent direction of Christopher Hinton, one of those responsible for the production of explosives during the war. Like Rickover in the United States, he was an expert at smoothing out difficulties between research scientists and construction engineers, ensuring that each passed mutually indispensable information and data to the other.

During this period, many British scientists were still working on the Canadian project at Chalk River; but it was no longer the joint enterprise it had been during the war, for the British, determined to build nuclear reactors and plants on their home territory, had refused a Canadian proposal immediately after the end of hostilities to integrate mainly in Canada the two nuclear programs. Like the British, the Canadian program was based on natural uranium reactors with heavy water as the moderator. This resulted from wartime Franco-British research and development work and led to the operation of two large experimental reactors that for some years were among the most advanced in the world.

The First French Developments

The work of the French Commissariat à l'Energie Atomique (CEA) in its earliest years was essentially scientific. The turning point occurred during 1952 when, thanks to the adoption by parliament of Felix Gaillard's proposals for a 5-year national nuclear plan, the CEA adopted an industrial program.

Prospecting under way in the Limousin mountains of France was well advanced and indicated that an annual production of some hundreds of tons of uranium could be expected by 1957. It was therefore possible to envisage construction of several plutonium-producing reactors, and it was decided to build two large ones, moderated by graphite, on a new site at Marcoule near Avignon. Graphite was chosen because French industry was able to produce it in very pure form, but could not yet fabricate heavy water.

The two reactors, from which no recovery of energy was initially planned, were intended to produce plutonium, the extraction of which would require a reprocessing plant for the irradiated fuel. This was to be built on the same site so as to provide a complete production plant.

The achievement of this phase of the French nuclear program was made possible thanks to the CEA's success in mobilizing French industry. Interest and enthusiasm for the nation's nuclear development were often of greater significance than technical competence in the choice of firms employed. At the time, I was in charge of the chemistry division of the CEA, and was thus called upon to select the firm to build the first French plutonium extraction plant at Marcoule. The choice lay between Rhône-Poulenc, better equipped from the viewpoint of technical specialization but not very anxious to work for the State, and Saint-Gobain, a less-advanced firm but one passionately eager to launch itself in this very modern field. I proposed the latter, and the choice turned out to be satisfactory to the CEA.

As we have seen, military considerations were never far from the minds of those responsible for the 5-year plan. However, the plan's explanatory notes recalled that France had always been short of coal in sufficient quantities and particularly in appropriate qualities; that she had fallen behind in the highly competitive race for oil; and finally that her hydroelectric generation installations, which had absorbed considerable investments, were necessarily limited in capacity. It was therefore indispensable for France to possess enough plutonium for her first ventures — power stations and motors — in the utilization of the new form of energy.

During this same year of 1952, a second heavy water research reactor with a power of two thermal megawatts [2 MW(t)] came into use at the Saclay center. Its fuel rods of uranium metal were cooled by compressed carbon dioxide; hence anticipating the use of this cooling method by the British. This was in fact the first use of the method in the world, soon to be adopted industrially in both the United Kingdom and France.

This cooling process was indeed so successful that it was proposed that the projected graphite reactors of the 5-year plan be modified so as to include provision for the experimental production of electricity. From the end of 1952 it had been intended that the first reactor should be cooled by air at atmospheric pressure, as were those at Brookhaven and Windscale, but that the second should use carbon dioxide under pressure. It was finally decided, toward the end of 1953, to add to both reactors a small electricity production unit. France thus took the road toward dual-purpose reactor prototypes. In this, France was two to three years behind Britain, who was in pursuit of the same goal.

From the point of view of foreign relations and policies, it is to be regretted that, throughout the period from 1948 to 1954, France, the principal country of Europe not bound by commitments to secrecy and in possession of research piles and reactors, was unable to take the lead in promoting a broad European collaboration — the role left unfulfilled by the United Kingdom due to its obligations toward the United States.

Norwegian Interlude

The French were by no means immune from the unfortunate effects of the secrecy policy nor from trying to turn this to profit. The failure of a Norwegian proposal for collaboration provides an example.

At the beginning of 1950, the Norwegian physicist Gunnar Randers found himself in a dilemma. Responsible for the construction of one of the first atomic reactors built outside the Anglo-Saxon world, he was unable to obtain the necessary uranium.

Shortly after the war, the Norwegian government, proud of its national production of heavy water and under pressure from Randers, had voted the funds necessary for building a low power heavy water research reactor — assuming that the essential uranium requirement would no doubt be met from discoveries within Norwegian territory during the course of the several years the reactor would take to build.

By 1950, construction of the reactor near Oslo had made great progress, but the searches for uranium had proved disappointing. Randers first made a vain attempt to exchange with the British government a quantity of heavy water for uranium, but London could not act in the face of American opposition.

Randers next turned to France, at this time the most advanced of the western powers not bound by secrecy agreements. Moreover, it was to Norway that France was indebted for the role in atomic development she had been able to play during the war, thanks to the 185 kilograms of heavy water supplied in 1940. Further, purchases of heavy water from Norway since the war had enabled the first two French research reactors to be built; in view of which the French had made available to the Norwegian scientists the principal data concerning the first French experimental reactor, ZOE, completed at the end of 1948.

Randers therefore traveled to Paris to seek the uranium he needed from Joliot-Curie, France having had more success than Norway in her initial prospecting for ores.

Joliot-Curie, convinced that no one but himself was in a position to resolve his Norwegian colleague's dilemma, decided to drive a hard bargain. He agreed to provide the uranium, but without the technical data for its essential purification and preparation in metallic form. He also asked that the reactor be considered a joint Franco-Norwegian project.

Randers thought these conditions too draconian and refused. He had previously warned Joliot-Curie that his government, for political reasons, was hesitant about collaboration with a Communist scientist, and had mentioned that he had another possible solution to his problem; in this he was thought to be bluffing. He was not and shortly afterward an agreement was concluded between Norway and the Netherlands, which had a stock of some 10 tons of uranium that had been purchased in 1939 on the advice of a far-seeing university professor. The uranium had been hidden during the war and its existence kept secret until then.

Recognizing their error, the French proposed to the newly created Netherlands-Norwegian association that they should become its third partner, this time offering generous conditions for information transfer and for the purification and metallic preparation of the uranium for the reactor.

But it was too late. The Anglo-American community, learning of the proposition, was clearly opposed to any Franco-Dutch-Norwegian collaboration in which France must inevitably have played a predominant role. Washington advised against the project, and the United Kingdom now offered to provide some of the needs and services that Randers had sought from Joliot-Curie, notably carrying out for him the purification and metallic preparation of the impure Dutch oxide. The Netherlands-Norwegian reactor was completed in 1951, at the Kjeller Centre near Oslo, which became the first nuclear establishment to throw open its doors to scientists and technicians from other countries.

The joint Netherlands-Norwegian collaboration was the first international civil atomic project and not only brought together research teams from the two countries, but also resulted in intermarriage between the nuclear work in those countries. This continued until 1960. However, destiny had decided — in giving to Norway large hydroelectric resources, and natural gas to the Netherlands, followed more recently by the discovery of North Sea oil — that neither country need hurry to produce nuclear electricity. The long-term importance of their initial collaboration was therefore less than it might otherwise have been.

France in turn became a partner in the last major atomic project of the period. This was in Sweden in 1954, at an underground site near Stockholm, where a research reactor similar to that in Norway was built. The required uranium was extracted from bituminous shales — ores of low grade but found in Sweden in abundance. The transformation into metal was carried out in France while Sweden awaited the development of an indigenous uranium metal industry.

The design and building of this reactor was due to the work of the physicist Sigvard Eklund, later to become a prominent international nuclear statesman at the head of the International Atomic Energy Agency.

In August 1953 in Oslo, at the suggestion of Randers, the world's first international atomic conference was held. It was devoted to heavy water reactors and was a great success. During the meeting Randers proposed the creation of a European Atomic Energy Society, which was in fact formally constituted the following year as a "discreet club" for European atomic organizations, with the aim of encouraging voluntary exchanges of information and data and of developing liaisons between the organizations' leaders. The society played a noteworthy part in breaking down the psychological barriers that were the legacy of 15 years of secrecy.

Thus did continental western Europe show a determination to overcome its nuclear underdevelopment; certain events were soon to facilitate the achievement of this objective.

Act Two

EUPHORIA
1954 to 1964

Panorama of the Decade

The British accession to nuclear arms in 1952, the Russian thermonuclear explosion the following year, and, less spectacular but nevertheless significant, the first French and even Scandinavian civil atomic achievements together provided final confirmation of the failure of the secrecy policy.

By their own legislation, the United States was prevented from giving any form of help in nuclear matters to other countries, so the field was wide open to the Russians. There was, in fact, a serious risk that the Soviet Union might offer to other countries — not only the nonaligned but even those of western Europe — invaluable atomic assistance that would naturally carry with it a degree of political influence.

Perhaps fortunately, the Soviet government at the time probably had neither the means nor the will to do this, for it was not until 1968 that the first evidence appeared of Soviet commercial competition with the Americans, precisely in the field that until then had been a U.S. monopoly, that of uranium enrichment.

The situation was retrieved for the Americans by Eisenhower's famous December 1953 "Atoms for Peace" speech to the United Nations, and more particularly by the radical modification, eight months later, of the McMahon Act. Thanks to this reversal of policy, the obstacles that until then had blocked international exchanges of data, materials, and equipment were all but removed.

This change to an open policy coincided precisely with the moment when the military and civil pursuits of atomic power in the most advanced countries were ceasing to overlap and beginning to go their separate ways with the construction of the first large power stations for generating nuclear electricity.

The rules of the new political game were laid down by the United States, which continued to dominate the atomic scene thanks to their technical lead and their political influence. Thus, as has already been explained, countries receiving nuclear materials had to undertake not to use them for military purposes, and in addition had to accept verification by foreign inspectors of their adhesion to this undertaking.

This decade saw the last years of the Cold War that had followed the Second World War. Already, Eisenhower's America and Khrushchev's Russia were coming together over certain issues: in 1954 at the Geneva Disarmament Conference; in 1955 at the first U.N. "Atoms for Peace" Conference, also in Geneva; in 1956 when the two Great Powers joined efforts to put an abrupt end to the Franco-British action over Suez; and in 1958 when negotiations began for the banning of nuclear tests. However, complete détente had to wait until after the Cuban crisis at the end of 1962, this episode marking both the climax and the end of the Cold War.

In Europe the decade began, after the failure of the European Defense Community, with the rearmament of the Federal Republic of Germany within the Atlantic Alliance, and later with the European economic revival leading to the Common Market and Euratom treaties. For France, these were the years of decolonization, of the return to power of General Charles de Gaulle, and of the end of the Algerian war.

From the viewpoint of energy supplies, the mid-1950s saw oil beginning to replace coal, gradually to become the main source of primary energy throughout the Western World. The Suez crisis of 1956 gave an early warning, unfortunately quickly forgotten, of the role the Middle East states were to play less than 20 years later in the geopolitics of energy. The proportion of electricity in world energy production was constantly increasing, and electricity consumption in the industrialized countries was doubling approximately every 10 years.

During this decade, the United States was producing electricity at a cost that, although it varied according to the area, was much lower than in Europe. The United States had immense coal reserves and important oil resources within its home boundaries and natural gas was becoming increasingly important as a source of energy. All these fuels were far more expensive in western Europe, such as in France where primary energy in the form of heat, cost on the average double the price of that in the United States. This of course meant that nuclear power would become commercially competitive in western Europe more rapidly than in America.

In 1954, there existed only one nuclear submarine engine and one nuclear station in the world, with power outputs of some 5 to 10 megawatts. Ten years later, dozens of nuclear submarines and several American nuclear-powered warships were in service on the high seas, together with a Russian icebreaker and a demonstration American merchant ship. Prototype nuclear power plants were to achieve outputs of 100 to 200 MW(e) by 1964, and units of 400 to 500 MW(e) were already being built. Several advanced countries had launched complete programs for nuclear electricity production, based on the hope that it would be competitive with electricity generated by conventional thermal power stations by the mid-1960s.

The fuel requirements of these programs, in natural or enriched uranium, remained remarkably small compared with those of the military programs; it was these latter that set the terms of the uranium market in the Western World, a market well on its way to overabundance as the American military demand began to be met solely from the country's national resources.

This period from 1954 to 1964, which began with the lifting of secrecy that provoked an interval of euphoria, led to and ended in sober reappraisal. Already complications were developing, technical questions were becoming inextricably tangled with politics, and antinuclear protest movements were beginning to appear.

The following pages analyze four different aspects of the decade: first, the evolution of nuclear technology; then the worldwide introduction of safeguards and controls, with the first repercussion being the politicizing of the natural uranium market; then the European problems resulting from dependence on American enriched uranium; and finally the developing trade in research reactors and nuclear power plants, characterized by competition between the champions of the main reactor types.

1. Technological Progress

The Geneva Conference

Toward the end of 1954, as a result of amendments made to the McMahon Act, the transfer from the United States to friendly countries of nuclear materials and data relating to their civil applications was authorized subject to the agreement of the Joint Committee for Atomic Energy (JCAE) of the U.S. Congress. In practice, only "declassified" information, i.e., information released from secret classification and subsequently published, was allowed to pass to foreign countries. Hence, a considerable amount of information was declassified for the U.N. Conference on the Peaceful Uses of Atomic Energy that was proposed by the United States and decided by the General Assembly.

Preparation for the conference was entrusted to U.N. Secretary General Dag Hammarskjöld, who was assisted by a small scientific committee drawn from seven countries: the United States, Soviet Union, United Kingdom, France, Canada, India, and Brazil. I personally represented France on this committee, which later became known as the U.N. Scientific Advisory Committee normally meeting about twice a year. Until the tragic death of Hammarskjöld late in 1961 this committee was able to make an appreciable contribution to atomic relations between East and West, for its Soviet Union representative from 1956 onward was the head of the Soviet atomic organization.

The U.N. conference, which was not only the largest international scientific meeting ever held, but also a noteworthy political event, took place in August 1955 in Geneva under the presidency of Homi J. Bhabha, chairman of the Indian Atomic Energy Commission. In his opening address, Bhabha emphasized the benefits that nuclear electricity production could bring, both to industrialized countries and to the underdeveloped areas of the world where, in his words, there was no energy more expensive than no energy.

He pointed out that 80 percent of India's energy was at that time still obtained by burning cow dung, which globally had only recently been replaced by oil as the second most used fuel after coal. In the same speech, looking to the more distant future, he mentioned the possible peaceful utilization of fusion and the hydrogen isotope thermonuclear condensation reaction, which was the basis of the H-bomb.

The Geneva conference was a remarkable success and confirmed the thaw in international nuclear relations. A date limit had been set for the receipt of summaries of papers to be presented. Four days before this date, 100 important Soviet papers arrived at the U.N. Secretariat in New York. The main American papers, which had been held back until that moment, were then also sent.

Fifteen hundred delegates, including many scientists from the East and West, met each other for the first time and presented some 1,000 papers, abolishing completely the secrecy that until then had cloaked many areas of atomic work.

But there were exceptions. Nothing was disclosed about national uranium resources and production, nor anything beyond the already well-known basic principles concerning methods for separating the uranium isotopes. On the other hand, technical data relevant to all other stages of what is now called the "nuclear fuel cycle" were made freely available to the assembled scientists and technicians from all over the world.

Papers disclosed methods for chemically processing the various uranium ores, as well as for the preparation of uranium oxide and uranium metal of "nuclear purity," i.e., purified to a very high degree from contamination by neutron-absorbing substances and thus suitable for use as nuclear fuel.

The preparation of moderating media, heavy water and high purity graphite, and of the special metals used for "cladding" uranium fuel rods, such as pure magnesium and zirconium, were also described.

The development of all this specialized industrial chemistry related to nuclear energy had been one of the greatest tasks of the past years. Publication of the data represented, for the newcomers in the atomic race, an invaluable savings in time and money.

But a new form of secrecy was appearing, for the practical know-how of making fuel rods and of their behavior under the operating conditions in a reactor — one of the most delicate aspects of the new technology — was now shrouded by industrial secrecy.

These questions of materials behavior were by no means limited to the conventional action of temperature; new phenomena concerning the effects of irradiation had to be considered. Under irradiation, even simple materials, such as graphite or metals, suffer structural changes that can be detrimental to their mechanical behavior and physical properties. Transmutational effects, such as the disappearance of certain atoms and their replacement by others, also contribute to changes in solid materials, which may make them unacceptable for their intended purpose. Knowledge of these phenomena is of capital importance because they affect the behavior of reactor materials and component parts and hence both the operational life of a reactor and the frequency with which its irradiated fuel rods require replacement. These are major factors in determining the cost of nuclear electricity.

If the 1940s had been the years of the search for nuclear purity in reactor materials, the 1950s were devoted to studying the behavior of these materials under irradiation and at high temperatures, while the 1960s were to be the years of research into nuclear power plant reliability, based on experience with the prototypes designed and built in the late 1950s.

All countries participating in the conference agreed on the importance to be given to the last stage of the fuel cycle, that of reprocessing irradiated fuels so as to recover the plutonium that has been formed together with the uranium that is now less rich in its 235 isotope than when it was first placed in the reactor, as well as the radioactive by-products of fission. Recovery of plutonium was then considered to be vital for the future of atomic energy, with a view to a much more efficient use of uranium in fast-neutron breeder reactors.

In the course of this Geneva conference, France took the initiative to become the first country to publish its method of plutonium extraction, hence obliging the other nuclear powers to follow suit and likewise lift their secrecy for in fact they were using the same method. The chemical steps involved in the production of one of the two materials needed for making atomic bombs were thus revealed, and no one at the time protested about the consequences of weapons proliferation. The extraction process, still the only one in use at the beginning of the 1980s, was a variant of the method first developed by my team in Canada during the war. Much of this development work (research reactors, reactor prototypes, nuclear power plants, and plants for reprocessing irradiated fuels) was technically very complex. Furthermore, because intense radioactivity prevented access to many vital parts of an installation, operations had to be conducted by remote control from behind thick reinforced concrete walls, repair work being subjected to the same precautions and carried out in working conditions never before encountered in industry.

All of this gives some idea of the scale and difficulty of the technological and industrial developments already under way in the advanced countries, on which the 1955 Geneva Conference threw the spotlight of actuality to emphasize a future full of promise.

In fact, production of nuclear electricity was still very much in its infancy, and the only purely civil power station in operation at this time was a 5-MW(e) plant that the Soviet Union had commissioned in 1954 at Obninsk, some 60 miles from Moscow. This plant, fueled by enriched uranium, graphite moderated and water cooled, had been given much publicity by the Russians at the Geneva Conference as a "world first," although it was preceded by more than a year by the American prototype submarine engine producing mechanical energy.

Only a few other nuclear plants of similar or higher power were under construction at this stage. They were the natural uranium, graphite-moderated and carbon-dioxide-cooled dual-purpose installations being built in England at Calder Hall and in France at Marcoule, each intended to produce some tens of megawatts electric; and the 60-MW(e) Shippingport, Pennsylvania, plant. However, a great many more prototypes were being studied and developed, particularly by the Americans.

Except in power plants using boiling water reactors, the coolant — whether gas, a liquid metal, an organic solution, or even water under pressure — transferred its energy to the turbine working fluid (water/steam) in a heat exchanger, a delicate and complicated piece of equipment whereby the external water was vaporized to produce the steam that drove the turbine, and thence the power station's alternator. In the case of boiling water reactors, the steam was generated in the reactor itself and it was this steam, slightly radioactive, that drove the turbine directly. For this reason the turbine in a boiling water plant was provided with radiation shielding.

It was at the Geneva Conference that there first appeared competition between two main classes of reactor: those fueled with natural uranium, adopted by the United Kingdom and France, and later by Canada and

Sweden; and those using enriched uranium, chosen by the United States and the Soviet Union. There was no way of being sure, at this stage, which type would prove best for the future. Only the two great nuclear powers at that time had sufficient production capacity for enriched uranium to enable them to divert a fraction for civil purposes. Its use, because it resulted in an installation of reduced size, required a lower initial investment than was needed for a natural uranium plant of the same output. However, the overall fuel costs during a plant's operational lifetime were certain to be higher in the case of enriched uranium fuel.

The advocates of natural uranium had scored a point in this emerging competition when, in February 1955, a few months before the Geneva Conference, the British government took the political and industrial world by surprise by becoming the first to launch a full-fledged program for nuclear electricity production. The British were then forecasting that, by 1963, the cost of a nuclear kilowatt-hour would be competitive with that of a kilowatt-hour from conventional sources.

The British announcement had a considerable effect, heightened by the success of the Geneva Conference. The breathtaking revelations at that meeting, at the end of eight years of atomic secrecy, and the enthusiasm and hopes they raised, helped momentarily to conceal not only the political difficulties ahead, but also the technical and economic problems still to be solved.

Enthusiasm even gripped the general public, leading to a considerable amount of financial speculation. Uranium mining shares rocketed in value, and any big industrial company had only to announce the creation of a nuclear division for the price of its shares to leap ahead. Not long after these ''crazy years'' for atomic energy, the financial gains more often than not gave way to losses, not surprisingly leaving the victims with a certain sense of resentment against nuclear power.

Three years later, in 1958, the United Nations organized a second atomic conference in Geneva, this time under the presidency of the French high commissioner for atomic energy, Francis H. Perrin. The meeting was less successful than the first, as there were too many participants and papers, with fewer disclosures of information than in 1955. Nevertheless these disclosures finally cleared away the last clouds of political secrecy, only to show up more clearly the thicker mists of industrial secrecy.

There was one exception. Isotopic separation of uranium remained a last stronghold of the older secrecy policy, despite several French papers on the subject that attempted, without success, to repeat the precedent of 1955 on plutonium and persuade those already in possession of enrichment technology to publish their data.

However, information on national uranium resources and production in the Western World was presented for the first time, confirming openly that supplies of this element were in fact abundant.

The United States and the Soviet Union gave full details of research they were doing on controlled fusion, that is to say, their attempts to reproduce in the laboratory the H-bomb reaction. Their reports made it clear that it was premature to hope that this reaction could be harnessed and put to industrial use in the near future, work being still at the stage of fundamental research. This served to emphasize the importance that all the advanced countries now attached to solving the problems of breeding, so as to achieve the fullest possible use of the energy contained in uranium.

Finally, and above all, the papers at this second Geneva Conference made it clear that the previous forecasts of the date when nuclear electricity would become competitive should be revised, due to the difficulties all the most advanced countries were having over construction of both prototypes and plants of higher power outputs.

The National Nuclear Programs

Only six countries took part in the race to build the first nuclear power stations. They were the first four members of the military club, plus Canada and Sweden; all entered the competition independently of each other and without any external aid. All other countries were in due course to turn to one or another of these six pioneers (or to the American-based technology of West Germany's nuclear industry) for the construction of their first power reactors.

In 1964, worldwide installed nuclear electricity capacity totalled some 5,000 MW(e). The most advanced civil reactors, 15 of which were in operation or had just been completed, had a capacity of 100 to 200 MW(e). Eight of these reactors (six in the United Kingdom, one in France, and one exported to Italy by British industry) used natural uranium and were graphite moderated and cooled by carbon dioxide gas. The other seven (three in the United States, two exported to Italy by American industry, and the other two in the Soviet Union) used enriched uranium and were both moderated and cooled by ordinary (light) water.

These plants already represented a considerable investment — between $20 million and $40 million each — and stations of greater capacity were being built. Contrary to the hopes of the mid-1950s, the nuclear kilowatt-hour was not yet competitive with electricity from conventional thermal sources, although its cost was gradually dropping. Efforts to reduce its cost further were leading to reactor units of increased size, which had lower investment costs per installed kilowatt.

It quickly became clear that nothing was truly conventional in a nuclear power plant, where even components completely outside the reactor unit, such as heat exchangers or turbines, operated under unusual conditions of temperature and pressure.

It also became clear that increasing the size of a nuclear plant was a more complicated operation than it appeared at first sight; delays and cost overruns were commonplace on construction projects, although the stage had not yet been reached when, as one Soviet nuclear leader was to remark, "every large nuclear project costs twice as much and takes twice as long as expected to build." But that stage seemed not far off, although sometimes there were cases of "lead times" of only four years between planning approval and first reactor operation.

There were also human problems for, in all the major countries, the arrival of nuclear power meant that normal relationships between electric utilities and the power plant construction industry were complicated by intervention of a powerful third partner, the national atomic energy organization, which was responsible for the new technology. As a result the general public, fascinated by this new triumph in the application of scientific progress, started to believe that the dawning nuclear age would be an age of unlimited energy supplies at virtually zero cost.

Not only was there scarcely any opposition to the construction of the first prototype nuclear plants or the announcement of the first large-scale national programs for nuclear electricity production, but there was public impatience in the face of delays, any and every incident being exaggerated by the media into a focal point of disappointment. Normal problems of startup, which would have passed unnoticed by the public in the case of a conventional power station, were highlighted and reported in detail so that the smallest incident could easily appear as a serious accident. Here, undoubtedly, was one of the root causes of the public anxiety that was to develop later vis-à-vis nuclear installations.

For governments and public alike, the problems of proliferation were limited to known or suspected nuclear armament programs; for this reason, these problems did not concern the civil field. The first reactor exports, mainly to Italy, Japan, and West Germany, caused no anxiety. Safeguards based on verification that only peaceful nuclear applications were being pursued were entirely sufficient both for governments and for the public in the supplier countries, among which a fierce commercial competition was beginning to appear.

The years of euphoria resulting from the lifting of secrecy lasted until

1958. Following the second Geneva Conference, the worldwide atomic situation began to change. The five years that followed saw both the outcome of the projects conceived during the euphoric period, and the slowing down of further programs while awaiting the technical and economic results of the operation of these first units. The slowing down was accompanied by an overproduction of uranium, together with a certain loss of interest both in industry and among the general public.

The panic that followed the Suez crisis led to a momentary feeling that energy must be produced at any price. The idea was quickly abandoned. The discovery of petroleum under the Sahara encouraged the belief (soon to be proved illusory) that the Western World could free itself from dependence on Middle East oil. For a brief moment, in fact, falling freight charges together with a calmer attitude among the Arab producers led to temporary overproduction of oil and hence a drop in the price of conventionally produced electricity. This price drop was also brought about by technical progress leading to improved classical thermal power stations.

Naval and Power Reactors in the United States

Two initially distinct industrial ventures had nurtured the production of nuclear electricity in the United States. The first, entirely financed from the federal budget and based on a single type of reactor, was the development of submarine and naval propulsion by the U.S. AEC's Naval Division under Admiral Hyman G. Rickover. The second venture was a highly diversified search for the most promising reactor type, which involved several forms of industrial cooperation between the U.S. AEC and private industry. This second effort ultimately led to the choice of the same type of reactor as was being used by the navy, namely one fueled by enriched uranium, moderated and cooled by ordinary water, and therefore known as the "light water reactor" to differentiate it from reactors using heavy water. This chosen reactor type was thus able to benefit from all the experience acquired in building nuclear marine engines.

At the end of 1964 there were 50 nuclear submarines in the American navy, equipped for a variety of missions, together with three surface warships: a cruiser and a destroyer each driven by two reactors, and an aircraft carrier, the *Enterprise*, driven by eight reactors. Thus a total of over 60 atomic-powered engines, all giving highly satisfactory service, demonstrated the substantial success of this nuclear program, which had cost some three and one-half billion dollars.

The U.S. AEC had also financed the construction of the first nuclear-powered merchant ship, the *Savannah*. Completed in 1962 (two years after the first Russian icebreaker *Lenin*), this demonstration ship steamed some 30,000 sea miles without any technical incident; her short career, however, was subsequently marked by strikes and economic problems.

The U.S. AEC had also designed and built for defense purposes a number of small "prefabricated" low power [about 2 MW(e)] transportable nuclear units to provide electricity at remote sites with difficult access: an air force radar station at the top of a mountain in Wyoming, a military base in Greenland, and a naval station in the Antarctic. The "active" lives of these small units were in each case beset by technical incidents that cast doubt on initial predictions concerning the use of nuclear energy in the world's polar or desert regions.

The Americans had also undertaken a program to develop and build nuclear reactors for propulsion of military missiles and interplanetary rockets. The greatest importance was attached to this program, particularly since one of President John F. Kennedy's first decisions had been to scrap the costly project for the development of a nuclear aircraft engine.

The U.S. AEC had been engaged since 1953 in a 5-year program to build experimental reactors for electricity production. A wide range of possibilities was explored, and at no time did any of these projects appear to be influenced by the problem of nonproliferation. The aim was to develop an economic and reliable source of electricity. There was no evidence of concern over whether one or another of the possible solutions was more likely to lead to the diversion, for military purposes, of the materials employed or formed as by-products.

In its official report for 1957, the AEC even emphasized the eventual importance of reusing the plutonium produced in a reactor to "enrich" natural uranium or uranium with a depleted content of uranium-235. At the time, it was believed that this "recycling" operation would allow the production of three to four times more energy from a given amount of uranium, and would therefore be of special interest to foreign countries having supplies of natural uranium but no enrichment plant.

The U.S. AEC's problems were not only technical; it was also necessary to persuade electric utilities to use the new form of energy. Ever since the economic depression of the 1930s, electricity production had been essentially in the hands of numerous private companies, varying considerably in size and importance. Only the largest could bear the financial risks associated with the as-yet noncompetitive nuclear power plants.

Before the depression, the electric power industry had been concentrated in the hands of a few very large utilities. When some of these collapsed financially, the federal government had split up the industry state by state, and this division persists to the present day. It is one of the causes of the relative weakness of this branch of national activity in which the government can, at either the federal or state level, have an influence by providing financial help and tax concessions.

The modified McMahon Act had allowed industry, since the end of 1954, to operate in the nuclear electric field: to build and own nuclear power stations and to have access to the necessary fuels to run them. However, the only such station under construction at the time was the U.S. AEC-owned Shippingport plant. In 1956, further legislation authorized electric utilities to join together to build nuclear plants, sharing both the financial risks and the technical benefits without falling foul of the famous antitrust laws. In addition, Congress passed legislation to protect private industry, up to $500 million, against the financial liability a serious nuclear accident could entail (Price-Anderson Act).

In 1955, the U.S. AEC launched a "Power Reactor Demonstration Program," offering assistance to private industry in the form of research aid and the provision without charge of the necessary fuels during the first years of operation of the plants. Despite this, because the electric utilities would in fact still be taking almost all the risks, they remained hesitant. The U.S. AEC therefore made a further offer, comprised of government finance for all or part of a prototype, and this led to some new proposals for joint undertakings.

Within the JCAE, the Democrats wanted the federal government to finance completely the various power reactors, accepting all the risks involved. The U.S. AEC chairman, Admiral Lewis L. Strauss, on the other hand, backed by the Republican administration, wanted as little government participation as possible. As for the private sector electric utilities, although some were ready to seek federal assistance, the vast majority were totally opposed to this, seeing in it a first step toward nationalization, which of course would have been welcomed by the trade unions.

This was the first appearance of conflict between the champions of free enterprise and those who sought federal intervention at various stages in the industrialization of atomic power; a conflict destined to cause many delays and complications in the American nuclear program.

The early 1960s saw three types of administrative structure in that program. There were prototype and experimental power reactors entirely subsidized by the U.S. AEC; there were demonstration power reactors being built in collaborative projects between the U.S. AEC, private industry, and private electric utilities; and third, the private sector was building power reactors without any direct government participation, but benefiting from many financial advantages such as guaranteed nuclear fuel supplies and reimbursement of research costs.

The U.S. AEC policy was in favor of free competition, generally making available all knowledge acquired from government-funded research by a given firm, so that the only individual advantages (by no means negligible) retained by the firm were its trained specialists and its first-hand practical experience of the application of the new knowledge.

A 1962 report to President Kennedy by Glenn T. Seaborg, then chairman of the U.S. AEC, emphasized that nuclear electricity would soon be important and one day vital in meeting the country's long-term energy requirements. Seaborg proposed a 10-year program and recommended continuation of government aid to both private and public electric utilities beyond the $1,300 million already contributed by the U.S. AEC in the civil sector. The objectives would include improvement of existing reactors, preservation of American leadership over the rest of the world, and, of course, development of fast breeder reactors.

It was in fact the construction of the first experimental fast breeder plant near Detroit, on behalf of a group of private electric utilities, that triggered the first manifestation of feeling against the potential danger it would represent for the neighboring population; though in this case the protest came from an organization — the powerful United Automobile Workers Union — whose main concern was to attack the monopoly of private electricity producers. In 1961 the U.S. Supreme Court, to which the matter had been referred, gave judgment in favor of private industry, supported in this case by the U.S. AEC. Construction of the plant went ahead.

In 1963 there were only three nuclear power stations in the 100 to 200 MW(e) range in the United States and none under construction. Two of the three had been privately financed, the third resulted from the demonstration program. The powerful American electromechanical industry, which had built both nuclear submarine engines and prototype power reactors, accordingly found itself with a very insufficient market. It therefore turned its attention to exports, particularly to the European market.

At the very same moment in Europe, and in complete contrast, the budding nuclear industry (essentially in the United Kingdom and France) was about to benefit from unification and nationalization of electricity production, for the respective governments were the first to set the precedent of launching their nationalized utilities into important programs for nuclear power plant constructions.

The British Plan

The first U.K. civil program for nuclear electricity production had been decided as early as February 1955, even before the first dual-purpose (civil and military) natural uranium, graphite-moderated, gas-cooled reactor was completed. The program was designed to achieve an output capacity of 2,000 MW(e) by 1965, using a series of progressively more powerful plants normally comprised of two identical reactors each, and it was intended that the electricity produced at that date would be competitive in price with that produced by burning coal or oil.

The first of these dual-purpose reactor plants, at Calder Hall in the north of England, was officially inaugurated by Queen Elizabeth in October 1956. At that time it was by far the most powerful generator of nuclear electricity in the world. Its 35 MW(e) were sufficient for the needs of an English town of some 100,000 population and it had more than a year's lead over its American enriched uranium rival in Shippingport, Pennsylvania. Seven more units of the same type were to be brought into service over the next three years, in two groups of four at two different sites; they were operated by the recently created British nuclear organization, the U.K. Atomic Energy Authority (U.K. AEA), a Crown company that had been established in 1954 by the Conservative government with a statute similar to that of the Commissariat à l'Energie Atomique (CEA) in France. The eight reactors produced simultaneously plutonium for the military program and 360 MW(e) for the national electricity grid. Their successful operation over nearly a quarter of a century bears witness to their excellent design and construction.

In 1957, following the Suez crisis, the British energy minister announced the tripling of his country's civil nuclear power program, the new objective being to produce by 1965 nearly one-quarter of all British electric power [i.e., between 5,000 and 6,000 MW(e)] from nuclear sources. Only three years later this burst of enthusiasm was to be brought back to earth and the program reduced to its former size. By 1964 only three civil power stations were in operation, each having two units of some 150 MW(e). The 5,000-MW(e) goal was not to be achieved until 1971 when there were a number of plants scattered throughout the country each with two reactors and with a power per unit that increased with successive installations.

Three other prototype reactors were being built in the United Kingdom in the late 1950s. One was a fast-neutron breeder reactor built at Dounreay in Scotland and designed to allow recycling of plutonium. The other two were graphite-moderated, gas-cooled reactors of a more advanced type. In a different field, the British — unlike the French who were obliged to proceed differently — in 1958 abandoned the idea of building their first submarine engine unaided. Instead, it was eventually built by Rolls Royce under license from Westinghouse, and was commissioned in 1963.

In 1956 in the early stages of the British nuclear program, Sir Edwin Plowden — one of the "Three Wise Men" of the Marshall Plan, and now in charge of atomic energy in the United Kingdom — predicted that within five years the export of nuclear power plants would have become the most important source of hard currency for the delicate British balance of payments. The first two export sales of power reactors in the world, to Italy and Japan, were indeed a success for British industry, although achieved at considerable financial cost. But it was a very short-lived success, for these two were the only power plants ever exported from the United Kingdom.

Plowden had communicated his hopes and illusions to the leading British electromechanical firms, inviting them to combine in setting up several large and distinct industrial groups. He had convinced himself that this was a necessary step toward fulfilling future orders for nuclear power plants.

The reasoning behind the decision was sound. Considering the number of possible client countries, it was reasonable to give to private industry, better equipped to deal with export trade, the overall responsibility for construction. Until then, the responsibility for building the dual-purpose units of the military program had been assumed by the Industrial Division of the U.K. AEA, which was thus able to keep close control of the fuel cycle. The reactors of the civil program belonged to the national electric utility, the Central Electricity Council, which gave the contracts for the various units to each of the five industrial groups that had been formed.

Unfortunately, Plowden's plan proved quite unreasonable in terms of size. With no further exports beyond those of the first two power stations, the national program, brought back to its initial development rate, was totally inadequate for filling the order books of the five industrial groups. A progressive reduction in the number of groups was brought about by a series of painful mergers: first to three and then, following the disbanding in 1965 of one of the remaining two, to the essential minimum of a single group. This was one of the major causes of the United Kingdom's nuclear decline, a decline that was all the more surprising since it followed many notable successes in the research and prototype stages.

The French Program

In France, a less ambitious but otherwise similar program than that of the British followed a few years behind. In March 1955, a month after publication of the U.K. plan, the characteristics of the second French plutonium-producing reactor, G2, were finalized. Like the Calder Hall units, it was to be a natural uranium, graphite-moderated reactor, cooled by gas under pressure.

As had been the case in the United States and the United Kingdom, the transition to the industrial stage was successfully accomplished thanks to the technical and human qualities of a specialist engineer from a leading branch of industry, in this case, the oil industry. Pierre Taranger, first industrial director of the CEA, like Rickover in the United States and Christopher Hinton in the United Kingdom, provided the essential working liaison between the laboratories and industry, succeeding in making each understand the problems and needs of the other. He was also responsible for the choice of prestressed concrete for the G2 reactor pressure vessel, in preference to the steel vessel used by the British. This successful innovation was subsequently widely adopted.

In 1955 the two top men in the CEA, Pierre Guillaumat and Francis Perrin, believed they had reached the stage where first steps should be undertaken in concordance with the national utility Electricité de France (EDF) with a view to launching a French program for civil nuclear power over the coming 10 to 20 years.

A consultative commission for nuclear energy production was there-fore set up. Known as the PEON (Production d'Electricité d'Origine Nucléaire) Commission, it included representatives from EDF, the CEA, and industry. By the summer of 1955 it had approved a program, proposed by EDF and the CEA, to develop a series of prototype power stations starting with a 70-MW(e) plant to be known as EDF1. All the units in the series would be based on the technology of the Marcoule reactors, each being more powerful than its predecessors. An interval of 18 months was planned between each construction, in the somewhat optimistic hope that a given reactor could then benefit from experience acquired in the construc-tion and operation of the one before last. In this way, it was foreseen that by 1965 generating capacity would reach at least 800 MW(e), obtained from three or four "first series" power stations.

In presenting this program, EDF Director Pierre Ailleret declared that "there can be no hesitation over the type of reactor to adopt for France's first atomic power stations: we can only consider graphite reactors of the type built at Marcoule. In doing this we shall of course be doing the same as the British."

There could in fact be no question, in 1955, of turning to enriched uranium reactors, for French research into isotopic separation of uranium by gaseous diffusion was still in its infancy, and the American government had not yet authorized the export of uranium-235 other than for research reactors.

The long-term French planning at this time foresaw three stages: first, power plants using natural uranium; then plants using fuel enriched with plutonium produced in the first plants; and last, plutonium-burning and -producing breeder reactors.

The minimum objective of 800-MW(e) installed nuclear capacity in EDF plants in 1965, first proposed in 1955 and confirmed in 1957, was not in fact achieved (and then exceeded) until early 1969. Only EDF1 was in operation in 1964, with construction of EDF2 just completed. Both plants were sited on the Loire River near Chinon.

In the course of this program, various far-reaching decisions had to be made. First of all, it was necessary to establish the respective responsibilities of the CEA and EDF in building the first purely civil power station. There was ample room for conflict between the ordinance of October 18, 1945, which set up the CEA "to develop industrial-scale atomic energy generating devices," and the nationalization law of April 8, 1946, under which EDF was to "design, build, and operate the means of generating electricity."

For EDF1, it was decided early in 1956 to adopt a symmetrical arrangement to that used for G2. With overall responsibility in the hands of EDF, the CEA was to prepare the basic plans for the project, design and manufacture the specifically nuclear items, and supply the graphite and fuel. For subsequent power stations, the CEA's role was more limited and by 1958 its contribution to EDF3 was specified as "conception and supply of fuel elements, provision of test results, nuclear and other technological information, checks and various tests."

The basic design for EDF1, as proposed by the CEA in 1956, was for a reactor with horizontal fuel and cooling channels (as in the G2 pile whose construction had just begun) contained within a prestressed concrete pressure vessel. However EDF, in its capacity as industrial architect for this as for subsequent plants, was seeking wider technological experience and, preferring the design adopted by the British, decided in favor of vertical channels (a choice that proved satisfactory and was maintained for the later power stations). Also like the British, EDF chose a steel pressure vessel, partly because the design pressure of the coolant gas was now to be higher than in G2, but no doubt also because severe problems were then being experienced with conduit pipes through prestressed concrete.

Early in 1959, a serious crack developed in the EDF1 pressure vessel as it was being assembled, drawing attention to the very difficult problem of welding thick steel plating in the large sizes required by the reactor. These problems took a long time to overcome, delaying by some three years the completion of both EDF1 and EDF2 [200 MW(e)]. Meanwhile, the excellent performance of the prestressed concrete vessels used for G2 and G3, which were quickly built and proved highly satisfactory from the moment of first operation, led EDF in 1960 to change to concrete vessels for the 400-MW(e) EDF3 and subsequent units. A year later the British made the same change.

Various difficulties resulted from the mutual relations between the three bodies responsible for the French nuclear program. As in Britain, a weak and relatively dispersed industry found itself faced with two established and powerful nationalized organizations, the CEA and EDF, whose complementary roles were imperfectly defined and whose tasks were insufficiently coordinated. However, in complete contrast to the relationships in Britain between the Central Electricity Authority and the U.K. AEA, EDF, besides having full overall responsibility and authority for its power plants, also acted as industrial architect; though in the case of CEA projects, this latter task was entrusted to leading private firms. This last system had the advantage of preparing industry for international competition in export markets.

Despite these differences in approach, the postwar technological gap that separated France and the United Kingdom from the United States in the civil nuclear field continued to narrow during the decade from 1954 to 1964: British nuclear electricity production even overtook that in the United States.

Soviet Achievements

During this same period, the Soviet Union, richly endowed with natural gas and oil, was running an essentially qualitative civil nuclear power program, based on two types of enriched uranium reactor. Both were cooled by ordinary water under pressure, one being moderated by graphite and the other by ordinary water (similar to the American reactors built by Westinghouse).

A series of 10 dual-purpose military units was also under construction in Siberia, each identical to an earlier unit, commissioned in 1958 and based on a 100-MW(e) natural uranium, graphite, plutonium-producing reactor cooled by pressurized water. But a program announced at the 1958 Geneva Conference, aimed at an on-line capacity of 2,000 MW(e), was far from being achieved by 1965.

The uranium resources available to the Soviet Union have always been kept secret, as they could have given some idea of the importance of the military atomic program. However, a less vigorous exploitation of uranium deposits in East Germany and Czechoslovakia from 1959 onward would seem due not only to reduced reserves in the mines, but also to discovery by the Russians of sufficient resources within their own territories. Only one Russian uranium mine has ever been opened to foreign visitors from the West. Situated in the iron and steel region of Krivoy Rog in the Ukraine, this mine alone, when I visited it in 1961, seemed capable of an annual production of some 1,500 tons, equivalent to the total French uranium production at that time.

Research into peaceful applications of nuclear power was carried out under the aegis of a state committee placed under the direct authority of the Council of Ministers via the minister for atomic energy, Vassili Emelyanov, an extremely able metallurgist who had distinguished himself during the war in the construction of turrets for tanks. He was to preside over the third Geneva Conference in 1964.

The most notable Soviet success in the late 1950s was the construction, two years in advance of the United States, of a first civil nuclear-powered ship, the 16,000 ton atomic icebreaker *Lenin*. Laid down in 1956, she was undergoing her first sea trials in late 1959, and by spring of 1960, she was at work in the Arctic.

In early autumn 1961, using only two of her three 22,000 horsepower reactors, *Lenin* completed a journey of some 6,000 miles in a month, three-quarters being through the thick polar icepack, which no ship in the world had previously dared to attack. This demonstrated the remarkable advantages that *Lenin*, despite her high cost, enjoys by being able to steam for very long distances without having to carry large quantities of fuel oil. Moreover, and in contrast to conventional vessels whose load decreases as fuel is used up, *Lenin*'s water displacement remains unchanged, which adds constant icebreaking power to her unequaled autonomy. Thanks to her use, the annual period for Arctic navigation has been considerably lengthened over long distances.

Also during this period the Soviet Union, like the United States, had built several small [1 to 2 MW(e)] "prefabricated" nuclear power plants, designed for use at isolated bases in Siberia. Again like the Americans, the Russians were beginning to study the use of nuclear heat for desalination, with a view to building plants producing both electricity and fresh water in desert regions (the Americans had Sinai in mind).

This review of the independent national nuclear energy programs in the early 1960s would not be complete without mention of the Canadian and

Swedish achievements. Both countries were developing, although on a more modest scale, natural uranium reactors, though this time moderated by heavy water. Following satisfactory operation of a demonstration unit, the first Canadian industrial power station, named "Candu," with an output of 200 MW(e) was to be completed in 1965, while a less powerful unit was to be used on the outskirts of Stockholm to produce both electricity [10 MW(e)] and domestic heating for some 10,000 urban apartments.

The Birth of the Antinuclear Movement

From the earliest moments of the American wartime nuclear effort, no one responsible for an atomic energy organization or program was unaware of the vital importance of full protection against the biological effects of radiation. Never before was a new industry developed with such concern for the health of the workers involved and the general population.

Over the previous half century, no more than a few pounds of radium had been isolated in the world. This radium was distributed gram by gram either to hospitals for treatment of cancer or to the metallurgical industry for use in radiography. The core of an operating nuclear reactor is in fact equivalent, in terms of radioactivity, to hundreds or even thousands of tons of radium. This intense radioactivity persists for some time after the reactor has been shut down, then decreases slowly until, after several centuries, it is again at the level found in the uranium ores from which the whole process began.

For many decades radioactivity, enshrined in the legend of the discovery of radium and the success of radiotherapy, had been regarded as beneficial. Before the war, the radium content of mineral waters was cited in publicity, and so-called "emitters" were on the market — these were small capsules containing an insoluble radium salt, intended to be placed in the family drinking water to make it radioactive. I can personally recall at this time seeing tantalizing posters advertising a beauty cream called "Thoradia" (hopefully containing neither thorium nor radium!), which had been concocted by a Dr. Curie, no connection of course with the famous scientist family, who were quite upset by the activities of their namesake.

This beneficial regard for radium was changed by Hiroshima with its instant devastation and its long-term effects stretching from 1945 into the future; followed in 1954 by the H-bomb test and its unfortunate effects on the Japanese fishermen; and finally the airborne tests and the resulting radioactive fallout, and the campaign for their prohibition.

The conditions were favorable to nurture a growing public anxiety about the dangers of the intense radioactivity produced in a chain reaction. Nevertheless, throughout the decade from 1954 to 1964, there was no general concern over prototype reactors or the first nuclear power stations coming into operation; they were even sought after as "prestige installations." Opposition was confined to a few particular cases.

Although there had been two fatal accidents resulting from incorrect operation of two experimental research reactors, one in Yugoslavia in 1959, the other in the United States in 1961, both cases were due to flagrant errors by the operators, who were the sole victims. There was also an accident in 1957 when one of the first two British plutonium-producing reactors at Windscale (not in operation at the time) caught fire as the graphite in its core was being reheated to relieve internal stresses caused by irradiation. The neighboring countryside was contaminated by fission products, and consumption of locally produced milk was prohibited for several weeks until the radioactivity had died away.*

Accidents such as these all occurred with prototypes but they were considered virtually impossible with the first full-scale power stations, which were equipped with multiple security devices for protection against any technical failure or human error.

The first public opposition to full-scale nuclear power stations appeared in the United States in the early 1960s. It began with the proposed construction of a unit in the region of the famous San Andreas seismic fault near San Francisco. In 1964, after four years of dispute, the project was abandoned.

Meanwhile, in 1962 New York City's principal electrical utility, Consolidated Edison, had announced its intention to build a 700-MW(e) nuclear plant — the most powerful to date — on the East River opposite the U.N. Headquarters. Both the city council and the local population were extremely perturbed.

The then chairman of the U.S. AEC, Seaborg, was in favor of the project and announced that he would be unhesitatingly prepared to live next to it. On the other hand, David E. Lilienthal, who had been the first chairman of the U.S. AEC but was now spokesman for the coal producers, stated he would not think of it for a second, and even declared to Congress

---

*Mention should also be made of the rumor, neither confirmed nor denied by the Soviet authorities, that a serious accident had occurred in the Ural mountains in 1958. According to the (possibly exaggerated) accounts given 20 years later by two Soviet dissidents, an explosion in a Russian military center had scattered fission products over a wide area, causing health risks for the population and a resulting prohibition on residency in the area.

that the national atomic program was deceptive, costly and dangerous. In the end, in early 1965 this project was also abandoned.

In winning this early victory, the budding opposition was undoubtedly helped by the circumstances of the case, for at this stage of nuclear development there were many to whom the building of such a power plant, in the heart of one of the world's greatest cities, seemed unnecessarily provocative. That would seem all the more so today, following the Three Mile Island accident.

In each of these early cases of opposition, exceptional circumstances affected the issue and the failure of the two projects was in no way an indication of general hostility on a nationwide scale to the new form of energy. By contrast, the American people and their government were then much more concerned with the need to limit its development abroad to peaceful applications.

2. Safeguarding Nuclear Energy

Origins

The first 10 years of atomic energy were thereby marked by considerable technical progress, by the construction of the first large-scale industrial nuclear power stations, and by the launching of a growing number of autonomous programs for the production of nuclear electricity.

During this same decade, political problems were sufficiently resolved to encourage the start of international nuclear trade and to allow some new countries to join in the industrial atomic adventure.

From as early as 1956, the U.S. Congress, together with the U.S. Atomic Energy Commission (U.S. AEC) and the State Department, had insisted that any transfer of nuclear materials to foreign countries should be conditional on acceptance by the importing state of the U.S. right to verify, by inspection, the uses to which the materials were put. Until the establishment of an international safeguards system, as already envisaged in successive draft statutes for the International Atomic Energy Agency (IAEA), the inspections were to be carried out by Americans.

The safeguards clause was a remarkably original condition. For the first time, materials that were physically and chemically identical were to possess different political properties. Their use could be either free or restricted to peaceful purposes by verified international commitments. As already noted, such safeguards were intially accepted without complaint by the nations concluding agreements with the United States.

On the other hand, by imposing these safeguards, the American government was putting the country at a disadvantage for future international competition as long as imposition of a similar condition was not generally and equally insisted upon by all nuclear suppliers. For with other conditions being equal, any importing country, even without military ambitions, would obviously opt for supplies over which it would have full control rather than materials subject to international inspection.

It was therefore very much in the United States' interest to secure an extension of this policy to the entire international market, by persuading as many countries as possible to apply the same export regulations as those imposed by Congress on American sales. At this stage only the United Kingdom and Canada, in accordance with the spirit, if not the letter, of the Quebec Agreement, were more or less obliged to follow suit.

Winning over the other countries therefore became a major objective for Washington, not only from the viewpoint of nonproliferation but also to overcome the commercial handicap resulting from the peaceful utilization clause. This was not to be the last time that the American government, to avoid becoming the victim of conditions imposed by its own Congress, tenaciously sought to make other countries apply these same conditions.

But the Americans at that time held a major trump card: the monopoly of enriched uranium supplies in the Western World. Only the Soviet Union could have countered this monopoly, and some 15 years were to pass before any such competition appeared. However, Washington's position over natural uranium was less strong and was soon to become even weaker, for from 1960 onward the U.S. AEC budget no longer enabled it to buy up the whole of the Western World's production.

Hence the adoption of the peaceful use condition by the exporters of natural uranium became a major objective of American policy, the more so because the future competitors of the United States in the markets for research reactors and power plants were to come from countries working with natural uranium.

Among the countries able to offer this reactor type, the only one not bound by the Quebec Agreement was France, who found herself in a double difficulty over the American policies: first, as a potential buyer of uranium on the world market for her civil (and perhaps also military) programs, and second as a potential exporter of natural uranium reactors. Thus, in the early days of the development of international safeguards policies, Paris was to find itself frequently in conflict with Washington.

Birth of the IAEA

As mentioned earlier (see p. 118), the first signs of dissatisfaction with this American policy appeared in 1956 during the preparation of the IAEA statutes; in particular, the clauses relating to the two "extremities" of the fuel cycle — safeguards on natural uranium and the future of plutonium produced in the reactors.

Until then, natural uranium had in fact been the only area in which the United States and the United Kingdom had shown interest in international collaboration, through their postwar extraterritorial prospecting programs for that mineral. For these ventures, no restrictions on its subsequent use limiting it to peaceful projects had been proposed. Indeed, since neither the United States nor the United Kingdom would have accepted such a restriction affecting their own uranium imports, they could hardly have imposed it on the producing countries.

This is why the financial help given for uranium prospecting in South Africa and Canada in the early 1950s, as well as for the unrewarding searches in Morocco (see p. 108), did not involve restrictions on the use of any share of the uranium found within their borders that those countries might retain. Such uranium therefore remained free for any use by themselves or by future importing countries.

The question of safeguards on natural uranium was raised during the preparatory conference for setting up the IAEA, held in February 1956 in Washington. The 12 countries* taking part included advanced industrial powers, uranium producers, and leaders of the Third World. During the month-long conference, the United States, determined that the new agency should have a technical character, with some difficulty obtained agreement that the most technically advanced countries together with those producing uranium — later to form themselves into a suppliers' group — should have a dominant role in the organization's administrative council or "Board of Governors." The initial composition of this board (23 members) involved bitter discussions with the Third World delegates, who considered their countries insufficiently represented.

Nonetheless, the decision to invest the agency with powers of safeguards and inspection was generally welcomed. These powers were to be

---

*Australia, Belgium, Brazil, Canada, Czechoslovakia, France, India, Portugal, South Africa, the Soviet Union, the United Kingdom, and the United States.

applied in all agreements for the supply of materials in the agency's possession, as well as, and above all, in bilateral or multilateral agreements whose civil nature the partners sought to guarantee through the agency's safeguards system, and also in cases where member countries might unilaterally request safeguards applications to their own programs. Despite opposition from the Soviet side, which wanted the countries concerned to meet the costs of the safeguards themselves, it was decided that the agency should accept the financial burden, the safeguards being considered a contribution to the protection of world peace by reducing mutual suspicion and distrust among nations whose atomic programs would involve militarily significant quantities of fissile materials.

While making no objection to safeguards on the "special fissile materials" — enriched uranium and plutonium — the Indian delegation was opposed to their application to natural uranium. The delegation felt that such a far-reaching system would inevitably divide the countries of the world into two categories. On the one hand, there would be those having no indigenous uranium. Because they would now be unable to secure supplies through the normal commercial market, these countries would be condemned forever to safeguards — even in the civil field, which was the only one where they could hope to develop. On the other hand, there would be the uranium-rich countries and the Great Powers with their military programs. For these countries there would always be supplies of nonsafeguarded, freely usable materials, which they could make available for their civil development, so protecting this from the risks of industrial espionage that are made possible by international inspection.

This was the first hint of a complex and obstinate Indian policy in atomic affairs, continually in the forefront of the campaign for nuclear disarmament and for an end to test explosions, and simultaneously championing nuclear independence for the nonaligned countries of the Third World. Of these last, India was at the time by far the most advanced in the atomic energy field.

The U.S. delegation, widely supported by most of the other states represented at the Washington meeting, rejected the Indian view. The Americans found it particularly unwelcome because, with an overabundance of uranium beginning to appear, they foresaw a time when they could no longer afford to renew their purchase contracts with Canada and South Africa and when uranium supplies would be available on the world market. In fact, they were beginning to put pressure on these two countries to make all future sales of uranium (except of course to the United States or to the United Kingdom) conditional on a guarantee of peaceful use and on the acceptance of international safeguards. Because these uranium-producing countries still hoped to maintain some of their sales to the United States, the American pressure was all the more effective.

The New York Conference

This was the situation when, in October 1956, representatives of 81 countries assembled in New York for the final negotiations on the IAEA statutes. It was, 10 years after the failure of the Baruch Plan, the first time that countries from all over the world had met at U.N. Headquarters to discuss ways for developing nuclear energy without the risk of encouraging its military applications.

During the previous summer, the French government had learned that Canada, although anxious to find new outlets for its uranium, could not supply it to France under "nonrestricted" conditions, i.e., under the same conditions that applied to the United States and the United Kingdom, who could use it for military purposes as and when they wished. Prior to this, Paris had in fact attempted, without success, to persuade the American government not to oppose a possible supply of Canadian uranium to France without restrictions on its use.

Under these circumstances the French delegation to the New York conference, led by myself, had been instructed to accept the principle of safeguards on enriched uranium and plutonium — both being substances of military significance prepared only with considerable industrial effort — but to oppose safeguards on natural uranium, in view of its wide geographical distribution and the discriminatory nature of any such controls.

In presenting this view to the meeting, I also called more generally for a "reasonable application" of the safeguards policy. In my opinion, the IAEA's slogan at that time should surely have been "once a client, always a client," rather than "once safeguarded, always safeguarded."

My intervention, completely contrary to present nonproliferation policies, whose object is precisely to keep the entire nuclear activities of nonnuclear weapons states under IAEA control, was considered by the Washington government to be deplorable. French Ambassador Hervé Alphand was summoned by U.S. Deputy Secretary of State Robert Murphy and I was obliged, in a second intervention, to tone down my initial declaration. The new American control philosophy was already a virtual dogma that must be accepted in full without discussion.

The French position was, in fact, one of the first expressions of dissatisfaction over the inevitably discriminatory nature of any effective nonproliferation policy. The French attempt to exempt natural uranium from safeguards and to restrict guarantees of peaceful utilization to the fissile material concentrates, enriched uranium and plutonium, had no technical justification. This attempt reflected France's wish for independence, if not her further intention to avoid control of natural uranium reactors in view of possible military requirements. In the end the attempt failed and the IAEA statutes made no distinction between natural and enriched uranium from the viewpoint of safeguards.

The French attitude had been shared by the developing countries, led by India, and by all the eastern countries except the Soviet Union, which at the beginning of the conference appeared to support the other countries already in possession of fissile materials. However, by the end of the conference, which was very animated, the Russians had joined ranks with those opposing controls.

India's position was clearly described by Homi J. Bhabha, whose personal prestige was considerable. He particularly attacked the perpetuation of safeguards through the successive transmutations of fissile substances, a vital matter for his own country where source materials existed but where external help was needed for the early stages of the program. He pointed out the illusory nature of strict safeguards, emphasizing that any aid in the atomic field, whether in personnel training or in the supply of nuclear materials, was potential military aid since it was bound to lead to the acquisition of new knowledge and could make available indigenous materials for potential military use. Bhabha proposed that IAEA aid should be given only to countries with no military program, a philosophy close to that advocated today by those who want to see assistance given solely to nations whose entire nuclear activities are subject to safeguards.

Last, the point on which the Indian delegate declared he would never give way was his total opposition to a clause that would have given the IAEA, in all projects subject to its control, the right to decide the future use of all the resulting plutonium, and to fix the quantity that each country should be allowed to keep for its safeguarded civil uses, the rest being deposited in the safe keeping of the IAEA itself. To endow the agency with such discretionary powers would be to risk giving it too great a hold over any country with an economy based on nuclear electricity produced through a program to which the agency would have contributed nothing beyond its earliest stages.

Negotiations between the American and Indian delegations continued throughout the conference. The Americans refused any substantive modification of their position, being totally opposed to the holding of plutonium stocks by nonnuclear weapons states, even if these stocks were subject to

IAEA control. This was to be the thesis readopted, 20 years later, by President Jimmy Carter when it became fundamental to the new American nonproliferation policy.

On October 19, 1956, the day the conference was due to end with a vote on the safeguards clause, the Soviet Union, whose position until then had remained uncertain, joined the other socialist countries in opting clearly for the Indian position. Seeing that the vote was likely to lead either to deadlock or to the adoption of the American proposals by only a small majority, the Swiss delegate, Ambassador Auguste Lindt, and myself decided to put forward a conciliatory amendment. This amendment gave countries the right to retain, from the plutonium they produced, the amounts they needed for their research programs and for their existing reactors and those under construction.

The American delegation requested 48 hours to consider this proposal. The matter was put before the secretary of state, John Foster Dulles, and even the president. After lengthy discussions, the three Anglo-Saxon delegations accepted the Franco-Swiss proposal, to which the Indian delegation also agreed.

Therefore, the conference ended in an atmosphere of unanimity, though failure had been only narrowly avoided. Finally the last obstacle had been overcome, and the way was clear for the creation of the IAEA and its inspection system, a system that would allow countries supplying nuclear aid to verify that it was used for purely peaceful purposes.

It is interesting to note that the much discussed clause concerning the agency's rights over the storage of excess plutonium has never been implemented. The clause has nonetheless acquired new relevance, twenty years later, as the evolution of nonproliferation policies has called for a legal basis for the international storage of plutonium extracted from spent fuels, while awaiting reuse in appropriate power reactors, notably in fast breeder plants.

France, although ready to submit her civil nuclear activities to the future control system of Euratom, had considerable reservations over the IAEA's safeguards, which she did not accept for her national projects until 20 years later, and then only in certain well-defined cases. In fact, in July 1957, during the debate in the French parliament on ratification of the international agreement establishing the IAEA, the government was requested not to make any request for aid from the new organization. It was considered vital that no part of the national effort should be submitted to international safeguards in which inspectors from unfriendly powers could participate. This condition was tacitly accepted by the government.

Safeguards on Natural Uranium

During the few months between the eventful adoption in New York of the IAEA statutes and this debate in the French parliament, a new dimension had been added to the problem of safeguards when France pursued negotiations with Canada for the purchase of uranium.

A decision had to be made on the source of part of the fuel for Electricité de France's nuclear power stations, as there appeared to be a choice between opening a new mine in Central France, which would lead to costly investments, or buying the necessary uranium from Canada. Should the latter solution be adopted, the French government was prepared to consider international controls restricting the Canadian fuel to peaceful uses, the country's indigenous resources remaining available in case of military need.

The Canadian negotiations were entrusted to me. They took place in Ottawa, in March 1957, through a renewable purchase contract for some 50 to 100 tons of uranium. Agreement was reached on quantities and on the safeguards, which were to be applied by the Canadians themselves for the French government considered IAEA controls inappropriate in a transaction between two North Atlantic Treaty Organization allies. There remained the question of price, for which I had express instructions. The Canadians suggested the same average price as paid to them by the Americans, which was $11 per pound of oxide. To their surprise, I replied that we could only accept the same price for the same material (in other words, if the uranium were similarly free of restriction) and that an imposition of the clause on safeguarded peaceful utilization should entail a substantial discount. As the French government had expected, the negotiations broke down on this point, to the great regret of the Canadians who, in view of their obligations vis-à-vis the United States, could sell to France neither nonsafeguarded nor cut-price uranium.

This was one of the first brief appearances of dual pricing, resulting from safeguard policies, in the world market for nuclear materials. Prices were dependent on whether the materials were sold free from restrictions or limited to purely civil applications.

Another example concerned heavy water. From 1955 onward, this substance was being offered for sale by the United States, restricted to peaceful uses and subject to safeguards, for less than half the price then charged by Norwegian industry. Despite this, Norway continued for some years to supply heavy water at the higher price both to the United Kingdom for its hydrogen bomb project and to France for a natural uranium submarine engine, a project pursued at this date but later abandoned.

However, as the IAEA safeguards became more generally accepted, and with the saturation of demand for military programs as such, the incentive behind this dual pricing diminished and virtually disappeared. Twenty years were to pass before the issue was raised again by the concept of "prior consent" from nuclear suppliers concerning their materials and the main stages of the uranium cycle.

In 1960, the United States replaced Canada as the Western World's leading producer of uranium. Shortly afterward the U.S. AEC was obliged to cease entering into new contracts for uranium purchase on the internal market, and at the same time decided not to renew existing import agreements.

The American military budget, huge as it was, did have its limits, as did the capacity of the country's uranium purification plants. So, toward the end of 1959 (as had been predicted), the U.S. government, obliged to give preference to its national producers, was forced to decide against renewal of its Belgian, Canadian, and South African purchase contracts, all of which were coming to an end in the early 1960s.

The Americans, in abandoning the markets they had themselves helped to create, gave up one of the essentials of their previous atomic policy: the stranglehold they had maintained on the principal Western sources of uranium through the Combined Development Trust created in 1943, whose purpose had now disappeared. No advocate of nonproliferation could at this stage have persuaded the U.S. Congress to vote the necessary fund for purchasing all of the now overabundant uranium. On the contrary, Congress was much nearer to prohibiting imports of foreign uranium into the country. Such a prohibition did in fact come into force in 1964.

The discontinuance of these foreign contracts was particularly serious for Canadian industry, which had been earning $300 million per year from uranium exports, the third largest source of external revenue for the country after wheat and wood pulp.

Canada and South Africa were able to soften to some extent the economic consequences of the American decision by spreading out the last deliveries of uranium that were due under the remaining contracts, which lasted until 1966 for Canada and 1970 for South Africa, and by exploiting only their richest mines. Nevertheless, a town in Ontario, Blind River, which had been built to house 25,000 people engaged in the large local uranium mining project, was reduced to a ghost town.

On the other hand, Belgium suffered hardly at all from the nonrenewal of the U.S. contracts for Congo uranium, for their principal mine was virtually worked out when its exploitation ceased in 1960, shortly before the

Congo achieved independence. Between 1944 and 1960 the mine had supplied the United States and the United Kingdom with over 30,000 tons of uranium, a quantity on the same order as the entire production of the Western World in that same year 1960. In terms of finished products, this corresponded to a turnover of about $1 billion with uranium taking third place, after copper and aluminum, in the nonferrous metal world trade.

During these times of overabundance, France was still engaged in her campaign against safeguards clauses in contracts for the supply of natural uranium. In 1959 France sold some tens of tons to Denmark and Sweden, modest sales that, at the request of the purchasing countries, were free of restrictions on use. The Euratom commission, drawing inspiration from the American philosophy of safeguards, called for these transactions to be made conditional on a peaceful utilization clause, quoting an article of the Euratom treaty that instructed it to ensure that beneficiaries of such deliveries "gave full guarantees that the general interests of the community would be respected."

The French foreign affairs minister, Maurice Couve de Murville, ignored this, considering the matter to be outside the field of community policy and therefore uniquely within his government's competence. Such conflicts of opinion persist even today, for the nuclear states of Euratom regard nonproliferation policies as coming within their foreign and defense policies, and therefore no concern of the commission, which takes an opposite view.

As the American nonproliferation policy evolved, a meeting was held the following year at the U.S. embassy in London, attended by the main Western industrial powers. The objective was to secure agreement among suppliers that they would not furnish essential nuclear materials or equipment to any nonnuclear weapons state, except under guarantees of safeguarded peaceful use. The American proposal was viewed with reserve by several countries on the grounds of uncertainty concerning the Soviet attitude. France was more forthright: She refused to join a common front of export suppliers as long as she was treated, by countries such as Canada, differently from the United States and the United Kingdom over sales of uranium.

Meetings of this sort were held without success every year until 1965. They usually took one day, and the French participants used to call them "Cashmere meetings" because, after voicing their disinclination to cooperate, they would leave and go to buy Scotch pullovers. After another 10 years, the London meetings were reinstituted, in different circumstances, with Soviet participation and French goodwill. They then achieved some measure of success.

In 1963, France was at last able to obtain an objective pursued since 1956: the conclusion of a substantial contract for the purchase of nonsafeguarded uranium. Where she had failed with Canada, she succeeded with South Africa, whose mining industry was also suffering from the nonrenewal of American purchases.

When I was invited in 1963 to make a tour of the South African nuclear installations, I learned with great satisfaction that their government had decided not to discriminate between the three Western nuclear powers and was ready to authorize its uranium producers to conclude a multiannual sales contract with the Commissariat à l'Energie Atomique (CEA). The contract, signed a few months later, covered a tonnage equivalent to nearly two-thirds of the annual French production at that time, at a price of about one-fifth that quoted in a limited offer by the Belgians to the CEA in 1955 (see p. 292) and about one-third that quoted by the Canadians during the abortive 1957 negotiations. Overabundance was beginning to make itself felt.

The existence of this purchase agreement with South Africa remained secret for a long time. But it was known to the United States, who was extremely irritated over such a restriction free sale, although it was made under the same conditions as sales to themselves or to the United Kingdom. Some three years later, Washington made its displeasure felt in Pretoria by refusing until the last moment to renew the 10-year cooperation treaty under which highly enriched uranium was to be supplied for the South African national research reactor, despite the fact that the latter was of American origin and under IAEA control. Had it been necessary, France would have been prepared to find, from her military production that was just beginning in her Pierrelatte plant, the few kilograms of uranium-235 that were needed by her South African uranium suppliers for their experimental reactor. Again, 15 years later, in 1981, the same problem arose but this time Pretoria was able to start supplying the required amount of sufficiently enriched uranium from its own national isotopic separation plant.

Thus, throughout the years the American policy of international safeguards was gradually put into effect, France had tried to mitigate the effects of such a policy on her own program. Over the same period, France was also confronted by Washington's influence in the negotiations setting up the European Atomic Energy Community (Euratom). Having failed to persuade all the future members of Euratom to renounce the military option, the United States nevertheless played an important part in the organization's foundation and early operation. They had two distinct objectives: European integration, and the promotion of American-type nuclear plants, fueled with enriched uranium and subject of course, to safeguards to guarantee their peaceful utilization.

At this point, we need to go back for a moment to the European scene in 1955, to examine the origins of nuclear collaboration between the western European nations.

3. European Collaboration and Euratom

Louis Armand's Two "Sauces"

The lifting of atomic secrecy led to two separate attempts between 1955 and 1958 at European multilateral collaboration: the European Atomic Energy Community (Euratom), involving the 6 member states of the European Coal and Steel Community, and the European Nuclear Energy Agency, involving the 17 members of the Organization for European Economic Cooperation (OEEC). The first of these was the more ambitious and politically oriented. The second, based on an organization where all decisions required unanimity, was backed by the United Kingdom following its hasty retreat, as an observer, from the 1955 negotiations in Brussels for the relaunching of European cooperation. The OEEC envisaged setting up nuclear joint undertakings with optional participation open to all its member countries.

A veritable race developed in the setting up of the two organizations. Louis Armand, chairman of the French National Railway Company, who in 1951 had initially accepted the leadership of the Commissariat à l'Energie Atomique (CEA) in succession to Raoul Dautry and then at the last minute turned it down, was convinced of the need for a nuclear Europe. So in 1955 he actively supported the moves to create the two European bodies. In a reference to Jean Monnet, the inspiration behind Euratom, and to the Château de la Muette in Paris, where OEEC's nuclear agency was being born, Armand often told me that what mattered was the objective, and that the chosen "sauce," whether Monnet or Muette, was of little significance. Later, however, he gave his main support to Euratom.

The race ended in February 1958 with the inauguration of the OEEC's agency, one month after that of Euratom.

The European Nuclear Energy Agency's tasks, which were technical rather than political, took immediate concrete form with the creation of three joint undertakings. Two of these were devoted to reactors:

- the Halden reactor in Norway, using boiling heavy water as both moderator and primary coolant, was initially built to supply steam to a paper pulp factory; the operation costs of this reactor were to be shared between the participant countries interested in research on the behavior of fuel elements and on safety in nuclear plants; and

- the Dragon Project, a high temperature gas-cooled reactor, jointly built and operated at Winfrith in the United Kingdom until 1976 when it was shut down; it was for a long time one of the foremost reactors of its type, providing invaluable experience and understanding of a reactor concept, which despite certain problems still has many protagonists; it could have a future as a source of high temperatures for the chemical or metallurgical industries.

The third joint undertaking, Eurochemic, was devoted to the acquisition of industrial technology and experience in the reprocessing of spent nuclear fuels, a field in which France was the only one of the 13 participant countries with any substantial practical achievements, the United Kingdom having declined to join the project. The objectives of Eurochemic were to build and operate a reprocessing plant and an associated laboratory, to develop reprocessing technology, and to train specialist engineers in this delicate area.

The joint sharing of this technology for the extraction of plutonium, today considered one of the most sensitive areas from the viewpoint of proliferation, seemed at the time a highly promising field for European cooperation and raised no political difficulties. Built at Mol in Belgium, thanks to an excellent collaboration between the main European chemical industries involved under the leadership of the French company Saint-Gobain, the Eurochemic plant operated satisfactorily for nearly 10 years until its closure in 1974 due to serious financial difficulties.

Euratom — Marriage of Convenience

Euratom was intended to be an instrument for pooling the scientific and industrial potential of the six participating countries in order to accelerate the development of nuclear industry in continental western Europe.

Apart from difficulties resulting from the military atomic ambitions of an important section of France's political elite, a considerable obstacle in the Euratom negotiations was the disparity in the nuclear achievements and policies of the six countries concerned. The British appreciated this, and in addition believed that their membership in the "Club of the Great" was incompatible with sharing their resources, their lead, and their secrets in the industrial field. Wishing to profit from the situation by exporting nuclear know-how and equipment, Great Britain saw among its neighbors some potential customers with whom it would prefer to negotiate individually. Great Britain also saw in France a future commercial competitor.

France was in fact following the same road as the United Kingdom, although some three years behind. Nevertheless, France had a considerable lead over her eventual partners in the Six. The French atomic budget for 1955 was over four times greater than the atomic budgets of the five other countries put together.

The only one of these five with an industrial capacity similar to that of France was the Federal Republic of Germany, a country that had, however, only just regained the right to proceed with civil atomic research. Furthermore, the very powerful German private industry was opposed to any form of state interference in the production of electricity, let alone intervention by a supranational body.

Of the other partners in the Six, Italy, with a nuclear organization still in its earliest stages, was proposing to buy reactors from the United States or the United Kingdom, while the Netherlands, despite their long-established association with Norway, had not yet succeeded in launching a truly national program. Last, Belgium had special agreements with the United Kingdom and the United States that limited its freedom over uranium in return for preferential treatment from those countries.

The modification of the McMahon Act in 1954, together with high hopes for the industrial development of atomic energy, had led the Belgians in 1955, when their uranium sales contract was renewed, to be more insistent than previously over the special treatment they had been promised. Under their contract, they should have at last been given access to secret American industrial information though without the right to pass this information on to their future partners in the "marriage basket" of Euratom, which was an unpromising augury for the future of that organization. In practice, the advantages turned out to be illusory because the information supplied to Belgium was the same as that offered to every buyer of an American research reactor or, later, every importer of an American nuclear power plant.

For the Congolese uranium, to which France attached an undenied interest and a symbolic value, things were no better. Under its new agreement, Belgium was allowed to retain certain quantities of uranium for its own use, these quantities being small at first but then increasing year by year. However, no uranium could be sold to other countries without prior consultation with the United States.

Before the war, uranium had been worth $1 per pound of oxide, if it could be sold at all. During the war the Belgians were paid $1.50 per pound. This price subsequently increased gradually until, in 1953, it reached $6 per pound or approximately half the price then paid to the Canadians.

The Belgians were, however, unprepared to treat one of their future Euratom partners with the same magnanimity. On the first day of the Brussels negotiations over the Euratom treaty, in July 1955, the president of the Union Minière du Haut Katanga approached the French delegation with an offer, presented as if it were a great favor, to sell some 20 tons of uranium oxide at $17 per pound — nearly three times the price being paid by the United States to the Belgians at that time. The offer, which could hardly have claimed to be made in any ''community spirit,'' was rejected in my presence by Pierre Guillaumat, who burst out laughing as he replied, ''At that price, I'm selling too!'' The treaty negotiations designed to bring about the pooling of the community's nuclear resources had not gotten off to a very promising start.

The idea of pooling the production of nuclear electricity seemed no more likely to succeed. Neither the Belgians nor the Germans wanted it since any supranational organization would have been unacceptable to the private industry and utilities in their countries where electricity was not nationalized.

Nor in fact was the French utility Electricité de France prepared to transfer its responsibilities, which had only recently been defined on a national basis, for building and running nuclear power stations to an international organization.

An American Memorandum

To specialists in the nuclear business, aware of these technical and political disparities between the six countries, the idea of bringing them together seemed ambitious and very difficult. But powerful political forces were at work in pursuit of European unity, and were to succeed in creating the proposed community organization.

The American government in particular was extremely anxious for the negotiations to succeed. In May 1956, the day before a conference opened in Venice between the foreign ministers of the six member countries at which the objectives of the future treaty were to be decided, Washington circulated to all six governments a memorandum (see p. 134) in which the question of nuclear armaments within the community was cautiously raised. At the same time the memorandum left no doubt as to what would be expected from Euratom.

This important document began by observing that some Europeans seemed unclear as to American methods and policies toward Euratom. Therefore, although the United States believed they should remain in the background at the Venice conference, the memorandum was circulated in order to eliminate misunderstandings by emphasizing once again some basic principles of American policy.

In reality, both the spirit and the letter of the document gave a clear indication that Washington had no intention whatever of ''remaining in the background'' during the negotiations. Another dogma had been revealed: that the United States would have no direct relationships with a multinational organization unless that organization had effective communal authority and could undertake responsibilities and duties similar to those of national governments, particularly with regard to safeguards. It was also made clear that such a community, integrated and linking Germany structurally with the western countries, would find the United States ready to make available substantially more important resources, and to adopt a more liberal attitude than they could show to the six nations separately.

This American dogma was set out with the utmost clarity in the matters of safeguards, ownership of nuclear fuels, and the monopoly that the future authority should have in the conclusion of supply contracts. According to the memorandum, if an integrated organization such as Euratom were given the necessary safeguarding powers, this would improve the chances of preventing diversion of nuclear materials in one of the major world areas where important atomic developments were soon to be expected. The Venice meeting would no doubt not be considering, at this stage, the important fundamental questions relating to ownership of nuclear fuels, nor the possibility that one of the member states might obtain such materials by purchases outside the Euratom framework. The U.S. government nevertheless wished to express its concern over the implications of a possible compromise on these matters.

If, in fact, Euratom wished to take seriously its task of communal responsibility, rather than becoming merely a coordinating mechanism with some duties in the field of safeguards, the United States believed it ought to have complete authority and control over nuclear fuels; as complete as if Euratom owned the fuels, even though this would not imply a legal right of ownership.

Any compromise in the Euratom statute by which one of the partner countries could, under certain conditions, obtain nuclear materials by special arrangements other than via Euratom would seem to the U.S. government to strike at the very heart of the concept of Euratom as the atomic community of the six countries.

The U.S. government wished these countries to know, at this moment when they were about to prepare their treaty, that for the reasons stated it would be unable to cooperate effectively with Euratom unless and until the problem of fuel ownership had been satisfactorily resolved on a European rather than a national basis.

This essentially European rather than national approach, as seen by Washington, drew much of its inspiration from the 1946 Acheson-Lilienthal Report, intended to apply to a worldwide atomic energy community. The report's influence can be detected throughout the Euratom treaty. Later on, the American government, becoming far more concerned over nuclear weapons proliferation than over European integration, was to suffer the consequences of its contribution to, and recognition of, the Euratom safeguards system. For in 1978 this regional system was to be used as an argument by member states that, after signing the IAEA's Non-Proliferation Treaty, tried to refuse the IAEA's controls or at least to restrict their field of application.

Shortly after the American memorandum had been received in Paris, Monnet summoned me and, without mentioning the document, asked me to tell Guillaumat and Francis Perrin that he was convinced the Euratom game would be "finished" unless the negotiations led to ownership (or something equivalent) of all nuclear materials by Euratom, with the six partners prohibited from concluding bilateral supply agreements. Nearly a quarter of a century later, in 1980, these were still the precise points on which the French government was attempting to have the practical application of the treaty modified.

Negotiating the Treaty

Following the Venice conference, at which the objectives of the future atomic community were approved in general terms, the current prime minister, Guy Mollet, who was aware of the CEA's reticence, somewhat provocatively decided to entrust final negotiations on the preparation of the treaty to Guillaumat, whose lack of enthusiasm for Euratom was known to him, but in whom he had trust as a disciplined state servant.

Up to this moment, Guillaumat, the CEA's administrator-general had shared my own skepticism as to whether the new organization, which we disrespectfully referred to as "Le Raton" (little rat*), could ever be set up. But overnight he became responsible for carrying it forward to the Brussels baptismal fonts. During the second half of 1956, he devoted himself completely to that task, chairing many long and tedious discussions, mainly concerning the extent of the supranational powers to be vested in the future organization.

It was during these negotiations that the French representatives, influenced by Monnet, set out the principles of community priority in the supply of nuclear materials, with equal access for all member states. This principle was established with a view to possible uranium shortages in the community, if not throughout the whole world, and in the hope that France would be able to gain access to some part of the Belgian production when the Anglo-Saxon sales contracts came to an end in 1960. Actually, neither the forecast shortages nor the foreseen Belgian supplies materialized: by 1960 the world market was in surplus, the Congo was no longer Belgian, and its uranium mine was virtually worked out. Soon the French were to be the first to regret the application, to themselves, of the clauses in the treaty that they had proposed when circumstances were different.

In June 1956, the French government initiated a wide-ranging debate on Euratom in the national assembly, at the end of which Prime Minister Mollet reaffirmed the country's freedom of action in the military field. The debate was especially noteworthy for the completely exceptional appearance at the dais of two nonparliamentarians who were called on to address the assembly: the high commissioner for atomic energy, Perrin, and Armand, who had chaired the early Euratom negotiations and who was to be the organization's first president.

The high commissioner's address, which dealt only with peaceful applications of atomic energy, nevertheless argued strongly for the maintenance of a national program. His address ended with the words: "European collaboration in the atomic energy field would be technically harmful if it were to result in reduced national programs. If, on the other hand, it can

---

*The pronunciation in French of "L'Euratom" and "Le Raton" is virtually identical.

help to stimulate these programs, and add further projects to them, it is desirable and will be profitable from this point of view."

Armand's address, in a much more political vein, was a discourse eulogizing the maxim "unity in strength." He emphasized the wide gap in every field between West European technology and that of the American and Soviet colossi, minimizing the achievements of European national programs and referring, among others, to the failure of the British Comet jet airliner compared with the success of the American Boeing. Speaking of future generations of French youth, he said: "We may well take them to the rue Vauquelin (where the Curie's laboratory was located) to show them the plaque recording that the French discovered radium, and they will say 'In that day and age, the size of the French nation alone was sufficient and appropriate for the time; but we should have been prepared to adopt the larger dimensions of our own century . . . by joining in association with other countries.' "

The Euratom treaty was signed in Rome in March 1957, at the same time as the treaty establishing the Common Market.

Under the Euratom statute, a council of ministers (those in charge of the atomic energy programs in each member state) had the responsibility of deciding the organization's policy, particularly the program and the budget. The commission, comprised of nationals from each of the six governments, was to implement the treaty according to the council's decisions.*

A supply agency was to ensure equal access by all members to uranium ore and fissile materials without any distinction between civil and miltary uses. This agency, set up in view of an expected shortage of uranium, was to have first option on all nuclear materials, as well as the monopoly of both supply and export contracts.

Euratom was to have ownership of all fissile materials other than natural uranium, i.e., plutonium and all grades of enriched uranium. However, should such materials be needed for fabrication of a military device, this (purely theoretical) ownership was to cease at the entrance to the military establishment concerned. The Euratom security control also stopped at this point.

This security control, as already noted, was in principle a verification of conformity and not of peaceful utilization; its purpose being to check that fissile materials, including ores, were not diverted from the uses declared by their owners, and to ensure that the supply regulations were respected.

---

*From 1965 onward, the same council of ministers and the same commission have carried out equivalent duties required under the three treaties of the European Coal and Steel Community, the Common Market, and Euratom.

The Euratom control system was based on that of the IAEA.

Finally, Euratom was empowered to negotiate and conclude agreements with third countries, and was given the necessary funds for a 5-year research program. This funding was on the same order as the national atomic budgets of the smaller member countries, though well below the budget of the French program.

With the Euratom treaty becoming effective in January 1958, and the inauguration a month later of the OEEC's European Nuclear Energy Agency, the European marriage market came to an end. It had led above all, though not without difficulty, to the creation of the atomic energy community through a marriage of convenience between six somewhat ill-matched partners. As a result, marital life among the partners didn't always run smoothly. But one important aim was undoubtedly achieved: The happy period, long since ended, when the prestige of the peaceful, civil atom was undisputed served to bring favorable publicity to another development that was to assume far greater importance — the Common Market.

Isotopic Separation Politics

Shortly before the start of the European negotiations, and for the second time as part of these negotiations, France attempted to break the American monopoly on uranium enrichment.

The process used by the United States, gaseous diffusion, had also been developed by the British, more or less independently for there had been no real wartime collaboration in this field. Therefore, at the end of 1954, the CEA began to negotiate with the British for the construction of a plant in France similar to that at Capenhurst, where the first enrichment stages were already in operation.

The French approach was at first warmly welcomed by the U.K. Atomic Energy Authority (U.K. AEA), which was hoping to put Britain in the vanguard of future nuclear technology exporters. However, a few weeks later the negotiations collapsed when, early in February 1955, Guillaumat and myself had a meeting in London with Sir Edwin Plowden and Sir John Cockcroft, administrative and scientific heads, respectively, of the U.K. AEA.

They began by showing us the preparatory work they had already done on our behalf, which included a list of all the possibilities for a gaseous diffusion plant. The plants were arranged according to the capacity with estimates of cost, construction time, and electricity consumption for each.

I hurriedly began to make a copy of this "menu," which gave us a lot of interesting information. Meanwhile, our British friends told us, not concealing their disappointment, that they were no longer able to proceed with the proposals because the United States, when consulted, had formally blocked the project by virtue of the 1943 Anglo-American agreements on atomic secrecy.

On the other hand, from the beginning in July 1955 of the Brussels meetings to prepare the Euratom treaty, the French delegation had proposed an isotopic separation plant as the main project to be jointly pursued. This priority was emphasized in the November 1955 report of the Euratom's negotiating group chaired by Armand, and again in the "Spaak Report," produced by the delegation heads and approved by the foreign affairs ministers of the Six. Following this, at the request of the French, Armand agreed to set up a working group of experts to study such a joint undertaking, without waiting for the Euratom treaty to become effective. He appointed me as chairman.

From the start of the group's meetings, in January 1956, the experts confirmed the future importance of uranium-235. At this time, French research scientists were the most technically advanced in this field among the Six. They were pursuing the now "conventional" gaseous diffusion process for which the Americans, like the British and the Russians, had refused to make public their own industrial technology at the 1955 Geneva Conference. Meanwhile, the Germans were investigating another type of gaseous-phase process, using large numbers of small nozzles rather than porous membranes.

In 1956, the group of experts was succeeded by a study syndicate which I also chaired in which Denmark, Sweden, and Switzerland then took part. Its task was to coordinate the progress of work in the various member states.

France was now pursuing gaseous diffusion research at an increasing pace. However, the Germans were veering away from their nozzle process and, in parallel with the Dutch, were investigating a method of isotopic separation by very high-speed centrifuge (ultracentrifugation). They had already been attracted to this method during the war and, despite its being far from industrially achievable, they had growing confidence in it as an economically viable proposition.

In 1956, the U.S. government decided to lift the secrecy requirement from data on enriched uranium power reactors and launched an export drive to sell these plants, backed by a guaranteed supply of uranium-235 at the American home-market price. This price, exempt from normal requirements of industrial profitability, was less than half the cost then expected from any future European enrichment plant.

The sale of uranium-235 by the United States was, of course, subject to safeguards guaranteeing peaceful application and prohibiting its use not only in weapons manufacture but also for the propulsion of warships.

In the face of the seemingly tempting American offer, French attempts to convince the study syndicate that it should build a European enrichment plant proved fruitless, despite the advanced state of the syndicate's work. Champions of the Euratom treaty who, like Armand, had initially seen no contradiction between close U.S.-European links and the proposed joint enrichment plant, which they had therefore supported, were by 1957 opposed to the enterprise. Rather than undertake a costly and as yet technically uncertain project, which would have yielded a European product at a much higher price than the imported equivalent, they preferred to see the countries of Europe buying uranium-235 from the Americans. This preference was also based on vague and distant hopes that the ultracentrifuge process would result in still lower priced uranium-235, so that building a "classical" gaseous diffusion plant could not be justified.

There was little concern about the constraints of safeguards; instead, there was far more worry over the political problems linked with a joint undertaking that might conceivably contribute to the manufacture of atomic bombs — a possibility that, of course, would be fundamentally resisted by Washington.

It is interesting to note that all forecasts were overoptimistic due to the many unresolved technical uncertainties concerning both gaseous diffusion and the centrifuge processes. The estimated cost of the plant proposed by the French experts proved to be about one-fifth of the real cost when the French eventually built it by themselves. Also it took almost 20 years to bring ultracentrifugation to the stage where its industrial use could be justified.

Thus it was that when the study syndicate met for the last time in May 1957, none of its members except France was in favor of an immediate start on building a plant. Soon afterward, the French parliament, during its debate on ratification of the Euratom treaty, approved a 5-year atomic energy development plan proposed by the government of Mollet. This plan included authorization to build a uranium enrichment plant as a national project, should it prove impossible as a joint European enterprise.

France continued bilateral negotiations, in late 1957 and early 1958, with Italy and Germany, respectively. Both of these countries seemed keen to collaborate, but not without awareness of the military implications and possibilities. Important financial contributions toward constructing a joint plant were envisaged at one stage by their defense ministers, but their colleagues responsible for atomic energy regarded the venture with disfavor. Early in 1958, as already mentioned (see p. 186), General de Gaulle opposed energetically this projected German-Italian participation in the French isotopic separation plant and the idea was abandoned.

It was not until 10 years later that a first group of European countries formally joined together in a project for the civil production of enriched uranium, and five years more before a second, separate joint enterprise followed. The two methods, ultracentrifuge and gaseous diffusion, were again competitors in these two rival ventures. In the meantime, France, by herself, had built an enrichment plant at Pierrelatte in the Rhône Valley. Completed in 1967, it was to be used mainly for military purposes.

Thanks to the failure of the 1957 project for a joint European plant, the United States continued to enjoy their monopoly of enriched uranium for civil purposes until, in the early 1970s, the Soviet Union appeared as a supplier in the Western markets, and 10 years later, European production attained a significant level. In the meantime, the American light water reactor type had captured the major part of the non-Communist world market.

The Report of the "Three Wise Men"

American nuclear export policies received a decisive boost in 1957 in the form of a report produced by three top European experts: Louis Armand from France; Franz Etzel from Germany, a member of the High Authority of the European Coal and Steel Community; and from Italy, Francesco Giordani, president of the Italian Nuclear Energy Committee. They had been asked by the six future Euratom governments to report "on the quantities of atomic energy that could be produced in the near future in the six countries, and on the means required to achieve this production."

Their conclusions were published in May 1957 (between the signature and the ratifications by the Six of the Euratom treaty) as a report entitled "A Target for Euratom." This surprisingly ambitious and enthusiastic report had a widespread and most important impact, and did much to facilitate approval of the community treaty by the parliaments of the Six.

With 20 years hindsight this "Report of the Three Wise Men" is reminiscent of school homework that is so often difficult to grade, for it contains both the best and the not-so-good, a mixture of brilliant original ideas with others obviously borrowed from elsewhere, the theoretical arguments being far more sound than the practical applications proposed.

The authors prophetically asserted that large-scale adoption of nuclear energy offered the only solution to the problem of Europe's energy dependence on the Middle East, whose dangers they clearly saw and described in the following words:

> As oil imports increase, so will the temptation to use them as a means for exerting political pressure. Not only would an interruption of our oil supplies during the next few years threaten us with economic catastrophy, but in more general terms it is evident that the dependence of highly industrialized countries on politically unstable areas could have serious worldwide repercussions. It is therefore vital that oil should remain only one of the factors in industrial expansion, and should not become a political weapon.
>
> Thus Europe, in order to protect her economy from risks of this kind, must find other sources of energy so that she can, if necessary, limit any further increases in her oil imports. This can only be achieved by making use of a new form of energy: nuclear power.

Writing while Europe was still reeling from the Suez crisis, the Three Wise Men could not foresee that it would be followed by 15 years of comparative quiet, with decreases in the prices of both oil and coal, which would severely hinder the development of energy from fission. The Wise Men were right, but they were right too early.

The report proposed an installed nuclear capacity of 15,000 MW(e) by 1967, i.e., approximately one-quarter of the total estimated electrical capacity of the six countries at that date. As soon as the report was published, the technicians criticized its conclusions as being far too optimistic in view of the current state of power station technology and of European industrial capacity. Installed nuclear capacity achieved by 1967 was in fact only one-tenth of that called for in the report. The called for capacity was not realized until some 10 years later when the European atomic industry was finally and fully under way.

The Wise Men had clearly underestimated both the time and the steps necessary to progress from the barely completed Calder Hall reactor in Cumbria, and from the submarine engine that had been in operation for only a few years, to full-fledged reliable and economic nuclear power stations. With hardly a mention of the French projects then under way, they seemed to consider natural and enriched uranium reactor types on the same footing; but they did underline the advantages that the United States was ready to grant to Euratom so as to promote their own light water enriched uranium reactor types in Europe, where the average cost of electricity was almost 50 percent higher than in the States.

Finally, the Three Wise Men delivered a fatal blow to the French project for building a European uranium enrichment plant. They believed it would be possible to obtain all the enriched uranium required for their program from the United States at one-half or one-third the price of the same material produced in a European plant. In their words:

> Construction by Euratom of an isotopic separation plant has been advocated, in order that the production of nuclear energy should not be dependent on materials obtainable only from another country. If the need to import large quantities of enriched uranium were to be permanent, this argument would have a certain weight: however, several years would have to pass before a Euratom gaseous diffusion plant could become operational, and even now the future of enriched uranium seems uncertain. Even without considering the prospects afforded by breeder reactors, it is important to take into account the plutonium inevitably produced in the first European reactors. It is highly probable that this plutonium can be used economically, thus reducing requirements for enriched uranium. Lastly, various improvements could even permit the use of reactors of all types fueled exclusively with natural uranium and recycled plutonium.

One could not at this time have found a better description and eulogy of a true "plutonium economy" than in this report, which drew its inspiration from the Washington administration and American industry that, 20 years later, were at loggerheads over the same question of plutonium utilization. It is curious to find the plutonium economy being quoted, at that time, as an insurance against prolonged total dependence on the United States for enrichment.

The construction of a European gaseous diffusion plant would admittedly, as the report pointed out, have "required considerable capital investment and consumed large quantities of electricity," but it would have given Europe an advance of 10 to 15 precious years along the road to independence in this field.

The report of the Three Wise Men and its consequences thereby strengthened the position of the Americans in the "reactor competition," in which they had already begun to take a solid hold of the markets, thanks to their policy of international assistance.

4. The Reactor Trade

Research Reactors

The "Atoms for Peace" policy had opened up new horizons for all countries. The most advanced countries saw export outlets as the fruits of the heavy financial sacrifices made in order to establish their nuclear industries. Other countries, because of restrictions on materials or of special political circumstances and in spite of being industrialized, had been unable to undertake an atomic program consistent with their position, and wished to make up for lost time. Finally, the least developed countries were impatient to gain access to the much vaunted benefits of this new form of energy.

Some countries fell into two of these categories. Thus, France was both an importer of materials she was not producing, such as enriched uranium, heavy water, and even natural uranium, while at the same time she was offering technical assistance and hoping to become a supplier of equipment and, if possible, to join in the competition to export experimental reactors or, better still, complete nuclear power stations.

In these early days of nuclear industry, such an export order was seen as a prestigious operation, both from the international point of view and for the exporting country's internal image. Such an order effectively demonstrated the competitivity of the operations, reassuring the national authorities in their concern over the amount of financial investment necessary for nuclear development. It enabled them to look forward to an important return on this investment, even if financial sacrifices were initially needed to enter the market.

This situation led to market competition that was often distorted by financial subsidies or by the offer of performance guarantees that were premature. The purchasing country initially benefited from this situation, sometimes being left, however, with insufficiently developed technology and heavy ''running-in'' costs. Finally, it was the reputation of atomic energy that suffered.

In support of these policies, the atomically advanced nations were also competing to train nuclear technicians from potential client countries. For there was no doubt that the first requirement of these countries was to acquire, as quickly as possible, a body of specialists able to prepare a nuclear program and to decide what imports would be needed. It was reasonable to hope that they would choose equipment and installations with which they were already familiar.

This led to the establishment of advanced training facilities at the American, British, and French atomic centers of Argonne, Illinois; Oak Ridge, Tennessee; Harwell, England; and Saclay, France. The courses included practical training with experimental reactors and in nuclear laboratories. The chemical technology of irradiated fuel reprocessing and of plutonium extraction, which was to become forbidden territory under the future American nonproliferation policy, was included in the practical courses, even in the United States.

Back in their home countries, the best way for the newly trained nuclear technicians to pursue work in their new specialized field was to do so with a research reactor. Such a reactor is a potential source of artificial radioelements, invaluable working tools for medical and agricultural research as well as industry and science. Later, with access to a more advanced research reactor, the trained specialists could take part in atomic energy development by working, for instance, on the behavior of materials under irradiation.

Finally, studies with power reactors could be undertaken, though only in the most industrialized countries having electricity grids that could be coupled to the most powerful plants — i.e., those that were the most economic.

These were the possibilities offered to countries willing to accept the constraints of a policy of "controlled assistance," meaning essentially that materials supplied, or used in imported installations, were subject to international safeguards.

However, in 1955 and 1956, the first two large research reactors to be exported had been sold subject to conditions of peaceful utilization but free from international safeguards, which were not yet operationally established. These two reactors used natural uranium and heavy water, as already mentioned; they had been sold by Canada and France to India and Israel, respectively.

Subsequently, and very quickly, the United States captured the major part of the market. From 1956 to 1959, they were able to conclude a series of bilateral civil agreements with some 40 "friendly countries." These countries had no objections to accepting safeguards effected by American inspectors, curiously called "bilateral" safeguards as opposed to the "international" safeguards of the IAEA that were to replace them later.

The majority of the bilateral agreements were so-called "research agreements," which allowed the signatory countries to acquire from the United States the few kilograms of uranium-235 needed for the operation of a research reactor.

These enriched uranium research reactors could be adapted to match the degree of advancement of the countries importing them. They were far less expensive than their natural uranium equivalents, costing in general between $500,000 and $1 million, with the purchasing country able to benefit from an American government subsidy of $350,000.

The moment for the handing over of the subsidy money was not always particularly well chosen. I thought General Franco's expression showed this when I saw him at the ceremonial inauguration of the Spanish reactor in Madrid in 1958. He received his check from the American ambassador in front of all the assembled foreign guests and Spanish dignitaries!

These research reactors were low power and in general had no military significance. For this reason, the safeguards operations comprised an annual verification of the fuel stock, which was either highly enriched uranium or, sometimes, uranium containing less than 20 percent of the 235 isotope, a concentration too low for any military use.

Agreements signed with the most advanced countries — the so-called "power agreements" — also covered the supply of uranium-235 for power reactors, which required a fuel charge of several hundred kilograms of this isotope in the form of slightly enriched uranium (approximately 3 percent).

The philosophy behind these agreements was basically liberal, for apart from the obligation to submit to safeguards, they contained no technological restrictions such as those linked to the extraction and use of plutonium, which were to become so important some 15 to 20 years later. Irradiated fuel reprocessing was accepted in principle: It was to be carried out either in the United States or in foreign plants acceptable to the Americans in order to ensure that the reprocessing installation used could be effectively safeguarded. Later, the first European prototype breeders were to be fueled with highly enriched uranium or even plutonium of American origin. The fact that these products, due to their high concentrations of fissile materials, could easily be used to make weapons caused no anxieties since the application of international safeguards was then considered entirely satisfactory.

However, enriched uranium supplied under a bilateral agreement could only be transferred to another country also receiving American assistance, and then only with the approval of the Washington government.

The USA-Euratom Agreement

Belgium, Germany, and Japan were the first countries to place orders with the United States for nuclear power plants, small units producing some 10 MW(e) that were demonstration rather than industrial installations. The first countries to import full-scale plants were the three former Axis powers, Italy, Japan, and Germany, whose atomic programs had been delayed as a consequence of their defeat in the last world conflict.

In 1958, the British, after signing an agreement with Italy on peaceful utilization under international safeguards, concluded the first sale in the world of an industrial nuclear power station to a foreign country. The plant was sold to an Italian electric utility set-up especially for this purpose. With a power of 200 MW(e), it was to be built at Latina near Rome under a private contract and would be a natural uranium, graphite-moderated installation cooled by carbon dioxide under pressure. The following year a second sale was made under the same political conditions but by a different British industrial consortium and this time to Japan. It was to be the Tokai-Mura power station, of the same type as the Italian one. French industry supplied the graphite. These two sales were early but unrepeated British successes.

General Electric also won their first international contract in 1958, with a group of companies operating in southern Italy, a region with high electricity costs. This was the first such sale by international bid. The competition for a 150-MW(e) boiling water plant had been organized by the World Bank, and the General Electric proposal was chosen from among several other offers. It was estimated that the cost of a kilowatt-hour from the plant would be about 10 percent higher than the current cost in the area where the station was to be built, at Garigliano between Rome and Naples.

The following year another consortium, from northwestern Italy, ordered a 240-MW(e) pressurized water power plant from Westinghouse. Thus Italy, equipped with three different types of stations, had a good start in the European nuclear electricity race. It then marked time for 15 years, limiting itself to these outdated prototypes.

Meanwhile, during the European Atomic Energy Community's (Euratom's) first operational year in 1958, the United States concluded an agreement giving Euratom some exceptional advantages — a token of American support for Euratom, which would also encourage the sale of American-type power plants in the member countries of the Six, where the cost of electricity was higher than in the United States. These American power plant types were reputedly "proven," although none was as yet in operation apart from the preprototype at Shippingport, Pennsylvania.

The so-called "USA-Euratom Agreement" was accompanied by substantial financial advantages and entailed a jointly financed research program worth $100 million. Through this program, American companies were able to benefit indirectly from subsidies that their government would have had difficulty giving to them directly within the United States.

Those companies benefiting the most from the agreement were the two industrial consortia, Westinghouse and General Electric, who in the early 1960s each had an atomic budget higher than that of any west European country except Germany, France, and the United Kingdom.

Overall, Europe was destined to serve as a testing ground for nuclear plants conceived and developed across the Atlantic, rather than for the truly European natural uranium, graphite-moderated, gas-cooled reactor that was at least as well proven, in England if not in France.

The agreement also included a guarantee to supply the uranium-235 needed to fuel the power plants and (particularly noteworthy) made Euratom responsible for ensuring the peaceful utilization of this enriched uranium. This was a token of trust and favor on the part of the United States for the European organization that was much resented by the American champions of International Atomic Energy Agency (IAEA) safeguards.

The agreement didn't contain any restrictions affecting transfers within the community. The American enriched uranium could thus circulate freely among the six member countries, and could even be reprocessed without prior American consent. Washington, in its nonproliferation policy some 20 years later, tried unilaterally to modify this position.

The agreement guaranteed that irradiated fuels would, if necessary, be reprocessed in the United States, under the same conditions offered to American clients. Repurchase by the United States of all plutonium produced in excess of the community's needs, which would always have first priority, was also guaranteed for the coming 10 years.

The USA-Euratom Agreement became effective in 1959, just as the era of euphoria was drawing to a close. The agreement was to suffer from these changing times, as well as from the gradual realization by European electric utilities that the American boiling water and pressurized water reactors were far from being as "proven" as they had been led to believe.

For two years, despite the many advantages offered, the only power station to be built in the framework of the agreement was the 150-MW(e) plant already ordered in southern Italy following the 1958 call for bids. Then in 1962, a group of private utilities in Belgium joined with Electricité de France (EDF) in a project to build a 240-MW(e) American-type pressurized water power station identical to the third Italian plant. It was to be built at Chooz in the French Ardennes on the Franco-Belgian border. This was the first new project to be launched under the USA-Euratom Agreement, and it gave EDF an initial chance to escape from technological dependence on the Commissariat à l'Energie Atomique (CEA), and to taste the foreign if not forbidden fruit of enriched uranium.

Until 1962, West Germany remained outside the electronuclear race. Contrary to expectations, its atomic program had not yet become part of the country's industrial expansion. Private industry had long been hostile to any provincial or federal subsidy, fearing this could be a first step toward dreaded nationalization. The state, through the federal ministry responsible for atomic affairs, had only a limited role to play since the main authority in many areas belonged to the provincial governments.

Nevertheless, on the European scale, only the Federal Republic of Germany possessed the considerable asset of an electromechanical industry capable of competing with the American giants. This was certainly true of the German company Siemens and to a lesser extent of AEG (Allgemeine Elektrizitäts Gesellschaft).

In early 1960, the CEA made several attempts to persuade Siemens to take up the natural uranium, graphite-moderated, gas-cooled reactor that CEA had been developing with EDF for some 10 years. These attempts failed for many reasons, principally from lack of a French industrial partner of a size comparable to the German enterprise and with the overall capability of building a nuclear steam supply system. In addition, the CEA was a state concern and there was a lack of encouragement from the European community that was, above all, anxious for the USA-Euratom Agreement to succeed. Perhaps, also, the negotiations were hindered by a certain German sense of inferiority in view of French successes in the nuclear weapons field, a field forbidden to the Federal Republic. Finally, the first interested German electric utilities showed a preference for American reactor types. It was the last of these that, perhaps more than any other, sounded the death knell of the natural uranium, graphite-moderated, gas-cooled reactor type.

Had the United Kingdom been a member of the European community at that time, joint Anglo-French pressure could probably have persuaded Euratom to show more interest in the reactor type being developed on both sides of the Channel.

The West German attitude must also have been influenced by the lack of indigenous uranium resources; the country would have been no less dependent on foreign fuel supplies had a natural uranium reactor type been chosen rather than one using enriched uranium.

So it was that, in 1962, American industry scored a major success by securing, in the face of British competition, a contract to build Germany's first nuclear power plant. The 240-MW(e) boiling water reactor was ordered from General Electric within the framework of the USA-Euratom Agreement by a consortium of German electric utilities. The plant was to be built at Gundremmingen in Bavaria. There is no doubt that the advantages offered by the USA-Euratom Agreement helped to tip the scales in favor of the light water reactor.

This was far from the six or seven nuclear power stations with a total capacity of some 1,000 MW(e) that the agreement had foreseen by 1963; and the target of 15,000 MW(e) by 1967, called for in the report of the Three Wise Men, seemed completely out of step with the developing situation. Nevertheless, several of the leading electromechanical construction groups in community countries had concluded licensing agreements with Westinghouse and General Electric that were to have great significance in the nuclear competition as circumstances improved toward the end of the 1960s.

The USA-Euratom Agreement did not prevent community member countries from having bilateral links with the U.S. Atomic Energy Commission. These relations, somewhat frowned upon by Euratom, resulted from earlier bilateral agreements. One such agreement, which France had concluded in 1956 for supply from the United States of enriched uranium for research reactors, was amended in 1957 to cover fuel supplies for possible future power stations.

French Policies

The United States thus supplied all the enriched uranium fuel for the first French prototype and advanced research reactors, in particular the most important of these, the first breeder reactor, Rapsodie, built in successful collaboration with Euratom.

It was about the supply of plutonium and highly enriched uranium for this reactor that in 1961 the president of Euratom, at the time a Frenchman, made an unsuccessful personal approach to the American authorities with a view to these materials being purchased through the USA-Euratom Agreement rather than through the bilateral Franco-American agreement preferred by the French government. This direct approach by the president of the commission to a non-community government was, of course, intended to strengthen Euratom's prerogatives in international relations as compared with those of the individual member states. It demonstrated a fundamental difference between Euratom's and the French government's concept of the respective roles of the community's member states and of the commission itself: The Americans made it clear that they were no less shocked by the Euratom chairman's "démarche" than were the French.

The incident led shortly afterward to opposition from General Charles de Gaulle, despite approval from the other five countries of the community, to the reappointment beyond 1962 of Étienne Hirsch, the second French president of Euratom who had taken over in 1959 when Louis Armand resigned for health reasons.

No such problem arose over the purchase of American highly enriched uranium for the prototype French submarine engine at Cadarache, which had been under construction since 1960 and was brought into operation in 1964. This supply was covered by a bilateral defense agreement, and there were special American safeguards to guarantee that the fissile material was not transferred from military to civil use, though of course also ensuring that it was not used for explosives.

French policies in these matters, somewhat outside the mainstream of current political trends, also appeared as reticent toward applying international safeguards to natural uranium. The French government, in fact, suspected that these safeguards were particularly aimed at its own activities and was much concerned by the continual growth and extension of international inspection networks. The result was a tendency to encourage opposition to these developments, which led in the early 1960s to the pursuit of two goals: the conclusion of a substantial purchase contract for nonrestricted uranium and the sale of a large and prestigious EDF-type natural uranium power station, subject to pledges of peaceful utilization but not to safeguards. The first goal was achieved in 1963 with the signing of a contract for the purchase of South African uranium, but pursuit of the second goal during the same year ended in failure.

The ideal potential customer for an exported French nuclear power plant seemed to be India. It had, in 1951, been the first country to conclude a nuclear agreement with France and had declared that it was particularly interested in natural uranium reactors. A champion of nuclear pacifism, India was nonetheless opposed to any form of control over its first nuclear power station.

However in 1962, during a conversation with Pierre Guillaumat — by now the French minister responsible for atomic energy — the chairman of the Indian Atomic Energy Commission, Homi J. Bhabha, in comparing the rival natural uranium and enriched uranium reactor types, jestingly remarked: ''The best one, of course, is the one you don't have to pay for!'' It turned out, however, to be no joke at all!

In July 1963, in fact, despite the French offer of a natural uranium graphite-moderated, gas-cooled power station without safeguards, the Indian government accepted an American bid, attracted by the extremely advantageous terms offered through the U.S.-controlled Export-Import Bank. Known as Eximbank, it offered an interest rate of 0.4 percent, with repayment over 40 years starting only in the 10th year. Under these conditions, the Indians ordered two enriched uranium boiling water reactors, each with a capacity of 190 MW(e), to be built at Tarapur, a site in the Bombay region. They also accepted initial American control to ensure peaceful utilization, with IAEA safeguards to follow as soon as they became operational.

This sale was a great victory for the American government. It was the first nuclear power station to be sold in the Third World, and was important because India, the very country that had been foremost in questioning the philosophy of international safeguards, had accepted those safeguards.

The ''Atoms for Peace'' policy, the decline of which India was to contribute to significantly 11 years later, gained all the more from this 1963 success because, only a few weeks earlier, the Soviet Union had come out openly in its support.

Up to that moment, the Soviet government had in no way committed itself to making its own exports of nuclear materials and techniques subject to safeguards on peaceful use, and even less to having such safeguards operated through the IAEA, which the Russians considered under too much American influence. They had, between 1955 and 1963, been selling two types of enriched uranium research reactors to the Eastern bloc without requiring any form of control; one type having the relatively low power of 2

MW(t) (thermal megawatts), the other type, sold to Yugoslavia and China, with five times that power. The only Russian sale outside their sphere of direct political influence was of one of the less powerful reactors, completed in 1961, to Egypt. In addition, they had undertaken to build a 70-MW(e) power station in East Germany.

None of these installations was subject to IAEA safeguards, and IAEA meetings often degenerated into bitter arguments between the Russians and Americans, following and reflecting the fluctuations in the Cold War.

But in June 1963, as negotiations between Washington and Moscow on the Nuclear Test Ban Treaty were nearing completion, the Soviet attitude toward the IAEA and its safeguards system underwent a sudden change. For the first time, the representative of the Soviet Union on the IAEA's board of governors voted in favor of the mechanisms of the safeguards, this time in their application to power reactors.

Until then, the United States had been obliged to operate the safeguards specified in their cooperation agreements. The sole exception was in their agreement with Euratom, where they were satisfied that the community's own system was adequate. Now they could transfer these inspections to the IAEA.

The safeguarding of atomic energy was about to become truly international, a vital step forward toward widening the world of nuclear trade and toward the opening of a new chapter in the story of controlled nuclear combustion.

Act Three

INDUSTRIAL EXPANSION
1964 to 1974

In the necessarily rather artificial and subjective division of this story into successive periods, the 10 years from 1964 to 1974 stand out as the decade of stability in world nuclear politics, the decade of great industrial expansion and achievement and of international collaboration. It was a period of confidence and trust among the nations concerned, whatever their human, material, and technological resources, or their state of development. It was also the golden age of technical freedom. The various stages of the nuclear fuel cycle seemed unencumbered by political obstacles and open to all nations, in particular those willing to accept international safeguards for their installations.

The decade began and ended with significant events in both nuclear and energy policies. It began as the Soviet Union rallied to the support of international safeguards, and as nuclear electricity became competitive; it ended with the shock of the rise in oil prices in the autumn of 1973, and the Indian nuclear explosion the following spring.

In world geopolitics, the period from the mid-1960s to the mid-1970s saw the beginnings of détente between America and Russia, which, despite the Soviet military intervention in Czechoslovakia in 1968, led to the conclusion of the first Strategic Arms Limitation Treaty (SALT I) for nuclear weapons. The decade was also marked by the United States becoming entangled in the Vietnam War, leading to the most powerful nation in the world suffering a first traumatic defeat. China, for its part, was to fall victim to its own cultural revolution, the distant and distorted echoes of which were not without effect on the events of May 1968 in France, one year before the departure of General Charles de Gaulle.

During the same period, the Middle East was twice shaken by hostile outbreaks between Israel and its neighbors: the Six-Day War of 1967 and the short-lived oil embargo that followed and the Yom Kippur conflict of 1973, which was followed by near-crippling rises in oil prices. Moreover, these events occurred at a time when the Western World's dependence on oil was continually increasing: French oil consumption (almost all imported) doubled between 1964 and 1974, rising in round figures from 40 to 50 percent of the country's total primary energy consumption, which itself rose by some 60 percent.

Despite a continuing fall in the prices of oil and coal (until the 1973 Middle East War) both in the United States and in Europe, electricity from nuclear power was claimed to be competitive with electricity from conventional thermal sources. This competitiveness was achieved by increasing the generating capacities of the nuclear stations to levels between 1,200 to 1,300 MW(e), thereby exceeding the 900 MW(e) per unit of the most powerful conventional thermal plants.

Nuclear electrification programs were no longer the sole prerogative of the leading industrial nations. Additional countries in Europe and several in the Third World were undergoing rapid development and launching into industrial production of the new energy form. There was fierce commercial competition for their orders.

There was also fierce competition between the various types of power reactors, the winner being decided by the harsh test of exportability. That test finally gave the accolade, in the early 1970s, to power plants based on reactors moderated and cooled by ordinary (light) water, either pressurized or boiling.

This was the decade when fuel cycle charges became more and more important in calculating the cost of the nuclear kilowatt-hour. There was a persistent glut of low cost uranium, but the price of enrichment was rising as "real values" in the United States replaced those of subsidized production due to use of plants originally constructed for military purposes. The enrichment scene was also changing due to the American loss of monopoly, the Soviet Union's entry into the market, and the establishment of joint

European ventures to exploit new technical processes, notably the "ultracentrifuge."

At the other end ("back end") of the fuel cycle, the full industrialization of irradiated fuel reprocessing and the final management of the radioactive by-products of fission were proving more complex than had been expected, both technically and from an economic viewpoint.

Although there was renewed interest in civil applications of nuclear energy other than for electricity production, such as nuclear propulsion for merchant ships and in space, nuclear heat for desalinization, industrial process heat, and peaceful applications of underground explosions, this did not lead to industrial development on any significant worldwide scale.

This decade also saw the appearance and growth of opposition to the development of nuclear power in the United States, an opposition that was later to spread to western Europe.

At the political level, as international safeguards for the peaceful use of atomic energy became commonplace, the Chinese nuclear explosion of 1964 was to prove the last for 10 years, the longest period during which no country carried out such an explosion. The climate was favorable for the negotiation, and subsequent conclusion in 1968, of the Non-Proliferation Treaty, a particularly important event of the decade. It seemed, in fact, that a solution to the nonproliferation problem had been found that, provided the international safeguards were generally accepted and further explosive uses of the atom were abandoned, would leave all countries complete freedom of industrial choice.

The account that follows will, as previously, deal in turn with technical then political aspects, with the national then the multinational scene. Technological evolution is treated first, followed by a survey of national programs with attention to the origins of antinuclear protests. Then multinational aspects of the fuel cycle are treated and, finally, the evolution of the political debate of nonproliferation.

1. Technological Evolution

The Third Geneva Conference

Exactly one-quarter of a century after the discovery of fission in the uranium atomic nucleus, the Third International Conference on the Peaceful Uses of Atomic Energy, held in Geneva in 1964 and sponsored by the United Nations, took stock of the technological situation and the options that seemed most promising for the future.

The euphoria of 1955, the less optimistic forecasts of 1958, and the skepticism that followed were replaced by the conviction — based this time on a degree of industrial experience — that the use of fission to produce electricity was on the threshold of real industrial success. No one disagreed. Nuclear energy had reached maturity.

Although in fact it was not until more than 10 years later, in 1975, that the title ''Nuclear Energy Maturity'' was given to an important international specialist conference held in Paris; both this and the 1964 Geneva Conference could justifiably claim to be devoted to the subject. The very long construction times required for large-scale atomic projects meant that installations coming into operation in 1975 had been designed mostly on the basis of scientific and industrial data available around 1964. Thus the 1964 conference heralded nuclear maturity, and the one in 1975 established this as a fact.

Uranium and thorium, the latter being studied very little due to the abundance of uranium, were still the only possible sources of energy produced from the atomic nucleus. Despite much research devoted to achieving a controlled thermonuclear reaction, that is, control of the fusion reaction used in the H-bomb, in 1964 it was impossible to say — as is still the case over 15 years later — if, when, how, and at what price unlimited amounts of electricity might be produced from the heavy isotope of hydrogen present throughout the water supplies of the world.

Improvements in the construction and operation of research reactors and prototypes had made possible considerable progress in understanding the behavior of materials under irradiation. By 1964, methods had been developed for making fuel rods that were able to withstand increasing levels of in-reactor radiation without distortion, this being a property of primary importance for the economics and reliability of future nuclear power plants.

There was still fierce competition between different reactor types, but for nuclear electricity production, American industry was offering a new and particularly attractive bargain. For the first time in the United States, a construction firm was offering an atomic power plant ready-for-use at a fixed "turnkey" price, to produce electricity at a lower cost than that obtainable from a conventional thermal power station on the same site.

However, this certainly did not stop the proponents of the various reactor types from confronting one another and prospective clients at the U.N. Geneva Conference with predicted relative costs for the nuclear kilowatt-hour. The fact that their calculations concerned reactors that had not yet been built and whose capacity exceeded anything so far completed and were based on fuel cycle costs unsupported by any valid economic data did nothing to lessen their convictions and enthusiasm! But there were also some sober communications presented at the meeting that made it possible to compare some first real figures from the satisfactory operation of six power stations with capacities between 100 and 200 MW(e).

At this point, the British were obliged to admit that contrary to previous forecasts, their Magnox type of reactor launched in 1955 with a blaze of publicity was not competitive. However, they had the remedy for this: the replacement of natural uranium metal fuel rods by rods of enriched uranium oxide, in a series of reactors operating at higher temperatures and with their coolant gas at higher pressures. The change to enriched uranium meant that, for the same power, a smaller reactor would be required, with a corresponding reduction in capital investment.

This left the Canadians and the French as the only proponents of natural uranium. The Canadians, using heavy water and awaiting the startup of their first prototype power reactor, had just sold one of these to India and a less powerful one to Pakistan.

The French for their part remained faithful to the graphite-moderated, gas-cooled reactor type, of which their most representative units were under construction in the Loire Valley south of Paris. They still had confidence in metallic natural uranium fuel, and were even forecasting that their natural uranium reactors would prove competitive thanks to an improved annular form of fuel expected to permit an increase in the power obtained from a given quantity of natural uranium metal. Finally, in the export field, French industry was at last on the verge of success with real hopes of selling to Spain a power station identical to the most powerful plant then under construction for Electricité de France (EDF).

The outcome of the competition between partisans of the American ordinary (or ''light'') water reactors and the Canadian heavy water reactors on the one hand, and the French and British promoters of graphite-moderated units cooled by pressurized gas on the other, still seemed uncertain. The British and French, although they had failed to win German industrial support for this system, nonetheless remained apparently formidable rivals. But in truth their graphite, gas-cooled system was living its last days of glory and the ''reactor war'' was nearly over. The British were about to lose their last chances in international competition, while in France construction and commissioning problems with the EDF power stations, and the resulting conflict between EDF and the Commissariat à l'Energie Atomique (CEA), were influential in the subsequent decision to abandon this national reactor type.

Accordingly, from 1970 onward only a very limited number of reactor types remained in the international commercial competition. All were moderated and cooled by water, either heavy or ordinary, and all used either natural or enriched uranium fuel in oxide form. The fuel elements, the essential basis of every reactor, were made up of clusters comprising from tens to hundreds of ''fuel pins,'' in which ceramic uranium oxide pellets approximately one centimetre in diameter were inserted one against another in thin tubes of stainless steel or a zirconium alloy, several metres in length. These clusters constituted the reactor core.

In heavy water power units, each pin cluster was placed horizontally in a second and stronger tube through which the pressurized coolant, heavy water in this case, was circulated, while the surrounding moderator (also heavy water) remained at normal pressure. The system of reinforced tubes made it possible to change fuel without shutting down the reactor.

In the light water reactors, the fuel elements were in the form of clusters between 100 and 200 pins each, containing several tens of tons of uranium enriched to a 3 percent content of uranium-235. They were supported vertically in either boiling water or water prevented from boiling by being maintained under high pressure, according to the type of reactor. The entire "core assembly" was contained in a reactor vessel of high-tensile-strength steel. Fuel elements could be unloaded and loaded by remote control after removal of the reactor vessel lid. In normal circumstances, about one-third of the core would be replaced in this way every year, a process necessitating long periods of reactor "downtime."

Between the Canadian-type heavy water power stations and the American-type light water plants, the scales were to tip in favor of the latter. But if this suggested that victory for the U.S. atomic industry was in sight, it could not hide the first clouds of nuclear opposition that were already gathering by the mid-1960s and that were later to spread across the whole of the nuclear world.

The first signs of opposition were not only various local actions against proposals to build nuclear power stations, but also objections from the coal producers. They disliked the granting of federal subsidies to such stations and underlined the dangers they claimed these plants would present in the case of an accident.

However, antinuclear protest movements had not as yet appeared on a nationwide scale in the United States, and had shown up still less in Europe, where the only evidence so far of disquiet had been some opposition in 1961 to the experimental immersion of slightly radioactive wastes in the Mediterranean Sea.

At that time, the use of plutonium, either by recycling in thermal-neutron reactors or as fuel in fast-neutron breeder reactors, was still considered to be the technical solution of the future that would enable nuclear energy to contribute substantially to the world economy. There was as yet no political opposition to this plutonium solution.

During the 1964 Geneva Conference, the Russians' announcement that they were beginning the construction of a 350-MW(e) breeder reactor power station had been one of the surprises of the meeting. But forecasts as to when breeders would be generally introduced still differed widely. One of the conclusions reached at Geneva was that the acceleration of research programs on fast-neutron reactors — which had already led to the Soviet prototype, to British plans for a 200-MW(e) unit, to advance studies for a French model (the future Phénix), and to advanced research in the United States — ought to make the date of entry into service of large 500- to

1,000-MW(e) industrial plants less than the previously forecast 25 years hence (late 1980s).

But meanwhile, while awaiting the industrial construction and use of these breeder reactors, it was imperative to find new sources of uranium. This despite the fact that during the decade beginning in the mid-1960s a temporary overabundance had discouraged further prospecting. Fortunately, important new deposits were soon to be found in Africa and Australia.

Although the production of electricity remained the most favored civil application of fission, the 1964 Geneva Conference gave positive attention to other civil applications that still seemed highly promising. In addition to the peaceful uses of underground explosions, discussed in Part One of this book, these applications included merchant ship propulsion (encouraged by the excellent performances of nuclear-propelled warships), nuclear rockets for interplanetary flight, and the use of nuclear heat in the production of fresh water by seawater desalinization. During the decade that followed, all of these related applications proved either too costly to develop or insufficiently economic to operate, and all were abandoned, at least temporarily. Only the nuclear production of electricity was to be pursued on a massive scale.

Nuclear Propulsion at Sea and in Space

Nuclear naval propulsion nonetheless continued to register apparent successes during the decade following the Geneva Conference. It was without doubt the most indisputable application of the controlled chain reaction. By the end of 1973, the nuclear fleet of the U.S. Navy was comprised of some hundred submarines, three cruisers, and the aircraft carrier *Enterprise,* powered by eight reactors.

The operation of the atomic engines in this fleet had been completely satisfactory, though the picture was marred by two tragedies. In 1963, the U.S. submarine *Thresher* was lost with all hands during a dive off the New England coast. Five years later, a second American submarine, the *Scorpion*, suffered a similar fate. But in neither case was the nuclear reactor or its associated propulsion equipment suspect.

The Soviets were also engaged in an important innovative program for their submarine fleet and the British were doing likewise though on a more modest scale.

In 1964, despite the earlier skepticism of Admiral Hyman G. Rickover, the French land-based prototype nuclear submarine engine was commissioned. France, unlike the United Kingdom, had not had the benefit of any American technological assistance in this field. Admiral Rickover, convinced that French engineers would be unable to build a submarine engine without help, had not blocked the supply of the highly enriched uranium needed for the prototype.

This prototype was built at the CEA's Cadarache establishment by a team under the direction of Naval Engineer Jacques Chevalier. Completed in less than five years, it was in full and correct operation by 1964. The first French missile-launching submarine underwent initial sea trials in 1969, only eighteen months after her British counterpart. By the end of 1973, three French submarines had been completed and two others were under construction.

The beginnings of a nuclear-propelled merchant fleet had been more modest and the results less encouraging. As already noted (see p. 265), by 1964 two civil nuclear ships were afloat: the American demonstration vessel *Savannah* and the Soviet icebreaker *Lenin*. Two improved versions of the latter were to be built in the 1970s: the *Arctic* with 75,000 horsepower from two reactors, which was commissioned in 1975, and the *Siberia*, which followed a few years later. The important role of these icebreakers was to keep the northern polar sea open for shipping for longer periods of time than was previously possible.

The American *Savannah*, after steaming some half million miles without any technical incident, was laid up in 1967 for economic reasons, following three years of service as a commercial cargo vessel.

The *Savannah* had her moment of glory during the 1964 Geneva Conference when the American government invited some 20 foreign representatives aboard for a short nuclear cruise in the strait separating Denmark from Sweden. From the traveler's point of view — or so it seemed to me during those few hours — an atomic ship differs from a conventional ship only in its greater acceleration power and in having no funnel; the difference seems no more than that between an electric and a steam train.

For several years after the Geneva Conference, the company responsible for the *Savannah*'s commercial management, American Export Line, which was operating at a loss, tried to obtain supplementary government aid toward the construction of three further nuclear-propelled American cargo vessels. The effort failed, despite its having the backing of the Joint Committee for Atomic Energy. Instead, the combined opposition of the all-powerful Admiral Rickover and the U.S. Atomic Energy Commission (U.S. AEC) won the day. Nuclear ship propulsion was not yet economic, nor was there yet a solution to the difficult problem of special insurance or that of securing entry into civil harbors and ports.

In 1963, the British Conservative government made known its intention to build a nuclear-powered merchant vessel. Two years later, no progress had been made, and the succeeding Labour government abandoned the project, considering it no more than a prestige-seeking operation.

Only two other countries launched into this field, West Germany and Japan, for whom the construction of submarines had been prohibited since the war.

The German nuclear-powered ore carrier, the *Otto Hahn*, was commissioned in 1969 and operated for a long time between Germany and Morocco. In 1979, having covered 650,000 nautical miles and having spent three-quarters of her time at sea, she was laid up.

The Japanese counterpart to the *Otto Hahn*, the *Mutsu*, had a career that was as short as it was dismal. Construction was begun in the port of Mutsu, whence came the name, and was not completed until 1974. Harbor trials of her nuclear engine were forbidden by the local municipal authorities, which had become hostile, so these trials had to be attempted at sea. After a few days of her inaugural cruise, a fault was found in her reactor's radiation shield. It could no doubt have been repaired in less than a year, but hostile press reporters and local officials dramatized the incident to such an extent that the ship was allowed to return to port using only her emergency diesels and only on the condition that her reactor was completely shut down and officially sealed. It was not until four years later, in 1978, that the *Mutsu* was moved to another dock where repairs could be started. These repairs took three years, during which a guarantee was given that the nuclear engine would not be operated. Trials therefore had to await the unlikely success of a search for a harbor where the authorities were prepared to allow them to be carried out.

These various difficulties encountered in the development of nuclear-powered merchant ships were in no way technical; the case of nuclear propulsion for space applications was very different.

In the western hemisphere, research into nuclear space propulsion has been confined to the United States. Started in the late 1950s, this research program continued until 1972 when it came to an end due to lack of funds. By then, about $1 billion had been spent on it, as had been the case with nuclear aviation.. The U.S. AEC had been responsible for research and for designing a suitable reactor, while the National Aeronautics and Space Administration was responsible for fitting the reactor into a rocket.

The aim was to build a very small nuclear reactor fueled with highly enriched uranium, designed to operate for a short time only but at high power and at a very high temperature, and cooled by compressed hydrogen.

Heated to several thousand degrees Celsius, this hydrogen would then be used to provide the required rocket thrust. Nuclear engines of this type were originally considered even as possible future replacements for the chemical rocket motors used in vehicles such as the interplanetary spacecraft *Saturn*; but the risk of such a reactor accidentally returning to earth while in operation made it necessary to restrict its possible uses to propulsion beyond the effective field of terrestrial gravity.

A series of incidents during the early development stages of the first reactors, and the considerably increased resulting costs, had on several occasions delayed the project, which led to even greater costs. In the end, it was financial difficulty rather than any real technical failure that caused the project to be abandoned, despite the achievement of some very interesting developments. It therefore remains entirely possible that the fuels used for the great interplanetary explorations of the 21st century will ultimately be uranium-235 or plutonium, in view of their extraordinary concentration of potential energy.

A second space application of nuclear power, this time completely successful, was found during the 1960s. Nuclear power can provide a long-lasting supply of electricity on board various types of rockets and satellites, particularly for the transmission of scientific and other data by radio or television. A first system, regularly adopted to provide a permanent power supply of a few tens of watts for interplanetary flights, was based on thermoelectric conversion into electricity of the decay heat from a compact but powerful source of long-lived radioactive elements, in particular, plutonium-238. A system based on the same principle and the same radioisotope is also used for nuclear-powered heart pacemakers.

A second system, this time for the provision of several tens of kilowatts, made use of small and compact nuclear reactors fueled with highly enriched uranium. Two variants were being developed, research on both of which was abandoned in 1973. It was no doubt such a reactor that was aboard the Soviet satellite that disintegrated in 1978 over the northern territories of Canada, exciting a certain amount of international emotion due to the recovery — fortunately from an entirely uninhabited area — of small amounts of radioactive debris. The affair caused a great outcry, and the Canadian government demanded $6 million from Moscow to cover the costs of the search for debris and of decontamination.

Nuclear Desalinization

Rather than nuclear propulsion, the ''star'' novelty of the 1964

Geneva Conference was the possibility of using part of the energy produced in nuclear electric power stations as heat for the production, by distillation from sea or brackish water, of fresh water for human consumption and agriculture. The United States, quickly followed by the Soviet Union, saw in this method at last a use for nuclear energy that could benefit the Third World countries. These countries had all had high hopes of atomic energy, only to face disappointment as it became evident that only the very largest nuclear power stations would be economic, at least in the early stages, for the production of electricity. Indeed, until the arrival of small commercial power plants with outputs between 100 and 200 MW(e), probably developed from advanced naval propulsion systems (a solution studied in France), the smallest industrial power plants available, with capacities of or above 440 MW(e) (Soviet reactors) were all too powerful for the needs or the electrical distribution grids of the vast majority of developing countries.

Israel, Tunisia, and Egypt showed immediate interest in the possibilities of nuclear desalinization. In 1964, President Lyndon B. Johnson offered Israel the chance to join in a collaborative project with the United States in conjunction with the International Atomic Energy Agency (IAEA), with the condition that the IAEA's safeguards for peaceful utilization would necessarily apply. Later, the IAEA set up a research project on a possible installation in southern California, which could have supplied fresh water and electricity to both Mexico and the United States. But finally, none of these projects materialized industrially.

The importance attached to this severe problem of meeting world requirements for fresh water was even reflected in a campaign, launched in 1966 by President Johnson under the slogan "Water for Peace," which was similar to and clearly inspired by the campaign so successfully launched by President Dwight D. Eisenhower 12 years earlier for nuclear energy.

However, here again premature hopes were dashed and research showed that the outstanding problem was not nuclear. The difficulties to be overcome were basically economic and concerned the industrial development of the conventional stages of desalinization, with the source of the necessary heat — whether nuclear or other — being only a secondary consideration. Ten years later, the world's only nuclear desalinization plant was in the Soviet Union, on the eastern shore of the Caspian Sea and linked to a fast-neutron power station that had been completed in 1972. The fresh water plant produced over 20 million gallons per day, sufficient for all the needs of the 100,000 inhabitants of the town of Shevchenko.

Desalinization is not the only industrial use foreseen for heat produced in the chain reaction. The greater part of the energy consumption of advanced countries is used in the form of heat: urban or industrial low temperature heating, industrial process heat, and high temperature heat for specialized uses such as in blast furnaces. The Germans have shown interest in reactors for producing gas at high temperatures, while urban nuclear heating has been pursued in several countries, particularly in northern and eastern Europe. France has specialized in a reactor project (known as ''Thermos'') designed to produce sufficient hot water for the needs of some 50,000 people. In the Soviet Union in the early 1980s, two reactors of this type for district heating were being built.

2. Nuclear Electricity

The Rush for Power Stations in the United States

In December 1963, a public utility in New Jersey issued an invitation to bid for a nuclear power station to be built at a site known as Oyster Creek. The ensuing competition between the two industrial giants, General Electric and Westinghouse, was won by the former with a proposal for a 500- to 600-MW(e) boiling water plant to be delivered at a fixed ''turnkey'' price. With the plant operating at its maximum capacity, the cost of a kilowatt-hour was estimated at slightly below that from an equivalent conventional power station on the same site. Shortly afterward, General Electric, accepting the risk of suffering initial losses in the hope that they would be quickly recovered as a result of series production, offered the same conditions for another proposed plant in the northeastern United States.

The Oyster Creek sale received wide publicity — President Lyndon B. Johnson was quick to call it an economic breakthrough — which brought about a sudden awareness of the viability of atomic power stations and a massive start up of nuclear electricity production, but also aroused the antagonism of the coal producers.

In the face of the sudden new competition, the reactions of the oil companies and the coal producers were, in fact, astonishingly different. The oil companies, aware that world reserves of their product were limited, decided to join the nuclear boom and, one after another, they launched into uranium prospecting and the manufacture of nuclear fuels. Some even began investing in the so-called high temperature plants.

On the other hand, American coal producers, with the world's greatest resources at their disposal, decided to resist the nuclear tide, attempting to stem it by legal actions. Their arguments were simple: If it were true that the kilowatt-hour from fission really cost less than that from conventional sources, then federal aid should be limited to breeder reactors only, and all such aid for these proven and competitive light water power plants (such as the renting of fuel, low cost enrichment services, guaranteed buy-back prices for all plutonium produced, and automatic government insurance of accident risks beyond $70 million) should be stopped. If, on the other hand, the figures quoted for Oyster Creek were propaganda rather than factual, all concerned should have the courage to admit the truth. The coal producers' lobby, the first important pressure group opposing nuclear power, did not hesitate to raise the specter of atomic catastrophe. But those backing the new energy retaliated by accusing conventional thermal power stations of polluting the air and being responsible for over 20,000 deaths in the United States every year from respiratory ailments.

Despite attempts from 1965 onward to reach a truce in this battle of criticisms and insults, the two sides continued to confront one another each year during hearings by the Joint Committee for Atomic Energy. This committee sidestepped the problem on the grounds that as yet there was insufficient experience in the operation of the new nuclear plants built from the mid-1960s onward. It was therefore not yet possible to decide whether or not these plants were economic, and consequently the various forms of aid would continue.

Toward the end of 1965, confidence in nuclear power spread from the plant constructors to the utilities (most of them private) responsible for the country's electricity supply. During 1966 and 1967, enthusiasm became really contagious and more new nuclear generating capacity was ordered than conventional thermal capacity. In all, decisions were made to build some 50 nuclear plants with outputs between 500 and 1,100 MW(e), representing a total of 40,000 MW(e) of new capacity.

All of this confidence and enthusiasm represented a substantial degree of trust in the future, for at the end of 1967 the only American-type power plants in the world with more than a year's operating experience were the few prototypes with capacities less than 250 MW(e) that had been ordered in the late 1950s when electric utilities were under U.S. Atomic Energy Commission (U.S. AEC) pressure to take an interest in the new power.

The nuclear construction market in 1966 and 1967, which was worth some $10 billion, was shared between General Electric and Westinghouse. The first nine units to be sold were all at fixed, turnkey prices, leading without a doubt to huge losses by their constructors; losses estimated at between $10 million and $20 million for each installation. Guaranteed fixed prices for a whole plant were dropped from subsequent contracts.

During these early years of the American "nuclear boom," some companies even considered abandoning coal and oil and burning only uranium in their new power plants. A particularly noteworthy event was an order placed by the government-owned Tennessee Valley Authority for one of the larger nuclear power stations to be built in the very heart of American coal-producing country.

Very full order books for power stations in 1966 and 1967 led to an inevitable slowing down during the following three years from 1968 to 1970 when orders averaged 10,000 MW(e) per year. Then, as the constructors became less overloaded, a new avalanche began with some 20,000 MW(e) ordered in 1971, doubling to over 40,000 MW(e) for each of the next two years, 1972 and 1973.

At the start of 1974, the total installed electrical generating capacity in the United States, which was expected to double over the following 10 to 12 years, was 430,000 MW(e). This was slightly more than twice the nuclear program then under way, a program that totalled nearly 200,000 MW(e) corresponding to some 200 nuclear power plants ordered, under construction, or already in service. This very substantial nuclear program corresponded to five times the *total* installed electrical capacity in France in early 1979.

However, more than half the American 200,000 MW(e) were tied to power plants on which building work had not yet started and which could therefore be postponed or even cancelled. Because of the long construction times involved and uncertainties over obtaining the necessary permits and licenses, the completion of the program was not expected before 1985, by which time nuclear power stations were to have been supplying over one-quarter of the U.S. electrical power consumption. In reality, subsequent delays and cancellations resulted in a substantial downward revision of these forecasts.

Among the early cancellations were several commercial size [between 800 and 1,200 MW(e)] high temperature gas-cooled reactors (HTGRs) systems that used enriched uranium and a graphite moderator and were cooled by helium at relatively high temperature. The system design was based on results from a prototype power reactor (undertaken as a part of the U.S. AEC's Power Reactor Demonstration Program) and the experimental ''Dragon'' reactor program in England, a joint international project of the Organization for European Economic Cooperation's European Nuclear Energy Agency (see p. 289).

At the time of the early cancellations, hundreds of millions of dollars had been invested in HTGR technology, since the concept was thought to offer greater thermodynamic efficiency and uranium utilization than water reactors; and final work was under way on the construction and startup of the 300-MW(e) Fort St. Vrain HTGR demonstration plant in the United States. Over the next several years, all eight commercial size HTGRs on order in the United States were cancelled due to the deteriorating financial positions of the utilities concerned, as well as a drop in the rate of growth of national electricity demand. This led to a costly setback for the two oil companies Gulf and Royal Dutch, who had jointly provided the majority of the investment capital in the venture.

Finally, the Fort St. Vrain reactor after many delays and difficulties achieved full commercial operation and HTGR design and development activity has continued in the United States as a joint government/industry activity.

As a result of these HTGR cancellations, the entire American nuclear power program became dependent on light water reactors; the pressurized water reactor by Westinghouse was adopted for approximately twice as many stations as the General Electric boiling water type. But although Westinghouse thus obtained the major share of the orders for the ''sub-type'' of water reactors it had pioneered, two new competitors offering the same type had appeared in the market in the late 1960s: Combustion Engineering and Babcock & Wilcox.

In 1972, Westinghouse took a further step, starting a project to develop ''offshore'' power stations on platforms floating in the sea a few miles from the coast.

Despite forecasts to the contrary and the resulting publicity campaigns, it soon became clear that the great avalanche of nuclear power plant orders would neither reduce the unit cost of installations nor shorten the building time required. On the contrary, construction times that in the early 1960s had been four or five years from the moment a power plant was ordered to when it came on line were to more than double during the years following the great Oyster Creek episode.

There were many reasons for this longer construction period. Some were technical, such as the difficulties of scaling up to a unit capacity five times that of previous plants; others were industrial, such as delays in the delivery of certain major components whose manufacturers could not keep pace with demand.

There were also losses of time due to administrative and social factors. The increasingly important part played by the trade unions and the strikes they called, and above all the ever-growing repercussions of problems linked with environmental questions and with the antinuclear movement. This movement led to a tightening of radiation protection standards, accompanied by frequent modifications to administrative regulations. The results included considerably longer delays in site selection and over obtaining the required planning permissions and subsequent operating licenses — all of which added to a plant's "lead time."

The considerable increases in the time required to build a nuclear power station were not restricted to such installations but also affected all large-scale industrial projects in the United States during that period. It was the result of a veritable upheaval within the American administrative machine at both state and federal levels. It was a remarkable contrast to the dynamic approach, the sense of priorities, and the exceptional methods of working that had characterized industrial operations during the past war and the years immediately following.

The increased construction delays, together with increases in the costs of borrowing money and the effects of inflation, had serious consequences for the cost of electricity produced. The "loss leader" prices of the first turnkey power stations ordered in 1966, which had to be kept down to only a little more than $100 per electrical kilowatt of installed capacity, were already up nearly 40 percent for comparable units ordered two years later. These prices had further quadrupled (in depreciated dollars) for power plants ordered 8 to 10 years later. As a result, by the end of 1973, the U.S. nuclear program represented a gigantic investment in the magnitude of $75 billion of the then current dollars.

In the early years of the boom, nuclear power stations had quickly matched and even exceeded the capacities of the largest conventional thermal units and had then stabilized at a level around 1,200 MW(e). By the end of 1973, the cost of such a station was in excess of half a billion dollars. This was higher than the cost of a conventional power station of similar capacity although, with fuel costs during the lifetime of the plant (estimated at approximately 30 years) clearly below those of the corresponding coal or fuel oil, the balance remained in favor of the nuclear kilowatt-hour.

However, due to the falling prices of coal and fuel oil throughout the 1960s, no doubt partly as a reaction to the nuclear breakthrough, the price difference between the nuclear and conventional thermal kilowatt-hour would probably have turned in favor of conventional energy. That this did not happen was due to the fourfold increase in oil prices in late 1973, which once again and indisputably tipped the scales in favor of electricity derived from the fission of uranium.

The Nuclear Opposition

The years of the nuclear boom in the United States also saw the first signs of a concerted opposition, moving forward from the field of local resistance efforts to embrace the more general theme of environmental protection. Later there was to be concentration on the risks to nuclear plant operators and to neighboring populations.

In 1967, the Democratic presidential candidate Senator Edmund S. Muskie was campaigning against pollution and attacked light water power stations because, for a given power output, they discharged more heat in their cooling water than conventional thermal stations and this might have some slight adverse effects on the ecology of the rivers where the water was discharged.

The first books attacking the industrial development of nuclear power appeared in 1969. They concentrated on two themes, that of possible catastrophic accidents and that of hidden dangers in low doses of radiation. Every attack was amplified by the media, which did not hesitate to emphasize, and often to distort every incident related to a nuclear station — even those having nothing to do with the reactor and which in a conventional plant would have attracted no attention whatsoever. Minor problems, inevitable during the commissioning of any power station, became accidents and the extremely rare real accidents became barely averted catastrophes.

A particular example of this type of exaggeration resulted in 1966 from the accidental melting of some fuel elements in the first industrial breeder plant, built in Detroit. The accident (very serious for the reactor, which was put out of action for an indefinite period) occurred during the first month of operation. The accident had no consequences for the health of the plant operators, and hence none for the neighboring population. Nevertheless, it was used several times as the basis for articles purporting to describe the effects on Detroit of a hypothetical reactor explosion ... despite the fact that this was technically impossible.

More serious were the public declarations made, from 1969 onward, by two scientists of the U.S. AEC who maintained that any additional dose of radiation above the normal, however minimal, would automatically cause an increase in cases of cancer and leukemia. This was because, according to them, there was no minimum threshold level for the detrimental effects of radiation, as was commonly believed by most other experts. Statistical evidence, in fact, seems to contradict the ''no threshold'' thesis, though without rigorously disproving it. Cancer and leukemia are not found more frequently among people who live in houses made of granite, which contain appreciable amounts of radium, or in towns at high altitudes, which receive greater amounts of natural cosmic radiation. And yet these people are subjected to higher doses than those living in wooden or concrete dwellings or at lower altitudes.

The widespread malaise due to the Vietnam War had nurtured protest movements in many fields: opponents of the ''consumer society,'' supporters of decentralized power and authority, champions of zero growth and of a ''return to nature.'' All of these groups quickly rallied to the antinuclear theme. They used the many channels provided by American legislation, tirelessly appealing against legal decisions in attempts to stop, at any and every stage, projects they were resisting on the grounds of refusal to accept an imposed risk, in contrast to voluntarily accepted risks such as those of tobacco, automobiles, or airplanes. Seeing the increasingly emotional nature of the nuclear controversy, and encouraged by the unexpected success and influence of their first efforts, they adopted the antinuclear theme as their main battle cry.

In 1971, nuclear opponents won their first big victory. The Washington District Appeal Court found in their favor and stopped the construction of a power station at Calvert Cliffs in Maryland. The U.S. AEC was condemned for nonrespect of a new law passed in 1970 on the protection of the environment because it granted a construction permit for the power station without giving due consideration to the risk of thermal pollution from the discharge of heat into river water. The court's decision caused an upheaval in the U.S. AEC's procedures and led to an 18-month moratorium on the issue of further construction permits for nuclear power plants.

The following year, the controversy assumed a more technical aspect when the U.S. AEC was obliged to admit that the emergency core cooling system (ECCS) in reactors, for use in the event of a sudden loss of the normal cooling fluid, was not absolutely fully proven and was the subject of internal argument between two of the commission's departments.

In fact, sudden loss of cooling fluid is a highly unlikely eventuality but, in the event of an equally unlikely simultaneous failure of the ECCS, it could lead to the occurrence of a so-called ''maximum credible accident.'' The consequences of such an accident could be tragic, for whereas a conventional thermal power plant will cool down quite rapidly as soon as fuel combustion is stopped, in a nuclear power reactor, the highly radioactive fission products in the core will continue to release important though decreasing quantities of energy (in the form of heat) after the chain reaction has been brought to a halt. Were it no longer possible to extract this energy due to loss of coolant or failure of coolant circulation, the reactor could eventually melt, which could lead to a release of radioactivity into the atmosphere. The molten core might even "melt its way" into the reactor foundations and the ground beneath, and so pollute the surrounding soil. In the United States, this hypothetical event, with the core sinking into the ground in the direction of antipodes and China (!) has been given the colorful name, "the China Syndrome," of which there will be more to say later in this book.

This was all that was needed to rally to the antinuclear cause the self-appointed champion and defender of consumer interests — Ralph Nader, who, with his powerful organization, was already well known for his victory over the American automobile industry. For his technical arguments, Nader turned to a group of dissident scientists founded in 1969 in an attempt to prevent America's great university research institutes from working for the national defense and the Vietnam War. Nader made a television appeal to millions of viewers and did not hesitate to accuse the government of allowing an industry to develop that could wipe the United States from the face of the earth.

Surprisingly, all of this agitation had not yet succeeded in shaking the electricity producers' faith in nuclear power. The orders for plants that had been placed in 1972 and 1973, the finest years of the nuclear boom, remained as proof. Nevertheless, the protest movement, which was directly responsible for the constant questioning of nuclear regulations and procedures and thus for increases in plant construction delays and costs, was soon to exhaust the patience of industry and cause considerable harm to the program for nuclear electricity production.

One of the first victims of this antinuclear protest movement was the powerful U.S. AEC itself. It was accused, not without reason, of being judge and jury in controversial matters; both promoter of nuclear power and controller responsible for its regulation. The days when the commission's chairman was, as Lewis L. Strauss had been, one of the most influential people in the country had long since gone.

In 1961, President John F. Kennedy had, for the first time, nominated a scientist to head this organization: Glenn T. Seaborg, the discoverer of plutonium and an ardent champion of the new energy source, both in his own country and throughout the world, which he traveled a great deal. He remained U.S. AEC chairman for some 10 years, during which time the considerable influence enjoyed in Washington by his predecessors, due to their being responsible for the early manufacture of nuclear arms for the greatest power on earth, gradually slipped from his grasp. His two last successors, James R. Schlesinger and the energetic and resolute biologist Dixie Lee Ray (the only woman to be appointed to the AEC's chairmanship) were, despite being more politically inclined, unable to regain lost ground. The U.S. AEC was destined to break up and disappear.

The American atomic organization was not the only one in the world to suffer the consequences of the accelerated industrialization of nuclear energy. During the same period, the roles and influence of several European atomic commissions were similarly challenged. On the other hand, the antinuclear movement took four or five years to spread across the Atlantic from the United States, appearing as a coordinated lobby only since 1974. The protest movement had been preceded by several years by a broad penetration in Europe of the light water reactor. Westinghouse and General Electric had succeeded in establishing this penetration on a large scale some 10 years after the report of the "Three Wise Men" and the conclusion of the USA-Euratom Agreement — one of whose objects had been precisely to facilitate such a penetration. But this success story was not without its drawbacks, for it bore within it the seeds of problems that were soon to confront the American nuclear industry. For, having until then enjoyed an undisputed monopoly on the light water reactor (LWR), that industry was granting licenses and so contributing to the creation of its own competitors in the markets of the world.

The Spread of the American Reactor Type

The enthusiastic confidence shown by the electric utilities of this great industrialized country in pressurized or boiling water reactors for cheap electricity production inevitably had repercussions outside the United States.

Countries preparing to undertake nuclear electricity production found themselves confronted by several available technologies, each promoted in high pressure sales campaigns. It was entirely logical that they should be attracted to the system chosen by American industry since, even if the reactor type were not intrinsically the best, the considerable operating experience acquired from many units of varying capacities throughout the United States would undoubtedly make it the most thoroughly proven system and hence the least risky to adopt.

So it was that, some two years after the start of the American nuclear boom, LWR fever spread through the rest of the world, particularly Europe and Japan. The main commercial beneficiaries from this were Westinghouse and General Electric, either through direct exports or through licensing agreements in many advanced nations.

Nevertheless, these two American giants did not have a total monopoly of their technologies. The Soviet Union was independently building pressurized water reactor (PWR) power stations and the Swedish company ASEA-Atom was building boiling water reactors (BWRs), while in Germany, as early as 1970, Siemens ceased to be a Westinghouse licensee and became a serious rival in the PWR market. As a result, about half of the German nuclear power plants and a slightly higher proportion of the Swedish market were lost to American industry, which nevertheless enjoyed, directly or through its licensees, some 80 percent of the construction contracts outside the United States for light water power plants during the decade from 1964 to 1974.

During this decade, essentially during its second half, about 50,000 MW(e) of LWR-type generating capacity were ordered in the Western World outside the United States. This total, four times that ordered for any other reactor type, involved 65 power stations. The first 15 of these, low capacity units with a total output of 7,000 MW(e), were already in operation by 1974 as the delays between the placing of orders and commissioning were less than in the United States.

American industry, which had won its first laurels in Germany, Italy, and France between 1959 and 1961 under the flag of the USA-Euratom program, returned in force to these countries with more powerful reactors in 1964, 1969, and 1970, respectively.

India was won over in 1963, then successively over the next few years Spain, Switzerland, and Japan; Belgium and Sweden in 1968, South Korea in 1969, and finally in this decade from 1964 to 1974, Brazil, Mexico, and Taiwan.

Two of the countries importing American power stations should have special mention: Sweden and Italy.

Sweden had initially chosen the heavy water, natural uranium reactor type, this being the basis of a national fuel cycle allowing the country to keep open the military option for a long time. That option had been the cause of much internal disagreement in each of the principal political parties. To avoid an open split, the most powerful party, the Social Democrats, had adopted a substantial civil program that allowed a decision on nuclear armament to be postponed.

Curiously enough, another factor in the early 1950s that helped nuclear electricity production in Sweden was the attitude of the ecologists, who at the time were opposing the construction of dams and hydro-power plants in the northern wilds of their country. Ten years later these same ecologists were among the most bitter enemies of nuclear power.

The heavy water system was finally abandoned in 1970, following completion at Marviken of a 100-MW(e) power station of this type. The plant was never commissioned, a reevaluation of its parameters having raised fears that serious instabilities might occur during operation. The plant had been conceived and planned by Sweden's national atomic organization and its construction, by the firm ASEA-Atom, had cost $100 million.

Swedish industry promptly switched to LWR power stations, with ASEA-Atom building its own boiling water version while another firm specialized in pressurized water units under license from Westinghouse.

Italy's role in the early days of industrial atomic energy had been substantial, involving the construction of three nuclear power stations of different types between 1958 and 1961. This promising program had been launched by Felice Ippolito, secretary general of the country's National Nuclear Committee. Ippolito's tremendous personal energy was equaled only by his lack of regard for administrative regulations. This fault was to prove his undoing, for he became implicated in a highly unpleasant trial in 1964 that was marked by violent political passions, in particular, those concerning the nationalization of electricity in 1962, which he had enthusiastically supported. Condemned to a lengthy prison sentence, he was pardoned after serving three years and was later completely whitewashed by Italy's president.*

The principal victim of this sad and unnecessary affair was the development of nuclear energy in Italy, which stagnated, reflecting the general difficulties the country was experiencing. From 1961 to 1974, only one decision was made to build another nuclear power plant. This was in 1969 and was for a General Electric 850-MW(e) BWR, which was brought into service in 1978.

The U.S. nuclear industry's first and main competitor was to be German industry, one of the last in the field whose early large-scale power plants (built under American license) were commissioned in 1966. In 1969, the two leading German electromechanical groups, AEG-Telefunken and Siemens, set up a special organization to bid for complete power stations incorporating either conventional or nuclear steam supply systems. This

---

*In 1979, Ippolito, who had shown some leaning toward the Italian Communist party, was elected to the European parliament.

organization was able and ready to undertake for clients all planning, industrial architecture, and general construction responsibilities, together with all necessary subcontracting. Because of the substantial financial commitment needed to follow the very rapid evolution in the size of rotating machinery, i.e., turbines and alternators, for electricity generation, particularly in a nuclear plant, these two German companies had decided to collaborate so as to share the work and rationalize the investment.

As far as nuclear steam systems were concerned, it was the new company Kraftwerk Union or KWU that from now on prepared all bids, both for Siemens-type pressurized water systems and AEG-General Electric boiling water units, including first and replacement fuel charges, as well as for a variant of the pressurized, heavy water, natural uranium reactor developed by Siemens. In 1968, following an Argentine invitation to bid, this German industrial consortium won its first overseas contract with an offer of this last type of installation.

Kraftwerk Union quickly made its mark on the international commercial nuclear scene. Indeed, in 1969, it won a contract in the Netherlands in the face of American competition. It was also first to enter the Austrian market in 1972, and obtained an order from Switzerland the following year.

As for Japan, despite an almost total lack of conventional energy resources, i.e., coal and oil, six years went by between the country's first nuclear power station order for a British-type natural uranium, graphite-moderated, gas-cooled reactor at Tokai-Mura (which reached full power operation after many incidents and difficulties) and the start of a substantial construction program for American-type units. Some 20 orders were then placed between 1965 and 1974.

Electricity production and distribution in Japan were shared between nine private companies, and three large industrial groups became involved in nuclear power station construction: Mitsubishi with a Westinghouse license, and Toshiba and Hitachi with licenses from General Electric.

The introduction of the American reactor types was carried out in such a way that major responsibility was quickly taken over by the Japanese constructor concerned. For every installation at each power level, the American firm was the main supplier, subcontracting some of the manufacturing work to its Japanese licensee. For all subsequent identical units this situation was reversed, with the proportion of American-made components being progressively reduced.

Similar procedures were adopted in other countries to allow the national industry, in so far as it was able, to assume an increasingly large part in the construction of foreign power stations. The sought-after objectives included a reduction in foreign exchange expenditures and a lesser degree of dependence on the supplier of the technology. In extreme cases, a country that had imported one installation could then duplicate it entirely unaided. This is what India sought to do with the Canadian heavy water, natural uranium reactor, which by the late 1960s was the only type still in a position to compete with the light water, enriched uranium reactor.

In the Western World, there were at this time three advanced nuclear countries — Canada, the United Kingdom, and France — where penetration of the LWR type was still being resisted. Canada continued to support heavy water systems, while the United Kingdom and France maintained their interest in graphite-moderated reactors cooled by pressurized carbon dioxide gas.

France alone decided, from 1970 onward, to switch to the American system. The Canadians remained faithful to their heavy water reactors, which they had been perfecting since the war and in which they could at last see their efforts beginning to bear fruit, both nationally and in the export market. Great Britain, suffering from the gradual disintegration of its nuclear background and accomplishments, seemed unable to organize any sort of atomic future and even less, in light of the problems of its indigenous reactor types, able to adopt another system.

The Communist World

The Russians had independently also adopted the pressurized LWR, together with a system that they alone were exploiting comprised of a graphite-moderated reactor fueled with enriched uranium and cooled by boiling water in pressure tubes. By late 1973, there were seven 1,000-MW(e) power plants of this latter type under construction in the Soviet Union, with another of similar power in operation and recently connected to the Leningrad electricity grid.

As for pressurized water reactors, by late 1973, four units of 200 to 400 MW(e) were in operation at Novovoronezh, with one further 1,000-MW(e) unit under construction at the same site. The 440-MW(e) unit was also intended as an export model.

In addition, a prototype fast-neutron breeder reactor, built at Shevchenko on the Caspian Sea, had been commissioned at the beginning of 1973. This was the first breeder reactor in the world in the several hundred megawatt electric power range to undergo practical testing.

Up to the end of 1973, the Russians had as yet made no attempt to transfer their nuclear technology to countries beyond their sphere of influence. Thus, there was little chance of competition between them and the Americans in the industrial field: Soviet exports were limited to the satellite states and Finland. There were, at that time, three Russian nuclear power plants in operation in East Germany with two more under construction there; two further units were being built in Bulgaria, and one each in Hungary, Czechoslovakia, and Finland. This last merits special mention.

In 1966, the Soviet Union had persuaded Finland to abandon a call for bids that was expected to lead to a final choice from the three American, German, and Canadian proposals that had been shortlisted. The Russians had then offered to supply one of their own 440-MW(e) power stations at a fixed all-inclusive turnkey price with an excellent payment plan. Construction of this plant began in 1970. Following this, the Russians raised no objection to the choice by Finland, in 1973, of a Swedish ASEA-Atom boiling water unit for its second nuclear power station.

Finally, in 1960, the Russians had broken off relations with China — now a hostile, inward-looking country in the throes of cultural revolution. In the nuclear field, China appeared to be exclusively concerned with its military program, the basis of which — research and the development of uranium resources — was later to serve it well in the inevitable progression toward the civil industrial phase.

From the political viewpoint, countries using Soviet reactors were entirely dependent on the Russians for their enriched uranium fuel. This was supplied to them in the form of assembled rods ready for use in the reactor; after use, the irradiated rods were returned to the Soviet Union. This really meant that the fuel was "rented" for the duration of its use in the power station — an ideal arrangement from the point of view of nonproliferation since the plutonium formed in the fuel could only be extracted in the Soviet Union. Hence, the Russians' "customer countries" had no real part in the nuclear fuel cycle as such, particularly in the sensitive stages of reprocessing and enrichment. In addition, they were apparently obliged to sell to their great protector any indigenous uranium they might find in their own territories.

This situation was reflected in a witty and penetrating remark by Vasily Emelyanov, chairman of the Soviet Union's State Committee for Atomic Energy, who had told us in the early 1960s: "Nonproliferation is not a problem, everyone should take care of his own."

In contrast to the Americans, the Russians, thanks to their system of renting out fuel elements, had no need for international safeguards, which in fact were applied to nuclear plants in Soviet allied countries only at a much later date, following their adherence to the Non-Proliferation Treaty (NPT).

If countries within the Soviet sphere of influence were thus completely dependent on Moscow for their nuclear energy supplies, they were no less so for their oil. It was this complete dependence that led to a remarkable incident in 1967, when the Commissariat à l'Energie Atomique's (CEA's) general administrator and myself each received a magnificent box of cigars from none other than the prime minister of Cuba, Fidel Castro. It seemed he was considering buying a nuclear power plant so as to become less dependent on the arrival in his country of a weekly oil supply via Soviet tanker, and he wished therefore to make contact with French specialists. The affair came to nothing, in particular, because it was unthinkable to provoke the anger of Washington, which since 1962 had been especially sensitive over any nuclear activity on Castro's island.

There remained the question of what to do about the gifts. The matter was the subject of many high level discussions. Finally, we were allowed to thank verbally the Cuban embassy in Paris and to keep the cigars, which for a long time were objects of admiration by connoisseurs. Eleven years later, Castro announced that he would be ordering the first Cuban nuclear power station from the Soviet Union.

The only Eastern country to attempt nondependence on the Soviet Union for its nuclear power program was Rumania, whose oil requirements happened to be partially met by indigenous resources. From the early 1970s it showed an interest in attempting an autonomous program based on natural uranium and the Canadian reactor system.

Canadian Perseverance

The Canadians, who since World War II had been faithfully persevering with no outside help toward perfecting their heavy water reactor, were at last to see their efforts rewarded both at home and in the export market. However, from the point of view of nonproliferation, this Canadian export trade inevitably presented a different aspect from that of the Russian or even the American sales abroad.

The first Canada deuterium uranium (CANDU) reactor, with a power of 200 MW(e), was brought on line at the beginning of 1967, considerably behind its planned schedule. It was to be used to develop, under the same name and under the dynamic direction of the physicist Wilfred Lewis, Canada's "national" reactor type. Work was in fact already under way near Toronto on four units of some 500 MW(e) each. These were commissioned between 1971 and 1973 under particularly efficient conditions; work on another four units, this time 750 MW(e) each, was begun in 1971 and 1972.

The success of the Canadian program was partly due to the very good relationships between the country's official nuclear organization, the Crown company Atomic Energy of Canada Limited, and its first customer, the electric utility Ontario Hydro, working together with a national industry which at that time was relatively weak.

The first Canadian exports were ordered in 1964 and 1965 before completion of the first prototype. Two reactors of the same power as the CANDU prototype were to be built in Rajasthan, India, and one lower power reactor was ordered by Pakistan. In 1973, both Argentina and South Korea decided to buy 600-MW(e) Canadian units for their second nuclear plants. In addition, Taiwan ordered a large experimental reactor of the same type.

All of these contracts were subject to IAEA safeguards. In the Indian case, however, the Canadian government was confronted with insistence on reciprocal controls. Canada was obliged to open its own partly completed CANDU plant to regular inspection by the Indians in exchange for similar Canadian control on the Indian plant. The Indian plant was sold thanks to this strategem by which the discriminatory aspect of safeguards being applied only to the importing country was eliminated. Ottawa subsequently insisted that this mutual control should be transferred to the IAEA.

Three years later, following General de Gaulle's call "Long live Free Quebec!," the first Franco-Canadian agreement was concluded at the federal level. This agreement nearly fell through because of the Indian-Canadian mutual inspection clause mentioned above.

The agreement concerned the sale of some hundred kilograms of plutonium from the CANDU reactor, which were intended for use in France's first fast-neutron reactor. Since concentrated fissile material was involved, Paris accepted a safeguards clause that was to be implemented by Euratom inspectors. When the contract was about to be signed, however, it was realized that Indian consent was indispensable since the plutonium had been produced in a Canadian reactor subject to Indian inspection under the mutual safeguards arrangement. In addition to the Euratom safeguards, which were not recognized by New Delhi, it was necessary to conclude a three-party agreement regarding safeguards to be applied in France by inspection teams from Ottawa that could include technicians from India. The complex nature of this affair, which after all concerned a minimal quantity of plutonium in relation to overall French requirements and production, illustrates the complicated and almost ridiculous consequences of inflexible application of a tangled network of international controls.

Strangely enough, these excessive Canadian stipulations were accompanied by a certain laxity elsewhere. Ottawa, wishing to promote its heavy water reactor, had agreed that the Indians should be allowed to build a series of installations identical to the two in Rajasthan, thus enabling Indian industry to free itself gradually from the need to buy components from Canada. The Canadians had accepted that these replica installations would be free from any safeguards, meaning that in practice the plutonium they produced could be used freely for any purpose.

The president of Atomic Energy of Canada, Lorne Gray, was an indefatigable world traveler in search of customers for his company's CANDU reactor. It is probable that many orders would never have been won had it not been for his efforts and dynamic approach; and it seems highly regrettable and unjust that, at the end of his career, he should have been accused in connection with his Argentinian and Korean contracts of under-the-table payments, made with the full agreement of his responsible minister.

Canadian-type reactors are unaffected by the controls applying to the supply of enriched uranium and their importers often have their own supplies of natural uranium to fuel their plants, or can hope to obtain this from elsewhere with no restrictions as to use. Moreover, the fuel can be loaded and unloaded while the reactor is operating, making it possible to produce only slightly irradiated plutonium (i.e., of weapons quality) without involving costly shutdowns such as are needed with LWRs. Finally, the countries that have imported Canadian reactors, such as India, Pakistan, and Argentina, have shown themselves most hostile to international safeguards and the NPT, or else, as with Taiwan and Korea, they have been in geographical areas of special political sensitivity. This was also true to some extent of Rumania, a country where the importation of CANDU-type power stations was being considered as early as 1968: a demanding and hesitant potential customer, patiently wooed by Ottawa.

This selection of customer countries is certainly not to be blamed on the Canadian government, which naturally would have preferred to find more buyers among nations that were closer, more stable, and better disposed toward safeguards. Yet Ottawa, while representing itself before the world as a champion of nonproliferation and of safeguards policies, could not have been unaware that the desire for nuclear independence, coupled with the use of natural uranium power plants, more often than not indicated military motives. It must have been no less obvious that those countries that had genuinely abandoned the nuclear weapons option preferred to buy LWRs.

It is legitimate to question whether the extreme positions adopted by Canada, particularly in matters relating to nonproliferation and the export of uranium, were not to some extent a kind of defensive reflex against criticisms that, from this point of view, could be leveled against the heavy water reactor and the conditions under which it has spread outside Canada.

The British Decline

In April 1964, the British government announced its second nuclear program. The program was still based on a graphite-moderated, gas-cooled type of reactor, but one using enriched uranium fuel. The government confirmed that building natural uranium "Magnox" reactors, the basis of the first program launched in 1955, was to be discontinued in 1971 when installed generating capacity would have reached a total of 5,000 MW(e). Although these Magnox stations had proved more costly than expected in terms of investment capital, their operation had been satisfactory from both the economic and technical points of view.

The new program's initial objective was a further installed capacity of 5,000 MW(e). In 1965, this goal was increased to 8,000 MW(e). The new stations were to come into service over a 5-year period from 1970 to 1975.

Construction of the first station, Dungeness "B," comprised of two 600-MW(e) reactors, was to take over 15 years and become symbolic of the decline of British nuclear development, which in the 1950s had earned an enviable reputation thanks to so many "world firsts" in both research and prototype units. Not all the stations of the second program suffered the long delays of Dungeness "B," but none was completed in under nine years and none was connected to the grid before 1976. There were numerous technical, administrative, and industrial reasons for this decline.

The new program was based on the satisfactory performance since 1963 of a 30-MW(e) experimental prototype reactor at Windscale, unhappily called "advanced." The epithet was adopted for subsequent much larger reactors of the type known as advanced gas-cooled reactors (AGRs). They used slightly enriched uranium oxide fuel in a stainless steel cladding.

The United Kingdom's basic mistake was to go in a single step from the experimental unit to one 20 times more powerful and to start building a series of such reactors without a proven prototype of significant size. This chancy move coincided with a problem that was developing in the earlier Magnox plants in which the hot pressurized carbon dioxide gas was evidently corroding some of the mild steel internal components more rapidly than had been expected. Reactors of the first program already in service therefore had to be operated at 20 percent less power, while the design and characteristics of units still under construction had to be revised.

A second problem was administrative. It concerned the working relationships between the main partners in the nuclear programs: the Atomic Energy Authority (AEA); three British plant-constructing industrial groups; and their two customers, the Central Electricity Generating Board (CEGB) and the South of Scotland Electricity Board (SSEB). These relationships had been deteriorating for some time, and had suffered particularly in 1963 when the CEGB did not choose the industrial consortium whose turn it was to build the final station of the first nuclear program, subsequently showing a degree of reticence toward the choice of the AGR. The CEGB would have preferred to adopt American light water technology, which among other things would have somewhat reduced the influence of the AEA.

The choice of the AGR for Dungeness "B," the first power station of the second program, was confirmed in mid-1965 following examination of bids from not only the British consortia but also from the three leading American construction groups. The decision was formally announced to the House of Commons by the energy minister who declared that the new British reactor type, besides its clear technical advantages, would show a savings of 10 percent in U.K. building costs compared with any of the other types proposed. Nevertheless, the minister added, these other proposals were not being discarded and might well be chosen for future units in the same program. Nothing of the sort happened. By the end of 1973, there were ten 650-MW(e) AGR units, grouped in pairs at five sites and all unfinished; work on them had been started progressively between 1966 and 1971.

Although all were AGRs, there were in reality three different models being built by the three industrial consortia still in the field in consultation with the industrial division of the AEA as the main source of technological data and wisdom.

No one at the time denied that the program was insufficient to keep all three consortia fully occupied. In addition, the groups were suffering from internal divisions and difficulties due to the weakness of some of their partners. Export prospects were virtually nil; one passing hope quickly died when Britain was momentarily offered the chance to supply a first nuclear power station to Greece, in exchange for . . . tobacco, unfortunately of a flavor unattractive to British smokers.

In the face of the difficulties encountered with the first AGR units, the British nuclear program and the industrial structure behind it became subjects of a series of multiple official inquiries, one conducted by a parliamentary committee, another by a body concerned with industrial reorganization, and a third by a senior official of the ministry responsible for atomic energy. Not surprisingly, all concluded that the industrial consortia should be concentrated first into two groups and subsequently into a single one. This concentration was effected in 1972, not without traumatic results for current projects. The reorganization had been preceded the previous year by radical changes in the AEA, notably ''detachment'' of the production branch to form an industrial and commercial company known as British Nuclear Fuels Limited.*

The single industrial group remaining, now known as the National Nuclear Corporation (NNC), was to be dominated by Sir Arnold Weinstock. He was a powerful, effective, and redoubtable industrialist who favored the American PWR. However, to further complicate matters, Sir John Hill, chairman of the AEA (which had a 15 percent holding in the NNC), was in favor of a new type of heavy water, enriched uranium reactor cooled by boiling light water, of which a 100-MW(e) prototype had been operating satisfactorily at one of the AEA's centers since 1967.

The general public, confused and unable to follow the details of the affair, was no longer convinced of the need for nuclear power now that gas and oil deposits had been found under the North Sea, at a time, moreover, when there was an evident overcapacity in the country for electricity production. In such circumstances, it was not surprising that 1972 and 1973 passed without a government decision on what type of reactor to order to complete its 8,000-MW(e) program.

The easiest solution, which had the benefit of trade union support, would be to continue with the AGR, which in the end would no doubt be perfected. Heavy investment in any new reactor type, for what could only be a limited number of power stations, certainly seemed hard to justify.

Nevertheless, the AEA and SSEB leaders maintained their support for the heavy water system (in which the Canadians were also interested) while the two chairmen of the NNC and the CEGB wanted to adopt the pressurized light water system — a course that might offer the possibility of association with the French program, and hence a gradual disengagement from the Westinghouse license.

The year 1973 ended with an inconclusive hearing before the Select Committee on Science and Technology of the House of Commons. Disagreeing with his SSEB colleague, who still favored the new nationally developed heavy water technology, the chairman of the CEGB publicly declared, for the first time, his preference for power stations using the truly proven LWR. He added that it should have been adopted years ago, but revealed that the CEGB had long been under pressure to forgo it, with all its

---

*The weapons branch of the AEA was similarly separated and attached to the Ministry of Defense in 1973.

advantages, in favor of the AGR. He described the current AGR program as catastrophic — a severe criticism but not an unjustified one.

Friction Between the CEA and EDF

Like their British counterparts, those responsible for France's nuclear program were slow to appreciate the full extent of the problems of the industrialization of nuclear energy. This new stage in the nation's atomic development was to provoke a crisis similar to that already gripping the United Kingdom. The causes were also similar: the evolution of the civil program, the relationship between Electricité de France (EDF) and the CEA, and the future role of the CEA in national nuclear affairs. Happily, the French had more success in overcoming the crisis than did their former atomic mentors across the channel. The price they paid included the abandonment of natural uranium reactors in favor of the light water, enriched uranium type, as well as some basic changes within the CEA.

In 1964, the only commercial nuclear power stations in operation in France were the first two EDF plants, with capacities of 70 and 200 MW(e). The next two, EDF 3 and EDF 4, with capacities of some 500 MW(e) each, were commissioned in 1966 and 1969. A special feature of EDF 4 was its integrated design, the reactor and heat exchangers comprised a combined unit inside a reinforced concrete containment.

It was characteristic of the relationship between the CEA and EDF that the fortunate decision to adopt this layout was made despite CEA reticence. The latter, wishing to avoid unproven design changes and hence the costs and time needed for plant installation, was opposed to EDF's tendency to make important technical advances from one project to the next in the hope of reaching competitivity more quickly. The CEA preferred the more cautious policy of duplicating existing units so as to keep down costs and construction delays.

During this same year of 1964, the startup and good performance of the land-based prototype French submarine engine demonstrated the CEA engineers' mastery of PWR technology. Also, EDF was acquiring this same technology by building, as a joint Franco-Belgian project, a Westinghouse-type PWR power station at Chooz, near the Belgian border. This station was commissioned in 1966.

Two other reactor types were being developed in France. The first, known as EL4, was a prototype CEA-designed heavy water, natural uranium unit, cooled by pressurized carbon dioxide. It was being built by EDF and the CEA in the Monts d'Arrée in Brittany; it became operational in

1967. The second was an experimental sodium-cooled fast-neutron reactor, "Rapsodie," which was completed in 1965 at the CEA's Cadarache research center in France.

The Pierrelatte enrichment plant was still under construction; it was not to be completed until 1967 when it began producing concentrated fissile materials for the first H-bombs and operational submarine engines, as well as the technology for a large-scale enrichment plant for civil purposes.

All of these projects originated in the 1957 5-year plan. In 1964, the Consultative Commission on the Nuclear Production of Electricity (CCNPE) proposed the start of work on a series of power stations with a total generating capacity of 2,500 to 4,000 MW(e). These stations were all to be based on natural uranium, graphite-moderated, gas-cooled reactors. There was no question at this stage of turning to enriched uranium as the British were doing, since this would have made the French nuclear electricity program dependent on American fuel supplies. Moreover, the CEA was proposing further development in naural uranium fuel that offered promise if used in larger reactors of from 400 to 500 MW(e). These new fuel developments offered better performances than with the British Magnox stations and even a hope of achieving economic competitivity with conventionally generated electricity.

In 1965, EDF decided to build two new nuclear power stations for entry into service in 1971. One was to be identical to EDF 4 and was to be built on the same site as this former plant at Saint-Laurent-des-Eaux on the Loire River. The second new station, with a 540-MW(e) capacity, was to be built at Bugey, near Lyons, and was to use the improved annular form of fuel with both external and internal cladding and cooling. This new fuel was intended to enable higher specific powers to be obtained by raising the coolant pressure; it was hoped that this could eventually lead to a new series of power reactor units producing up to 1,000 MW(e).

But other developments were to intervene. These two power stations, to which must be added the similar plant at Vandellós in Spain, were in fact the last of their type to be built. All three were brought into service in 1971 and 1972. It was ironic that the much sought-after export to Spain should prove to be the last power station to use what had come to be known as the "French type" of reactor.

It was also paradoxical that French industry had had to await the construction of this last plant, the Vandellós power station in Catalonia (on which work began in 1967), to escape for the first time from the state of fragmentation in which it had been kept as a result of EDF's role as national industrial architect. An ad hoc group of French construction companies was set up to build the Spanish plant, for which the fuel elements were to come

— as with all French power stations — from the CEA. Vandellós was to benefit from very advantageous financing arrangements, including EDF's provision of one-quarter of the capital investment in return for the same proportion of the electricity to be produced.

It was now — at the moment when the United States was caught up in a veritable nuclear frenzy; when Germany, helped forward by a powerful and united industry ready to compete with the Americans, was putting new effort into an expanding nuclear power program; with Japan doing the same and with Britain apparently pumping new life into its graphite-moderated, gas-cooled reactor; and with Canada beginning to appear in world markets as champion of the heavy water system — that French nuclear development seemed to be struck down by a paralysis that was to last for some five years.

It must be admitted that proponents of the French national reactor program were handicapped by a number of developments: the spectacular fall in the price of fuel oil, which dropped by almost half between 1964 and 1968; rising interest rates, which directly affected the already relatively high investment costs for natural uranium power stations; a general impression that the torrent of orders for American plants was bound to make the LWR the least risky investment and therefore the best in the world; and, finally, the hope that a supply of enriched uranium could eventually be available from a European plant modeled on the technology of Pierrelatte, if such a joint enterprise were set up.

All of this helped create uncertainty as to which direction should best be followed, an uncertainty that was quickly reinforced by very real concern following the poor initial performance of the three power stations completed in 1966, which represented the three reactor types considered to have reached the prototype stage in France: the American-type plant at Chooz, the EL4 heavy water plant, and especially EDF 3. This last station was suffering from failures in the detection system for damaged fuel cladding, from heat exchanger and turbine troubles, and to some extent even from poor fuel element performance.

As always in matters concerning nuclear energy, the press amplified these unfortunate incidents and the CEA, instead of supporting EDF in the face of public opinion, showed a tendency to criticize the national electricity undertaking in order to justify its own technical choices. This provoked EDF to blame its misfortunes on the technical advice given by the CEA. Industry was not left out of the dispute, and each of the three main contributors to the French nuclear program was accused by the other two of being responsible for the situation.

The CEA was criticized for its large size, for its military program, and especially for its choice of the natural uranium reactor type. At the same time, EDF was attacked for its audacity in risking too much technical innovation between one power plant and the next, for wanting to play the role of industrial architect and for having, in the interests of competition, initially spread its industrial orders too thin. As for industry itself, it was easy to criticize its lack of cohesion (compared with German industry), its dislike of taking risks, and its lack of care and strictness in dealing with orders. The fact that the CEA on the one hand, and EDF together with industry on the other, were the responsibilities of two separate government ministries did not ease the path of compromise and reconciliation.

The problem eventually reached the level of the head of state when it was raised during a restricted council meeting in December 1967. The moment was not very well chosen for EDF was just recovering from a difficult strike and its new director general, Marcel Boiteux, had only recently been appointed and therefore had played no part in the nuclear decisions and projects in dispute.

The two principal contestants were the minister for industry, Olivier Guichard, who was responsible for EDF, and Maurice Schuman, delegate to the presidency of the council, who was the CEA's guardian and protector. Both were close and faithful followers of General de Gaulle, but the latter had the advantage of having been with the general from the day he arrived in London in June 1940 to call the French people to fight on despite the invasion of their country. Hence, Schuman won the day. De Gaulle — after severely reprimanding EDF Chairman Pierre Massé, who defended his position with conviction — accepted the arguments for the national natural uranium, graphite-moderated, gas-cooled reactor and decided on the construction of two twin units with a greater capacity than the EDF 4. Construction was to start in 1968 at Fessenheim on the Rhine in Alsace.

However, EDF had obtained authorization to take a 50 percent share in the construction of a 900-MW(e) pressurized water power station at Tihange in Belgium, a Franco-Belgian project similar to that at Chooz. It was also expected at the time that EDF would participate in a Franco-Swiss boiling water plant to be built at Kaiseraugst, on the Rhine not far from Basel. Construction work on this had not yet begun. Determined to enter the LWR field, EDF was hoping to maintain a minimum of competition in the field by investing in these two types of American power reactors. In the traditional manner, this would enable EDF to keep a firmer hold on prices, thereby keeping them down.

General de Gaulle's choice of the graphite-moderated, gas-cooled reactor was, for the CEA and other proponents of the technology, a Pyrrhic victory. The Fessenheim power plants were not ordered until late 1970 and were — in the end — built with PWRs under the Westinghouse license, for

EDF had meanwhile been authorized to discontinue the graphite-moderated, gas-cooled system. During this time, the CEA went through a serious crisis that it barely survived, and not without losing some of the power it had held since its creation.

One of the unexpected results of this period of upheaval for the country's nuclear affairs was a gradual reconciliation of the EDF and CEA points of view. This occurred through the constitution of combined working parties and in the course of a joint study undertaken by their nuclear experts. This study concluded that a light water power station should be built in France without delay. The CCNPE also approved this approach; indeed, in April 1968, it estimated the cost of a kilowatt-hour from an LWR plant at 10 percent less than that from a "French-type," i.e., graphite-moderated, gas-cooled power station.

The CCNPE in fact recommended the early construction of an American-type nuclear plant, as well as a preproject for a heavy water power station, which it also found of great interest. The commission was of the opinion that, without prejudice to possible future developments, it would be inadvisable to undertake — at least before the end of 1970 — construction of any further graphite-moderated, gas-cooled units beyond those already "decided" for Fessenheim.

The student-incited political upheaval of May 1968, the financial difficulties that followed, and General de Gaulle's departure from office the following April all helped toward the postponement of construction of these Fessenheim units.

In October 1969, the director general of EDF presented the completed fourth power station of the French nuclear program, EDF 4, to the press. Built at Saint-Laurent-des-Eaux in the Loire Valley, it had a capacity of 500 MW(e).

The day of the unveiling was one of paradoxes, beginning with EDF Director General Boiteux declaring that, as things stood, nuclear power was not economically viable and could not compete with power from conventional plants burning fuel oil, which could now produce a unit of heat at 60 percent less cost than 10 years earlier. (He had no way of knowing that 11 months later the situation would change radically following Libya's approval of increased oil prices.) Nevertheless, the director general was convinced that nuclear power had an inevitable role to play in the long term, and that construction of atomic power plants in France must therefore be continued, for, in his words, French industry had to be given a chance to develop its "nuclear muscles" in preparation for the day when atomic power would be both economic and essential.

Referring specifically to EDF 4, Boiteux acclaimed the technical success it represented for a reactor system now thoroughly developed and proven. But at this very moment, when the latest graphite-moderated, gas-cooled power station was giving every satisfaction, he pronounced its death warrant by declaring unambiguously that EDF favored American-type light water plants.

Finally, and again paradoxically, on the very evening of EDF 4's inauguration, an inexcusable action by one of the plant operators, carried out despite an alarm light warning against it, caused a fuel rod to melt and put the installation out of action for a year. Its performance had been faultless up to that day of glory and of shame.

This event proved to be the last overpublicized breakdown* in a power plant of the graphite-moderated, gas-cooled type. These plants were now to slide quietly into the anonymity of invaluable if unspectacular operation throughout the 1970s, particularly from the moment when the 1973 oil crisis made economic comparisons between French and American reactor systems no longer significant.

Finally, a restricted council meeting under the new president of the French Republic, Georges Pompidou, held during the month following the inauguration of EDF 4, confirmed the discontinuation of the graphite-moderated, gas-cooled reactor. A communiqué from the Elysée stressed France's need to maintain her mastery in all branches of nuclear technology, thus guaranteeing in large degree her energy independence and her industrial power. The communiqué underlined the priority given to the development of the fast-neutron reactor and announced that as of 1970 EDF would undertake a program of diversification covering several types of power stations fueled by enriched uranium.

The page had been turned. The French government had crossed a threshold, avoided the British mistakes, and made the essential decision . . . to end the indecision.

It must be emphasized that the choice of the LWR was justified neither by a better performance of the prototypes (the commissioning of the first Franco-Belgian PWR was followed by a two-year shutdown) nor by any important economic superiority, but essentially by the advantage of benefiting from operational experience that was to reach worldwide proportions.

The Canadian heavy water, natural uranium reactor had for a time been considered as a possible alternative solution. In 1969, it had been supported by Robert Galley, the last minister for scientific research before this was placed under the aegis of the Ministry of Industry. A detailed

---

*In 1980 an accidental meltdown of a fuel element of a twin reactor of EDF 4 put the unit out of action for two years.

proposal for the construction and operation of such a nuclear power plant was prepared by French industry, but at the end of 1970 the CCNPE rejected the corresponding commercial proposition, not on the grounds of its slightly higher costs compared with a light water system, but rather due to lack of operational experience with the reactor type.

Thus the LWR had overcome its last obstacle in France, and EDF, having freed itself from the CEA's tutelage, was about to form a much closer relationship with its former guardian, whose technical support and advice were to be welcomed in the future as work proceeded on the American-type power stations.

Both the reactor war and the conflict between the CEA and EDF were over.

The New French Program

As early as 1970, EDF ordered its first pressurized water power station, with a generating capacity of 900 MW(e), from the Schneider Group. Their subsidiary, Framatome, held the Westinghouse license. At the end of that year, on the recommendation of the CCNPE, the decision was made to build more of these plants over five years to a total capacity of 8,000 MW(e). Two of them were ordered in late 1971 from the same constructor, despite considerable pressures in favor of the BWR. In 1972, Framatome was reorganized, giving Westinghouse a 45 percent holding and Creusot-Loire 55 percent.

Early in 1973, rising oil prices gave warning of oncoming trouble, and the new French nuclear program was accelerated for the first time. In the autumn, 17 years after the forecast in the report of Euratom's Three Wise Men, the oil crisis burst on the world.

Arriving on an already troubled international political scene and at a time of financial imbalance, the decision of the cartel of oil-producing states to steeply raise their prices, while those in the Middle East also reduced their production and began selecting their customers, reopened the question of economic choice between different sources of energy supply. For France, nuclear energy, now even more competitive, became the only possible resort. The prime minister immediately decided to increase the rate of ordering plants to between five and six 900-MW(e) units per year. With government support, a last attempt was made to push the boiling light water system. An initial order for two units, with the possibility of further options, was placed with the Compagnie Générale de l'Electricité (a licensee of General Electric) but the business fell through the following year. The official reason given for this failure was the considerable difference in price as compared with the Framatome PWR system.

Paradoxically, the choice of the American reactor system, despite its two variants, was to lead EDF to its most stable industrial policy of placing orders in series for the same unit of the same type with the same constructor. This was a far cry from their previous policy of ordering units of ever greater power from constantly changing suppliers, a policy that had been characteristic of the era of the graphite-moderated, gas-cooled reactor. At last, the industrial architecture of the country's nuclear heat supply was in the hands of a single national constructor. The way was open for France to take her place in the world nuclear competition.

In late 1973, EDF announced that it would build no more fuel oil- or coal-burning power stations, but would follow an "all nuclear" policy, with an annual rate of increase in generating capacity of some 5,000 MW(e) [total installed nuclear capacity at the time was some 2,000 MW(e)]. With each nuclear plant then taking about six years to build, the target date of 1985 for producing over 60 percent of all French electricity from uranium was ambitious. The nuclear tidal wave from across the Atlantic had arrived in France.

During this same year, the 250-MW(e) prototype breeder reactor "Phénix" was brought into operation at Marcoule. It had been built under the direction of Georges Vendryès, a physicist turned power station constructor and the inspiration behind the French fast-neutron reactor program. Though it was true that a prototype Soviet breeder reactor had been commissioned somewhat earlier, but its performance was still far from satisfactory, while a similar reactor in Britain, begun several years before the Phénix, was still unfinished.

In America, the decision to build a prototype breeder had only just been made. In fact, the U.S. AEC was still assembling a group of private industrialists willing to put up part of the necessary financial investment. A site had been chosen at Clinch River, near Oak Ridge, Tennessee, and the installation was to be built by Westinghouse. The Americans could not believe that others might be ahead of them. Their applied research program was accumulating basic technical data and they were convinced that this work, together with the great power of their nuclear industry, would put them ahead in the race when the moment came. They were nonetheless unhappy when the French refused to pass on to them, without due compensation, data from the successful operation of the experimental fast-neutron reactor Rapsodie, soon followed by similar data from Phénix. It was true that the Americans had published, without financial compensation, large amounts of noncommercial data on the LWR, but they had already been financially rewarded by the victorious spread of their technology and the exports of their industry.

The French lead in the fast-neutron reactor field had also resulted in requests for assistance from other countries interested in this technology: Japan, Italy, and India. India was at this time, apart from Spain with the Vandellós power station, the country where French nuclear industry was the most involved. Some years ago, in fact, in 1969 in Paris, it had been decided that French industry should build two heavy water production plants in India, employing an original process, and should also help the Indian Atomic Energy Commission with the construction near Madras of a research reactor similar to Rapsodie. This was to be fueled with mixed plutonium and highly enriched uranium, the latter supplied by France and the former by the Indians from their own fuel reprocessing plant that had been in operation since 1965.

Finally, following the successful commissioning and startup of the Phénix, EDF without further delay began canvassing the construction of a "precommercial prototype" at a site between Lyons and Geneva. The 1,200-MW(e) plant was to be known as "Super-Phénix" and to be built in association with a German electric utility, RWE, and the single nationalized Italian utility, ENEL. Such an international undertaking was a daring innovation for EDF; it required parliamentary approval.

The CEA Crisis

Satisfaction over the success of the Phénix and the reactor type it represented was all the greater for the CEA since, at the very moment when it seemed to have reached its zenith in 1968, it had only just escaped becoming the main victim of the reactor war.

Its growth record over the past 15 years had been impressive: 3,000 employees in 1954, 10,000 in 1958, and a maximum of 32,000 in 1968. The increase in its budget was no less spectacular, rising over the same period from some $30 million to $200 million (in current money of the years in question) and finally reaching nearly a billion dollars or 3.5 percent of the total national budget.

Covering the entire fuel cycle, from mineral prospecting to the manufacture of concentrated fissile materials, and possessing a wealth of special instruments and equipment for basic and applied research, mostly designed and built within its own laboratories, the CEA had been rewarded with numerous successes, whether in the satisfactory operation of prototype reactors and chemical plants for reprocessing and isotopic separation or in the performance of its submarine engines and various types of weapons.

Being a specialized organization and thus escaping from the inflexible rules of the traditional civil service to some extent, particularly in the financial field, the CEA excited both admiration and envy. Surrounded by a certain atmosphere of mystery and secrecy, to the outside world it could seem impenetrable, untouchable, and beyond all financial control. From this image to accusations that it was a state within the state was a simple step, frequently taken.

In conflicts with other ministries, the CEA had the advantage of being directly answerable to the prime minister, either through a minister of state or through a specially appointed ''minister-delegate.'' The task of such a representative was not without its difficulties, and sometimes the minister responsible for the CEA became all too well aware of the contrast between the power and wealth of the organization he was supposed to control, and the insufficient means at his disposal for exercising that control in so wide a field of complex activities.

As of 1962, the minister responsible for the CEA was also in charge of scientific research, and here too the contrasts between the budget allocations for this fundamental national activity and those for CEA research gave rise to some unhappy jealousies; there was a great temptation to put the nuclear research centers with all the other national laboratories under the same authority. These jealousies made it inevitable that disruptive forces should operate against the CEA and that there should be attempts to split up its ''empire.''

The first attempt at toppling the ''empire'' occurred in the early 1960s in the military field. The minister for the armed forces, Pierre Messmer, would have liked to absorb the CEA's Military Applications Division under his authority. This division's program, more costly than the CEA's civil program, was funded by transfers from the armed forces' budget. However the CEA, in using these funds, did not act as a subcontractor but as a partner with equal (or even greater) responsibility. The minister's attempt was unsuccessful because General de Gaulle was determined, especially during the delicate period when the H-bomb was being designed, to protect the organization he had created in 1945.

There were others who would have liked to see the CEA's industrial activities placed under EDF control and its applied research carried out by private industry under contract, a formula used in the United States between the AEC and many industrial firms. It was in this area that the worst tensions developed; the CEA-EDF conflict, brought before the head of state at the restricted council meeting in December 1967, and resolved in favor of the CEA, had left wounds that were slow to heal.

The agitation that occurred at the Saclay research center during ''the events'' of May 1968 presented opponents of the CEA's monolithic structure with new arguments for change. Then, the successful demonstration the following August of French accession to thermonuclear weapons removed the most powerful argument for maintaining the status quo. Finally, General de Gaulle's resignation eight months later, in April 1969, robbed the CEA of its paternal protector.

Two months later, during the formation of the first government under the Pompidou presidency and on the initiative of Guichard (who had been industry minister at the time of the stormy 1967 restricted presidential council on reactor types), the CEA and EDF were brought together under the latter's former ministry, which now became the Ministry for Industrial and Scientific Development.

At the beginning of 1970, the holder of the new ministerial office, Francois Xavier Ortoli, gave the CEA's general administrator, Robert Hirsch, the distressing task of reducing the organization's staff by almost 10 percent over two years. At the same time he called on Charles Cristofini, a senior civil servant with special knowledge of industrial affairs, to study and report on the role and future of the CEA and on its reorganization, possibly by developing forward management techniques with a view to increasing efficiency and improving the exploitation of results.

The Cristofini Report, produced in April 1970, rejected the idea of breaking up the CEA but advocated reform of its statute. As a result, in September 1970, the government modified the 1945 ordinance, ending the dual direction of Francis Perrin and Hirsch among other things. The scientist Perrin had brilliantly guided the scientific fortunes of the CEA for 20 years. The former prefect, Hirsch, had worked since 1963 for the maintenance of the organization's unity and cohesion, navigating its passge to the thermonuclear weapon and improving its world image and reputation. All of this was done, however, without holding back its growth too early, so as to avoid its becoming too complex internally and too unwieldy administratively.

In place of the dual direction, the general supervision of the CEA was entrusted to an ''administrator-delegate.'' The high commissioner became his scientific and technical adviser and was also responsible for the safety of the installations. This task was given to Jacques Yvon, a theoretical physicist whose unruffled competence had, over two decades, had a great influence within the CEA.

The 1970 reform replaced the principle of monopoly, which had dominated the 1945 ordinance, with the notion of sharing. The CEA was to become a partner in work of the sort where earlier it had been wholly responsible. In pursuit of this objective, the CEA was not only expected to coordinate government-funded research and development projects, involving industry in them at the earliest possible stage, but also to act as adviser to industry and as a research and testing organization, whose advice would be heeded, appreciated, and sought after by industry for the improvement of its products and the fruitful pursuit of its initiatives.

After a gap of almost 20 years, the CEA's destiny and its part in the country's nuclear industrialization were once again entrusted to a mining engineer and oil specialist. If Pierre Guillaumat had introduced French industry to nuclear energy and piloted the CEA to the height of its prestige, André Giraud, the new administrator-delegate, presented with an organization under attack from every quarter, had the task of persuading industry — despite its tendencies to avoid government involvement — to recognize the considerable value of the CEA. He was to succeed in this task, after first reestablishing good relations between the CEA and EDF.

However, the reform did not concern only external matters. The new management system had to be set up. The new posts were filled by men already working for the CEA, but tragically some of these posts again became vacant in early 1971 when seven members of the CEA's top management died on active service in a grievous aircraft accident. Among them were several of the organization's chief directors, in particular Jean Labussière, the director of finance; Jacques Mabile, the director of production — to whom France owed her place among the world's uranium producers; and Hubert de Laboulaye, the director of programs, who had worked with me for many years, always enthusiastically carrying out his heavy duties.

One of the CEA's unusual characteristics was that it brought together the most diverse disciplines to the mutual benefit of all concerned. A proposed dispersal of the organization's departments, advocated as a means of improving control of its main activities, would have meant the disappearance of this characteristic. Giraud avoided this by proposing, then carrying out, a program of ''peripheral decentralization,'' creating a whole new range of subsidiary companies around the essential nucleus of the organization, which was thus preserved. This central nucleus concentrated on research and development, not only in the nuclear field but also in a variety of projects that drew on the CEA's exceptional multidisciplinary potential that had been accumulated over a quarter of a century.

All nuclear fuel production activities, from mining to the production of concentrated fissile materials, were grouped together as early as 1976 in a CEA wholly-owned subsidiary known as the Compagnie Générale des Matières Nucléaires, which employed almost one-third of the parent organization's workforce. (The British AEA had carried out a similar operation in 1971). The CEA also bought back 30 percent of the 45 percent Westinghouse holding in Framatome, therefore becoming directly involved in the construction of commercial nuclear power stations.

Meanwhile, in 1973, the overall responsibility for the safety, authorization, and control of nuclear installations was entrusted to a "Central Service for Nuclear Installations" that, like the CEA and EDF, operated as part of the Ministry for Industry.

In 1975, the new structure was completed with the creation of an interministerial committee for nuclear safety and protection of the population against risks and nuisances due to nuclear installations and radioactive substances.

Finally, it was necessary for France to regain in her nuclear fuel supply the same independence that she had previously enjoyed for the graphite-moderated, gas-cooled reactor. Only thus could she avoid being subject to the American monopoly of enriched uranium.

This reinforcement of France's position in the nuclear fuel cycle was to be one of the major tasks of the early 1970s. It was to prove part of a scene in which, in any case, the United States' monopoly was beginning to disappear.

3. The Fuel Cycle

Uranium: Supply and Overabundance

Throughout the 1964 to 1974 period of rapid nuclear industrial expansion, decisions made in Washington and the attitudes of U.S. industry continued to have considerable influence on nuclear activities in all other Western countries.

The Americans were holding two major trump cards: They were owners of the industrial technology in nuclear power plants now installed, or about to be installed, in many parts of the world and also had a monopoly on the enriched uranium required to fuel those plants. Yet, unknowingly, they had already crossed the threshold of a series of events, some brought about by themselves, that was to reduce substantially the value of the second of these cards. Consequently, this denied them some of the fruits of their victory over the natural uranium reactor type. This was to mark the beginning of the end of American domination of world civil nuclear development and it gave advance warning of approaching national atomic decline.

The year 1960 had been the best yet for uranium production in the Western World. Production had reached a peak of some 30,000 tons with about one-third coming from the United States and another third from Canada.

Then, with the nonrenewal of the American purchase contracts with Canada and South Africa, overabundance suddenly became evident, followed by a consequent drop in production. By 1967, with the remaining U.S. contracts stretched out to give producers more time to adjust to the new situation, western production as a whole had fallen by half. The main victim was Canada, whose production fell by 80 percent. South Africa was affected less for its uranium industry was linked with gold mining, which was flourishing during this period of worldwide inflation.

Despite reduced prospecting in the Anglo-Saxon countries, several deposits of worldwide importance were nevertheless discovered during these years of uranium surplus. These included one in the Agadès zone of the Republic of Niger south of the Sahara, discovered in 1964 by the French prospecting team from the Commissariat à l'Energie Atomique (CEA) under the leadership of Jacques Mabile. Some four years later, very rich deposits were identified in the northern Australian territories, and large quantities were found in southwest Africa — the future Namibia — by Rio Tinto, a group that already owned uranium mines in Canada and Australia. The situation was not favorable for the early exploitation of these resources.

Meanwhile, it was becoming increasingly important for Canada, and to a lesser extent for South Africa, to find new buyers to replace the United States and Britain. France seemed an obvious possibility with her ambitious nuclear program, but her military activities raised complications despite the fact that her home production of uranium, from deposits discovered in the Massif Central and in Brittany during the 1950s, was more than sufficient for the needs of her nuclear armament. Because of this, any French purchase of uranium from another country could be considered as contributing only to her civil program, or possibly to future exports, for the CEA was also hoping, in due course, to be able to become a uranium supplier in the world market.

Nevertheless, the French were unwilling to accept the application of safeguards for peaceful utilization, by any potential supplier of their uranium. They instead expected to be treated in the same way as the two Anglo-Saxon nuclear powers.

South Africa, in its 1963 major uranium sales contract with France, had accepted this point of view and had decided against discrimination between the three western powers of the Atomic Club.

Canada on the other hand, the first country to have decided against producing nuclear weapons despite having the means to do so, adopted the opposite point of view to that of South Africa and tried to persuade the

French government to reconsider its position. The Canadians were unsuccessful and, rather than alter their policy on safeguards, they renounced a commercially and financially advantageous French contract.

The transaction would have been substantial, involving the purchase of some 50,000 tons of as yet unmined uranium from the Denison Mines Company for a total price of $750 million spread over 25 years. The Denison uranium mines were at that time, 1965, the most important in the world, and their continued exploitation seemed to depend on the outcome of the negotiations with France. The mines were situated in Ontario, in the parliamentary constituency of Prime Minister Lester Pearson, who had been persuaded that the purchase was essential for France and that she would therefore, in the end, accept safeguards. Conversely, the French government had been told of the importance for the Canadian prime minister of concluding the contract and of the probability that he would therefore drop the safeguards condition during final negotiations. Under these circumstances, failure was assured, though it was not confirmed until the very last moment when the Canadian minister responsible for the final negotiations, Mitchell Sharp, came to Paris to conclude the transaction with his French counterpart, Yvon Bourges. Sharp was so incredulous that the French would not accept any form of safeguards clause in the contract that Bourges had to telephone, in Sharp's presence, Prime Minister Georges Pompidou to prove that this was indeed the government's position.

Ottawa, with Washington's support, had also been emphasizing its commitment, resulting from Canadian participation in the Nuclear Test Ban Treaty, not to provide nuclear aid to any country likely to carry out aerial testing. For the second time in eight years, Franco-Canadian negotiations over uranium had ended in deadlock on a question of discrimination between France on the one side, and the United States and Britain on the other.

However, the affair did not end badly for everyone because, under pressure from the producers that had been deprived of a sale for political reasons, the Canadian government decided on a 5-year plan for uranium purchase and storage, so as to keep the mines open in the hope of better times to come. Meanwhile, at the political level, the Canadian prime minister, to avoid any further accusations of practicing discrimination, announced that all new uranium export contracts would in the future be subject to conditions of peaceful utilization and to international safeguards, even in the case of sales to the United States or the United Kingdom. He was thus adopting the same policy that Washington, in accordance with the McMahon Act, had followed in all cases of civil nuclear aid to foreign countries since 1954.

During this same year of 1965, the unconditional pursuit of the principle of safeguarded peaceful use of uranium resulted in the collapse of a project of smaller dimensions, proposed by the Swedish Atomic Energy Commission (AEC), where there was no possibility of military assistance being involved. The Swedish AEC was seeking a partner to provide half the finance, over a period of five years, for the operation of a pilot plant to process bituminous shales with low uranium content, which were plentiful in Sweden. The French CEA was interested in the ''industrial know-how'' involved, and incidentally would also have received its share of half of the production, amounting to 60 tons of uranium per year, 300 tons in all over the five years.

Although the Swedes had bought, some years previously, a total of nearly 40 tons of uranium from France with — at the request of Stockholm — no control conditions, the Swedish government had more recently become a convinced supporter of international safeguards, and now insisted that the International Atomic Energy Agency (IAEA) safeguards system should apply to the future 300 tons. This was a minimal quantity compared with France's own uranium production and current stocks, and the Swedes were of course unable to persuade the CEA to accept their conditions, so the project came to naught. This was all the more unfortunate since no one had realized that the uranium could have been left temporarily in Sweden, for a few years later Sweden bought 1,000 tons from France, of which the transport of 300 tons might thus have been avoided.

From the French point of view, these abortive transactions though interesting were by no means vital. Indeed, there was already an important and attractive alternative to the first of them: an initial project to exploit the uranium rich deposits discovered in Niger. This operation was arranged between Niamey and Paris following abandonment of the Canadian purchase. This time there was no question of conditions of utilization, since Niger was bound to France by agreements covering priority options and defense.

In the early 1970s, the uranium production from Niger was equivalent to a substantial fraction of the Western market, thereby contributing to the oversupply and enabling the CEA in turn to become a significant supplier.

But this production was by no means the only cause of imbalance between uranium supply and demand. The main responsibility lay with the Americans who, 15 years after setting in motion a veritable treasure hunt for uranium throughout the Western World, suddenly closed their frontiers to all foreign imports.

The American Embargo

The 1964 American-Soviet détente had a direct effect on military programs. Reduced production of fissile materials for weapons — by 40 percent for American uranium-235 — released these materials for civil enrichment as well as more natural uranium to add to the existing oversupply. The U.S. government, long under pressure from industry to relax its monopoly of concentrated fissile materials, now decided to authorize the private ownership of uranium and plutonium.

Under the new system, the operators of nuclear power stations, who had previously obtained enriched uranium for their fuel by leasing it from the U.S. Atomic Energy Commission, were given the right and eventually the obligation to own it. They were also permitted to negotiate their own enrichment contracts for this uranium. A period of transition allowed the previous system of leasing to be continued in parallel until 1973, at which time ownership became compulsory.

Foreign countries having agreements for cooperation with the United States could also profit from these arrangements, and could have their own uranium enriched in U.S. plants (so-called "toll enrichment") on the same financial conditions as the American power plant operators, i.e., at the officially announced price on the date of delivery. These prices were fixed unilaterally by the U.S. government and were modified from time to time to take account of inflation. Enrichment costs were relatively low, corresponding only to the costs of running the separation plants, which were considered to have been amortized long ago through the military program. This was one of the "subsidies" then attacked by the coal industry lobby, and later by the promoters of would-be rival enrichment enterprises.

Finally, to protect the national uranium producers, some of whom had also had their current purchase contracts spread over longer periods than originally foreseen, as a form of compensation for possible nonrenewal, it was decided that the toll-enrichment facilities for private American consumers would be restricted to nationally produced uranium. Therefore apart from the completion over the extended period of the government's existing contracts with Canada and South Africa, the new legislation instituted a complete embargo on the purchase of foreign uranium. This restriction was to be enforced until 1977, thereafter being gradually lifted with the complete liberalization of the market forecast for 1983.

The embargo was contrary to the regulations of the General Agreement on Tariffs and Trade, and the Canadian government complained bitterly to Washington about it on several occasions. Annual export figures for Canadian uranium had fallen to $25 million by 1969 compared to the $300 million of 10 years earlier, and the Canadian mining industry was operating at only one-quarter capacity.

One of the first results of the embargo was the maintenance of a relatively high price in the United States for uranium ore, while its value on the world market dropped dramatically. There were many other unfortunate consequences. To some extent even the American mining industry was harmed, in that it was encouraged to confine itself within the national frontiers and discouraged from starting new ventures in other parts of the world. Following the oil crisis of 1973, which in the course of a few months sent the price of natural uranium rocketing from about $8 to at least $40 per pound of oxide, the mining industry had to make up lost ground, being helped in this by the U.S. government, which suddenly found itself in favor of Western World coordination at the "front end" of the fuel cycle.

The embargo policy was disastrous not only for the Canadians but also for the other western producers, among them the Australians, the French, and the South Africans, as well as for the important British mining company Rio Tinto, established in several Commonwealth countries. Every one of the rare sales contracts gave rise to savage competition between the leading production companies, resulting in continually falling prices, accumulation of available stocks, and several producers being reduced to bankruptcy or forced to close their mines.

Even with the more profitable mines, prices were becoming insufficient to give the private producers the cash flow necessary for prospecting and discovery of new deposits; they were therefore unable to produce the mining reserves that were indispensable to meet the nuclear industry's future requirements. The situation was in fact characterized by a contradiction between the low levels of immediate requirements, which were well below production capacities, and forecast future consumption, which was tending to double every five years. The immediate plethora and the American embargo were together the root causes of a possible uranium shortage in the early 1980s, when the many nuclear power stations built in the boom years would be coming into operation, while others would require renewal of their fuel charges.

France alone, whose uranium mining industry was for the most part nationalized, had not slowed down her prospecting program, and the Niger project, which she had developed, was beginning to bear fruit — its production adding yet further to the general oversupply. In view of the importance of these deposits, France, together of course with the Niger government, had in 1969 begun inviting participation in the venture from German, Italian, and (sometime later) Japanese uranium producers. At a later stage, the French were also to sponsor American and Spanish partnerships again with the full agreement of the Niger government.

A Uranium Cartel

From the start of 1974, under the influence of the world oil crisis and very soon of the increased demand for uranium by Westinghouse (as we see shortly) and the electric utilities (who had all become accustomed to an oversupply and held no stocks), uranium prices dramatically increased fivefold. This sudden reversal of the supply situation brought to light some operations that until then had remained secret.

In 1975 it was discovered that the depression of uranium prices during the period of overproduction had been artificially aggravated, both inside and outside of the United States, by Westinghouse. In its competition with General Electric, Westinghouse had tried to gain advantage by including in the sales contracts for its nuclear power stations a proportion of the uranium necessary for their operational lifetimes, without however making arrangements with the producers to ensure these supplies.

The American firm had thus increased the market imbalance by selling to national consumers — it had done the same with some Swedish power station contracts — uranium that it was not buying and was not in a position to produce. The company had assumed it could always obtain the uranium at a later date, after the lifting of the embargo, at a still more favorable price and from the strength of potentially being the dominant buyer in the whole world market. The uncovered commitments of the company were estimated at more than $2 billion. The amount of uranium concerned was some 30,000 tons, or about one and one-half times the Western World's annual production at the time.

By its action Westinghouse had, to a certain extent, even nullified the embargo in the United States. The company was in effect offering American customers uranium that it did not have but which it intended to buy on the foreign market, thus speculating on the lifting of the embargo rather than concluding long-term contracts with national producers, as required by the current American legislation.

The fivefold increase in the price of uranium in 1974 was a crippling blow for Westinghouse, forcing the company to try to escape from its commitments on the pretext that the oil crisis could not reasonably have been foreseen, nor the sudden rise in uranium prices that, it declared, made its contracts ''commercially impracticable.''

But the electric utilities were unwilling to listen to Westinghouse's problems, and refused to renegotiate their contracts, taking the view that the company had speculated in falling prices and now had to face the consequences. One after another, in late 1975 they began legal proceedings against the company.

While the pursuit of these lawsuits was gradually leading toward a compromise that it was hoped might limit the losses suffered by this major electricity-producing company, a new drama took shape, following information supplied to the U.S. Justice Department. This information came from documents stolen from an Australian mining company by an employee who was also a member of an antinuclear protest group. The documents referred to arrangements made between 1972 and 1974 among non-American western uranium producers experiencing difficulties due to the slump in their markets. These arrangements, some of which had even been publicized in official communiqués, appeared to be of relevance to the American antitrust laws.

This development certainly made the already complicated situation even more so. While the U.S. Justice Department was setting up an antitrust inquiry before a grand jury, and Congress also was looking into the affair, Westinghouse seized its chance. Abandoning the initial excuse that the fivefold increase in uranium prices had been impossible to foresee, Westinghouse now declared that it had been the victim of an international uranium cartel determined to prevent the utility from buying supplies in the world market and to engineer price increases in that market.

Westinghouse called a large number of uranium producers as witnesses in the lawsuits it was fighting, and also sued them in the American courts, seeking damages at three times the value alleged to have been suffered, estimated at $2 billion.

At this point, in 1977, the Canadian government reacted vigorously, officially declaring that its policy had been to encourage and to participate in international market arrangements, from 1972 to 1975, so as to ensure the survival of its uranium industry toward which U.S. restrictive commercial practices had been prejudicial. Ottawa also emphasized that the prices adopted following the ''arrangements'' concerned sales to third countries and not those to the United States, South Africa, Australia, Canada, and France; also that the prices had always been below those on the internal American market.

Finally, the Canadian government decided to introduce legislation to forbid its nationals to divulge any information relating to agreements made from 1972 to 1975 between its own uranium producers and producers in the other countries concerned. These countries in turn forbade their citizens to give evidence before the American courts, or at least advised them against this. In the United Kingdom, the House of Lords itself authorized several directors of Rio Tinto to ignore the summons of an American judge.

The sequels to this affair of the so-called "uranium cartel" or "club" thus took an embarrassing turn for the U.S. administration, well aware of the risks of exciting Canadian or Australian hostility. For these two countries, besides being important uranium producers whose contributions would sooner or later be needed by the Americans, were also among the most faithful supporters of U.S. nonproliferation policies.

In the end the Justice Department, early in 1979, stopped the grand jury inquiry. At the same time, the Chicago judge who had been hearing the many cases brought by Westinghouse against the uranium producers, accusing them of unlawful conspiracy to bring about price increases, refused to deliver a final judgment. Instead he succeeded, between 1979 and 1981, in persuading the principal contestants in the affair to reach compromise settlements, under which the producers agreed to pay Westinghouse in cash what they would otherwise have continued to pay (if not rather more!) to their lawyers. At the same time the producers would supply certain quantities of uranium to Westinghouse at reduced prices, so enabling the company to meet at least part of its previously unsecured commitments.

The only real beneficiaries in the affair were the lawyers, whose fees reflected the astronomical sums the litigation had been about.

End of the American Enrichment Monopoly

The embargo on entry of foreign uranium into the United States, leading to profound difficulties for the western market in atomic energy's basic fuel, had been a corollary to the new American toll-enrichment policy. This policy was welcomed by the countries that had adopted enriched uranium reactors, for they could now fuel their nuclear power stations with their own uranium, enriched to order in American plants but still remaining their property.

Once enriched the uranium was, of course, subject to international safeguards, as was any plutonium produced during the chain reaction. Except for the European community, which enjoyed a privileged position under the USA-Euratom Agreement, both reexportation of the enriched uranium and its reprocessing after irradiation were subject to American consent. These restrictions, being then no more than something of a formality, caused no serious problems for the countries concerned.

Therefore all appeared set for a reasonably prolonged extension of the

American monopoly of enrichment. Yet once again, Washington's policies both at home and abroad were to turn against their authors, bringing about a whole series of events that in a few years had severely shaken the U.S. monopoly and the very foundations of the "Atoms for Peace" concept.

The Americans had fought the reactor war in the knowledge that, if they won, their victory would be reinforced by the trump card of their enrichment monopoly. They did indeed win the reactor war with their light water systems, but at the same time they destroyed their trump card.

As on many other occasions, their first difficulty was with their wartime ally. In 1965 the British, having launched their advanced gas-cooled reactor (AGR) program, turned to the United States for the necessary supply of enriched uranium, a solution preferred to that of reactivating and enlarging their own Capenhurst plant, which had been built to supply their defense needs. Most of this plant had been closed down since 1962, for under the 1959 Anglo-U.S. Defense Agreement the British could acquire American uranium-235 for their military program in exchange for some of their home-produced plutonium.

The U.S. Joint Committee for Atomic Energy (JCAE), whose peaceful right hand neither knew nor wished to know what its military left hand was doing, decided, under the pretext of nondiscrimination in the application of the Atomic Energy Act, to impose IAEA safeguards on any slightly enriched uranium supplied to the British for their civil power stations; this despite the fact that highly concentrated uranium-235 was already being provided for the nuclear warheads of Britain's Polaris missiles. This uranium-235 was, of course, exempt from safeguards but could not be "diverted" to peaceful uses!

The American congress, by imposing conditions based on a principle but lacking practical justification, had set in motion a series of events that were to lead to the end of the U.S. enrichment monopoly and the beginning of the American nuclear decline.

The Washington officials had — by no means for the last time — fallen victims of their ignorance of the technological and industrial progress taking place outside U.S. frontiers. Perhaps also, as a result of using the term "agreement for cooperation" to describe arrangements that in reality were no more than commercial contracts subject to certain political clauses, they were forgetting one of the most elementary rules of commerce: avoid creating unnecessary competition.

The British rejected the American safeguards conditions, and decided to modify without delay their Capenhurst plant to produce the slightly enriched uranium needed for their AGR power stations. They even considered enlarging the plant with a view to supplying foreign customers. This proposed expansion could have been the first breach of the U.S. monopoly.

However, the necessary capital had to be found to carry out this transformation, and in 1967 the British offered a proportion of the enriched uranium production from the proposed enlarged Capenhurst gaseous diffusion plant to several European countries, in particular West Germany, in exchange for their financial support. But the plant was nevertheless to remain technically out-of-bounds for the investors, for the United Kingdom was still bound by its 1945 agreements with the United States on atomic secrecy.

The idea of a possible European isotopic separation plant had been relaunched in late 1966 by a number of European nuclear industrialists. The following year, as a result of interest by the Germans and the Italians, the idea was raised with the European Atomic Energy Community (Euratom).

There were two main reasons for this renewed interest in European enrichment, one technical and the other political. Technically, it was feared that a shortage of U.S.-enriched uranium might develop due to the spectacular increases in American home orders for nuclear power stations. Politically, reaction was building up in the principal European powers against their total dependence on the United States for this fuel. In the face of pressure from Washington to persuade them to accept the Non-Proliferation Treaty (NPT), these countries did not want to remain solely dependent on an American product, and were even ready now, if necessary, to pay a higher price for an assured European supply.

The Germans, unable to accept the discrimination inherent in the British proposal, which they considered humiliating because they didn't have access to the technology of the plant they were paying for, approached the French CEA, which was free to do as it wished with the gaseous diffusion technology employed in its Pierrelatte plant, now completed and operating satisfactorily. The Germans would have been happy for France to supervise the construction of a European enrichment plant in which they would have been full partners. But the time for its acceptance by the head of the government, General Charles de Gaulle, was not yet ripe.

Urenco

At this point conversations again took place between Great Britain and Germany, but this time jointly with the Netherlands, and also on the subject of another process, ultracentrifugation, which all three countries had studied in secret and on which all appeared to have reached approximately the same stage of development.

The use of an ultrahigh speed centrifuge for isotopic separation had already been recommended in 1957 by the Germans and the Dutch in a research study for a possible European enrichment plant as part of the Euratom program.

The new approach was accepted by all concerned, and the announcement of the tripartite collaborative project made dramatic headlines, all the more so because of the contrast between the publicity surrounding the ensuing negotiations and the mystery surrounding the state of advancement of the development work already carried out by each of the three partners in the so-called "troika."

The method of ultracentrifugation, which had been explored during the war in both the United States and Germany, had at that time been abandoned because of the difficulty of building gas centrifuges to revolve safely at the very high speeds needed to "cream off" the small percentage of lighter-isotope uranium in the gaseous hexafluoride, the identical working substance to that used in the diffusion plants.

Research into the process had been taken up again during the 1950s in Germany, the Netherlands, the United States, and Britain, being stimulated in particular by the development of new structural materials better able to withstand the enormous stresses occurring at the necessary speeds.

In 1960 the United States obtained from the other three countries concerned a promise that their work on this method of isotopic separation would remain strictly secret since, should it prove successful, the method could be used by any industrialized country for the annual manufacture of one or two nuclear weapons, the essential explosive being produced in facilities much smaller than the huge gaseous diffusion plants. A little later, the American government in fact went as far as prohibiting any work on this process by its own private industry, unless under government contract.

From the political standpoint, in view of the French refusal at that time to consider British entry into the Common Market, the tripartite project emphasized that British technological achievements and abilities were nonetheless sought after by other European countries. The three partners gave an estimate that technically it would take two or three years of research and development, plus the construction of two pilot installations — one at Almelo in the Netherlands and the other at Capenhurst in England — to demonstrate, as they hoped, that the process could yield under European conditions (in particular because of its much lower requirements for electric power) isotopic separation costs well below those of gaseous diffusion. So far, however, the promised benefits once again turned out to be further off than expected.

A number of specialists, among them Robert Galley, French minister for scientific research and "father" of the Pierrelatte project, thought the matter more political than technical. They also doubted the possibility of series manufacture, at sufficiently low prices and for a single installation, of the hundreds of thousands of centrifuge machines that would have to operate for years at rotational speeds higher than had ever before been achieved. The same specialists still supported the idea of a European gaseous diffusion plant and wished to see France, with her rich experience from Pierrelatte and unhampered by any secrecy in this field, take both the initiative and the leadership for such a project.

General de Gaulle finally gave his agreement and Galley traveled to Germany to put to his German colleague Gerhard Stoltenberg a proposal to build a European gaseous diffusion enrichment plant. It would straddle the Rhine with the actual separation unit on the French side (secrecy being maintained on the more specialized components) and the necessary electric power station on the German side of the river, thus giving the Germans, if not all the "know-how," at least one of the essential keys to the plant's operation.

Fifteen days later, the cherished hope of German acceptance vanished with Stoltenberg's answer. It was negative. The Germans were determined to equip themselves with their own industrial technology.

Later, the Italians tried unsuccessfully to join the Anglo-Dutch-German "troika." They were told that it was already difficult enough for three partners to agree. This must have been true, for the final treaty creating "Urenco" (Uranium Enrichment Company) was not signed until March 5, 1970, on the very same day that the NPT became effective. It was as if Germany had wanted to show that, in ratifying this treaty, it had freed itself from Konrad Adenauer's pledge renouncing the manufacture on home territory of nuclear weapons or their vital components, defined as "plutonium or uranium containing over 2.1 percent of uranium-235."

Enrichment in the Soviet Union

Toward the end of the same year of 1968, another breach was made in the American enrichment monopoly. The Soviet representative on the IAEA board of governors announced that his country was prepared to provide nations party to the NPT with toll-enrichment services up to a uranium-235 content of 5 percent. The offer was entirely genuine and, surprisingly, even included a telephone number for the Moscow-based firm concerned! At the time, however, it was thought to be something of an act of propaganda, to encourage further adhesions to the NPT.

Meanwhile the United States, instead of preparing to counterattack these very real threats to its monopoly, was becoming embroiled in a long and sterile internal conflict. The cause was the conviction of the new administration, which had taken power in 1968 following the election to the presidency of Richard M. Nixon, that the country's enrichment plants should be freed from public ownership and transferred to private industry, which would then be responsible for any necessary modifications and extensions. The JCAE, however, with its eyes on the expanding national program for nuclear power and fearing a probable fuel shortage in the early 1980s, was far more concerned over rapidly increasing the capacities of the plants than over their denationalization. The JCAE therefore wanted priority approval of funds to improve production from the 15-year-old plants, and did not think it a good moment for fundamental changes in administrative structures.

The administration nevertheless remained determined that its own changes should precede any decision on increasing capacity. To give itself more time and defend itself against the fuel shortage argument, it launched a vast program of advance enrichment, using the available capacities to process the then substantial national stockpile of natural uranium, which was equivalent to some two years of the Western World production. In this way it contributed even more to stagnation in the uranium market outside the United States.

A new threat to the monopoly, though a more distant one, appeared during the summer of 1970 with an announcement by the South African prime minister that his country had secretly perfected a new enrichment process, superior to gaseous diffusion, and that partners would be sought for the construction of a large-scale civil plant as an international enterprise. The secrecy was meanwhile maintained until 1977, when the new principle was disclosed: a type of centrifuge but with the high speed rotating container replaced by a vortex formed in rapidly flowing uranium hexafluoride. However, the South Africans found no partners and were obliged to industrialize their method on their own.

Following the creation of Urenco, France and the United States were the only remaining users in the Western World of the gaseous diffusion process. This being so the French minister responsible for atomic affairs, Maurice Schuman, went to Washington in 1969 in an attempt to promote bilateral cooperation in this area. His proposals were very firmly rejected by the State Department. The following year the French government decided to test the Soviet enrichment offer of two years earlier, which no other country had yet taken up.

Thus, early in 1971, I took part in some extremely easy and straightforward negotiations. In view of France being a nuclear weapons power, already in possession of adequate national resources for her military program, the Soviet Union required no safeguards on the utilization of the material to be supplied, asking only for a promise that it would be for peaceful purposes.

This first contract for toll-enrichment placed by the CEA with the Soviet export organization Technabexport finally and completely ended the U.S. monopoly. The price of the operation, which was below the current American tariff, was agreed in advance — a real advantage when compared to U.S. contracts which were priced on the day of delivery, at a rate which varied with inflation, being fixed from time to time by the American administration.

In setting an example with this Soviet purchase, France had benefited from very favorable and almost promotional conditions. Subsequently, in the face of quite a number of Western orders, Moscow fell in line with Washington and also fixed prices on delivery day, simply charging five percent less than the Americans. However, the Soviets restricted their customers, among the nonnuclear weapons powers, to those that had ratified the NPT and were therefore subject to IAEA safeguards. As a small consolation, for several years the United States subsequently retained the monopoly of price fixing in this field, without doubt the only one in which they controlled the price of Soviet services!

The American counteroffensive came at last in the form of an offer from President Nixon for the transfer of technology to friendly countries. The conditions of this offer were made known in November 1971 at a large meeting organized in Washington by the U.S. government.

The American proposals reflected two kinds of preoccupation, one political and one commercial. That of nonproliferation was covered by requiring the technology to be used in a multinational plant subjected to IAEA safeguards. The technology, in fact, would be transferred only to a group of countries that had agreed to build a joint installation. But in addition, the proposed plant, in which American industry hoped to participate by supplying equipment and capital, would have to be "transparently open" and avoid competition with U.S. production.

In short, the American government was prepared to relinquish its monopoly in exchange for a tight control of the enriched uranium market in the Western World.

Eurodif

The American offer was given a very cool reception, to the substantial benefit of France. Early in 1971, the French had announced their intention to build, with or without foreign partners, a large civil uranium enrichment plant, for which they had been improving and testing the necessary technology at Pierrelatte since 1969. To ensure that this project and the French technology were taken seriously, a site survey in France had been entrusted — with a certain amount of publicity — to the important American engineering firm Bechtel.

In view of the effect of the size of the installation on its economic viability, it was very much in the French interest to find partners, for which purpose they needed to make known their as yet unrecognized technological achievements in the enrichment field. To do this, a study was launched with the Japanese into the technical and economic aspects of the French process. The CEA then made an offer to other European countries to participate in a study association for a European gaseous diffusion plant — ultimately to become the Eurodif Company.

Finally, a joint study was begun with the Australian Atomic Energy Commission on a possible project to make the most of that country's very large recently discovered uranium deposits, together with the high quality, low cost coal from New South Wales, which could provide the necessary energy for a local enrichment plant. This study, which began very well, suddenly came to an end following the Australian elections of late 1972. The victorious Labour Party had made opposition to the French nuclear tests in the Pacific one of its main electoral themes, and could hardly associate itself with France in the civil atomic field at the same time.

Disappointment over the conditions attached by Washington to the American offer led not only the interested Belgian and Italian firms, but also those in the ultracentrifuge partnership, toward the CEA's proposed study association. The members of the ultracentrifuge partnership, in fact, had at last consummated their marriage by revealing to one another the precise status of their development work. It was seen that the Dutch spouse, whose attractions had been flaunted the most temptingly before the others, was in reality able to contribute the least to the joint enterprise, the viability of which would not in any case be assessed before 1975.

For the moment, therefore, French technology enjoyed a relatively clear field. It was decided that the Eurodif study association should begin work in February 1972, representatives from Spanish and Swedish industry joining later. The first year was devoted to a general evaluation of the competitivity and usefulness of the proposed plant, and the second year to more specific aspects such as economic viability, choice of site, and forecasting likely orders.

The centrifuge partners withdrew from the syndicate in March 1973 at the start of the second phase, which was completed rather quickly. The nonparticipation of the ''troika,'' apparently a serious blow to the project, was largely counterbalanced in the autumn of 1973 by two major events that greatly stimulated and helped launch the Eurodif project. The first was the oil crisis triggered by the Organization of Petroleum Exporting Countries or ''OPEC,'' which led to an acceleration of the world's nuclear programs and thus favored the order book of the proposed enterprise. The second event was the American government's decision to make the conditions for toll-enrichment of uranium in the United States much more stringent: Customers now had to place an order, corresponding to a 10-year supply for a nuclear power plant, eight years before the plant became operational, and they had to make an advance payment that would be partially forfeited in case of cancellation. These conditions had been imposed in order to increase the attractiveness of the enrichment industry for potential private investment and they could not have come at a better time to help push the Eurodif project from the planning stage to concrete execution.

Shortly after, the Americans announced that they were temporarily obliged to stop accepting new enrichment orders for foreign power stations; they finally undermined the dogma they had done their best to publicize: the dogma of unlimited U.S. capacity for the provision of enrichment.

One last difficulty was still to come. An organization of European nuclear electricity producers attempted at the last moment to rearrange demand so as to stimulate competition between Eurodif and Urenco, with the threat of possible recourse to the American technology offer thrown in for good measure. This attempt was not successful.

Finally, in late 1973, the decision was made to set up a company to undertake the practical exploitation of the study association's work. The new enrichment plant was to be built at Tricastin, adjacent to Pierrelatte in the Rhône Valley, and the enterprise was to be doubly nuclear, for the necessary electric power would be supplied by four 900-MW(e) light water reactor (LWR) plants. Enriched uranium production would be sufficient to meet the fuel requirements of over 100 power stations of similar capacity.

The Eurodif company initially was comprised of firms from Belgium, Spain, and Italy, in addition to the CEA, which was majority shareholder.* Swedish companies that had been interested were obliged to withdraw at the last moment, in the face of a threatened moratorium on their country's nuclear program.

The projected plant, which was to be brought into service gradually

----

*In 1975 Iran took a 10 percent shareholding.

starting in 1979 and completed by early 1982, was to have a capacity equivalent to over half that then available in the United States and over a third of that foreseen for the same country in the mid-1980s. The project had been led throughout by André Giraud, the CEA's administrator-delegate.

French gaseous diffusion technology had won a vitally important round in its competition with the ultracentrifuge, and France was about to regain her independence in nuclear power, momentarily lost when she first adopted the LWR. But above all Europe, frustrated since 1956, was at last to have at its disposal a large autonomous uranium enrichment capacity.

The Plutonium Cycle

The overlapping between the civil and military applications of fission, between the associated technical and political questions, and even between the technological stages in the nuclear fuel cycle, makes the history of atomic energy extraordinarily complex. But to complicate still further the round of nuclear problems, it seems inevitable that the spotlight of current events and of competition is continually moving from one to another particular aspect of the panorama, leaving other aspects in the shadows before highlighting them in turn and emphasizing the importance they have always possessed.

This happened during the period of rapid industrial expansion between 1964 and 1974, when attention was focused mainly on the nuclear boom, development of the uranium market, enrichment and the end of the American monopoly, the introduction of international safeguards, and the NPT.

On the other hand, the reprocessing of irradiated nuclear fuel, the future of the extracted plutonium, the final disposal of the resulting radioactive wastes, and the breeder reactors themselves, which depended on the reprocessing operation, were questions that at that time gave rise to less emotion if not less interest. Nevertheless, the future of civil nuclear energy depended on also finding answers to these questions, albeit with less immediate urgency than some of the others. A few years later, these less urgent matters were to become the key problems on the political and technical scene.

At this time the only opponents of fast-neutron breeder reactors — and even these were indifferent rather than actively hostile — were the supporters of the Canadian natural uranium, heavy water reactor ("CANDU"). Among the various slow-neutron power reactors, CANDU was the one that used uranium the most efficiently, so much so that it was not considered necessary, at least initially, to extract plutonium from the irradiated fuel. The used fuel rods were therefore stored away. In 10 or 20 years time, according to the evolution of both the price and the availability of uranium and the considerably decreased radioactivity of the rods, it would be much easier to extract plutonium from them should this then be desirable.

Had the United States adopted and developed the Canadian reactor type, and had all other countries followed suit, one could have temporarily visualized a world without enrichment or reprocessing. The corresponding NPT could have been reduced to a few lines: "Until further notice there will be no nuclear power stations other than natural uranium, heavy water ones; the enrichment of uranium and the extraction of plutonium will be forbidden; and irradiated fuels will be placed in storage under international safeguards!"

But in reality it was the LWR that was now dominant, in a world where the earlier natural uranium, graphite-moderated, gas-cooled system still played a part through the power plants of the British and French programs. For both these reactor types, the recovery of plutonium was considered technically and economically necessary.

As a first possibility, the plutonium could be mixed with depleted uranium (i.e., uranium with a reduced content of the uranium-235 isotope) for making new fuel that could then be reused in reactors. This process, known as "plutonium recycling," enabled substantially more energy to be produced from a given quantity of uranium. As a second and later possibility, the plutonium could be used in breeder reactors, in which natural uranium could be burned several dozen times more efficiently.

For these reasons, both the Americans and the Germans were at the time deeply interested in plutonium recycling, hoping at first to reintroduce the plutonium into LWRs without modifying them, while also pursuing fast-neutron breeder reactors, on which the British and the French had concentrated their efforts in the first place.

From the moment the French had revealed their plutonium extraction process, during the first Geneva Conference in 1955, followed by the release of further technical data by the Americans and the British during the 1958 conference, it was generally accepted that it would be impossible to restrict the spread of reprocessing technology, which used only conventional equipment although, of course, with specialized remote-control operating systems made necessary by the intense radioactivity present.

It was under these circumstances that in 1957 work began in Mol in Belgium, under the aegis of the Organisation for European Economic Cooperation, on the "Eurochemic" project, the world's first multinational nuclear enterprise. The Americans, who were much in favor of international operations in any "sensitive" field of activity, gave their support and the plant became a technical success and a rich source of industrial know-how for the thirteen European participant countries.

The technology was largely attributable to the CEA for the overall process, and for the engineering detail to the French company Saint-Gobain. This company had built the plutonium plant at Marcoule during the 1950s, and 10 years later was to be responsible for the second French reprocessing plant, at Cap de la Hague near Cherbourg, constructed at the same time as the European plant.

The Eurochemic enterprise was a failure from the financial and commercial points of view, the shareholding governments and companies either refusing or getting impatient over payments to meet successive deficits, which had not been foreseen in the planning or the statutes. The plant was neither a pilot unit as the Italians had wanted, nor was it sufficiently large to meet commercial competition from British and French national reprocessing industries with surplus capacity, or from a privately owned facility commissioned in the United States in 1969.

At the time of Eurochemic's closure in 1974, no one could foresee that the cost of reprocessing irradiated fuels was far from stable, and would in fact increase over a few years to around 10 times its original figure as a result of technical problems and more elaborate protective measures needed to ensure the safety of such plants and their workforces.

In America, in contrast to the difficulties over transferring uranium enrichment to private industry, the transfer of the whole of the reprocessing stage of the fuel cycle raised no problems. However the consequences for the companies concerned were disastrous. By the end of 1973, the only civil reprocessing unit to have reached the operational stage was closed down while awaiting authorization to build an extension; this was never built and the shutdown became permanent. A second facility, still under construction by General Electric, was abandoned as soon as it was completed, an original variant of the conventional process having proved unworkable. A third plant to be built at Barnwell, South Carolina, then still at the drawing board stage, was also due to run into serious difficulties, as is seen later.

Paradoxically, although the existing installations and those under construction in 1970 represented substantial surplus capacity at the time, they would be totally inadequate to meet the reprocessing demands of power stations whose operation was expected in the 1980s.

In these circumstances, a commercial and technical agreement was established between the British, French, and Germans by which, should the British and French reprocessing plants reach saturation, the Germans would be able to accommodate any excess work. An association was formed under the name "United Reprecessors," and a complementary agreement was made covering information and data exchange between the three partners, the Germans contributing their experience from a small national plant based on Eurochemic technology that had been operating since 1970.

By the end of 1973, a second reprocessing plant was under construction in India, while Saint-Gobain was responsible for building a first Japanese plant and was carrying out final negotiations for smaller units in South Korea and Pakistan. These three facilities were to be subject to international safeguards and had not yet given rise to political difficulties.

However, the first signs were appearing at an international level that nonproliferation issues were soon to be raised in connection with this sensitive technology. The United States had decided unofficially not to provide industrial aid in this field of plutonium extraction. This decision became official in late 1973. No American aid could be provided to foreign nations for fuel reprocessing without a governmental agreement that would take into account, among other factors, the country's status regarding the NPT, as well as whether the proposed facilities would be placed under multinational management.

4. The International Organizations

The Role of Euratom

The early 1970s saw a number of successful multinational European projects, ranging from the joint exploitation of uranium mines in the Niger to the prototype breeder power station Super-Phénix, and including enterprises for fuel enrichment and reprocessing. The part played in these projects by the European Atomic Energy Community (Euratom), which had been born under the Treaty of Rome with such a prestigious flourish (greater than that initially accorded to the Common Market itself) was extremely limited and in some cases negligible. Its activities and influence had not in fact (in contrast to the Common Market) produced the hoped-for results among the Six — to become the Nine in January 1973 following the adhesion of Denmark, Ireland, and the United Kingdom. There was no technical or political field — research, industry, fuel supply, or external relations — in which Euratom could show truly positive results.

Euratom failed to instill, in research or in the industrial field, in external relations or in the policies of fuel supply, the spirit of community solidarity that had been expected. In science itself, the prestige of Europe and of the atom proved a combination of bad influences for the community, providing it from the start with excessive ''nondesignated'' financial resources, which led to the creation of a geographically dispersed ''Joint Research Center'' having a general interest in every aspect of nuclear energy and thus often duplicating work under way in the national laboratories. Coordination and concentration of effort in some particular area, as was the fortunate but exceptional case with thermonuclear fusion, would always have been much more fruitful.

The resulting ills were many. Funds were scattered widely and indiscriminately among projects with promising futures, and to others of little interest; research contracts were placed with university or private industrial laboratories, which sometimes were even subsidized, mainly under the pretense of ''pooling information.''

Euratom successively took over the main Italian nuclear research center at Ispra and several sections of the Belgian and Dutch nuclear centers (Mol and Petten), together with a project to build and operate a large-scale plutonium research laboratory at Karlsruhe in Germany. It also acted as an organization for the equitable distribution of subsidies to existing research organizations, rather than itself undertaking complementary activities in areas of special interest. Moreover, the painful annual bargaining over the distribution of research funds among the partners was often more related to their respective appetites than to rational allocation according to priorities and competences.

The success of a true atomic energy community in fact required not only the achievement of a common research policy, but above all a common policy in the field of industrial promotion. But the Euratom treaty had by-passed problems of industrial policy, apart from a timid notion of coordinated investment programs and objectives, and the possibility of so-called ''joint enterprises.'' The commission was initially content to attribute this status of joint enterprise to a few power stations of American origin in the USA-Euratom program, for the sole purpose of obtaining for them certain tax benefits, so making a mockery of a valid community concept.

From the point of view of external relations, for prestige reasons the commission frequently tried to introduce itself as an intermediary between community members and other countries. This led to the conclusion of a number of agreements comprising window dressing rather than concrete substance, but it also led to the USA-Euratom Agreement — an agreement that undoubtedly gave the community access to more generous supplies of enriched uranium and plutonium, but also gave American industry better access to community markets. In addition it included U.S. recognition of the Euratom safeguards system, which became the only system applied to materials of American origin following the nonrenewal in 1966 of the bilateral agreements between the United States and community countries. As a result, controls effected by American inspectors were discontinued.

Apart from the exemption of materials intended for the defense uses of nuclear weapons powers, Euratom safeguards now applied to the remainder of the community's nuclear activities. That this application was to good effect was evidenced by an affair, revealed only in 1977, which became a cause célèbre in the annals of nonproliferation and international control.

It concerned a shipment of 200 tons of uranium concentrates, produced in a Belgian refinery and sent in 1968 from a German port to Italy, where it never arrived. The ship into which it had been loaded returned after some time in the Mediterranean — without the precious cargo but with a new crew and different owners. Inevitably, the name of Israel was cited in connection with the matter.

The affair was incorrectly seen as an example of unsatisfactory operation of the Euratom safeguards system, when in fact it was the very opposite. The Euratom control, which did not operate outside the community, was alerted when it received no confirmation of the arrival of the sodium uranate in Italy. An irregularity was thus detected, which the national police authorities — the only ones responsible in this case — should have already discovered. The detection of this by the Euratom system in fact demonstrated that any member state attempting to divert nuclear materials for military purposes could not have done so without it being discovered, which was the sole purpose of the international control.

If the safeguards of Euratom operated satisfactorily, its supply agency did not. The agency had been given monopoly powers so as to ensure equal access for all member countries to uranium resources, and to pursue a common supply policy. These objectives were by no means achieved.

To start with, the Germans did not wish to be obliged to buy natural uranium or enrichment services from France when they could obtain the same from Canada or the United States at lower prices, which they naturally preferred. To them, the accompanying safeguards conditions, which would have been unacceptable for France, were no inconvenience.

Secondly, the French wanted to retain for their own use the nonrestricted uranium they had extracted from their home territory. What is more, they had noted that the community's right to ownership of plutonium and enriched uranium used for civil purposes had remained purely symbolic, the supply agency having never in reality exercised its right of purchase, being content to give automatic approval of contracts for the buying and selling of nuclear materials by community members.

It was specified in the Euratom treaty that, seven years after its entry into force, the ''modalities'' of Chapter VI of the treaty, covering the supply of nuclear materials, would be confirmed or revised. This chapter had been conceived during a period of shortage when state monopoly was often the rule and by 1965 it had become totally inadequate. None of the countries involved proposed confirmation, but no agreement could be reached over the commission's proposals for modification.

France considered these regulations no longer valid and ceased to conform to them; in particular, she stopped submitting her contracts for the approval of the supply agency. The commission appealed to the European court of justice, particularly when France delivered to Italy, without Euratom agreement, a quantity of enriched uranium required for research on a naval propulsion engine for a warship, a project that never materialized. In 1971 France was condemned by the court, which took the view that the original supply regulations — although unanimously rejected — were still legally valid until new measures had been officially agreed and adopted. The original articles covering the supply of nuclear materials therefore remained in force without being fully applied either by the community's member countries or by the commission itself.

In 1966 the supply agency had become the obligatory channel through which American fissile materials were obtained. The United States, as part of their policy of support for the commission, had decided, as was their right, not to renew expiring bilateral agreements with community countries. These countries were not favorably impressed by the decision, which forced them always to use the USA-Euratom channel, inseparable from the unwieldy Brussels administration. As a result bilateral nuclear relations with the United States, and particularly those of France, became momentarily rather strained.

The advantages that Euratom enjoyed under its agreement with the United States had been cited as a major argument in favor of this multilateral policy. However the Americans, as seen later, soon had to review the principles behind these advantages, and even had to seek their renegotiation because of new legislation that had certainly not been foreseen. The renegotiation was coupled with moves to give the International Atomic Energy Agency (IAEA) a predominant position in the application of the 1968 Non-Proliferation Treaty (NPT), which was part of the further evolution of U.S. policy following the 1974 nuclear explosion in India.

Reinforcing the IAEA

Whereas the NPT was a source of problems for Euratom, it was to consolidate the IAEA by giving it greater purpose and increased importance.

The beginnings of this world organization had been difficult for, constantly subject to the hazards of changeable Soviet-American relations, it had been unable to achieve the role of banker, or even broker, for fissile materials; nor that of "controller" of these materials as intended by its founding members. The continued pursuit by the United States and other leading powers of a policy of bilateral agreements, together with Soviet reticence toward an American-inspired organization and its safeguards system, had been the main reasons for these pains of the immediate postnatal period.

The circumstances led the IAEA to concentrate its early efforts on questions of technical assistance, international regulation, and the organization of conferences and seminars during which technicians from every part of the world came to Vienna to exchange their knowledge and theories. During these same years, the mechanics for a practically applicable safeguards system, at this stage no more than a theoretical vision, were discussed at length within the organization's administrative council, the board of governors.

I was a member of this board since the creation of the IAEA in 1958, and in 1980 I was nominated its chairman. From this position I at first made an effort to play down the excessive formality of the meetings, which were often marked by successions of lengthy monologues, and to speed up the work by avoiding exaggerated importance being given to secondary questions of operational management. But the general atmosphere of the three sessions of the board per year, their length progressively reduced to only a few days each, never lacked interest, reflecting during the 1960s the temperature of American-Soviet relations, and during the 1970s the ambitions of Third World countries to assert themselves.

In one particular respect the Soviet Union was quick to make use of the IAEA — to inflict public punishment on one of its citizens. Vyacheslav M. Molotov, during the final stages of his disgrace, was appointed for a year as deputy to the head of the Soviet delegation to the board of governors. Andrei A. Gromyko, Molotov's successor to the post of foreign affairs minister, in announcing this appointment to Vassili Emelyanov, the astonished Soviet governor, asked him on Nikita Khrushchev's instructions not to be absent from a single session, so that the ex-president of the council could never occupy the delegate's seat; this was in sadistic contrast to the famous negotiations that Molotov had directed at the height of his fame. The expression on his face when he took his place behind Emelyanov showed how well the torture had been calculated.

During this same year of 1961 the Soviet Union, supported by India, had vigorously opposed the appointment, for a 4-year term as IAEA director general, of the Swedish nuclear physicist Sigvard Eklund. His appointment had been proposed by the Western powers in succession to the first holder of the post, former American Congressman W. Sterling Cole, previously a chairman of the Joint Committee for Atomic Energy of Congress but not too familiar with the outside nuclear world.

Eklund, whose mandate was subsequently renewed four times — an almost unique occurrence in the annals of international organizations — guided the IAEA to its maturity with calm, energy, and great diplomacy.

There was an important discontinuity in the evolution of the IAEA when, in June 1963, the Soviet Union suddenly and unexpectedly rallied to the support of its safeguards system. Emelyanov declared, without any outward sign of embarrassment, that the organization's function of control and inspection was by far its most important task. This was in marked contrast to the same man's observations, some years earlier, when he had compared the international system to an American spider's web "paralyzing all science and all scientists in the world." From now on, however, with the NPT safeguards and inspections entrusted to the IAEA, the agency's important international position was confirmed.

A reaction was bound to take place, making itself particularly felt within the board of governors. The opposition was essentially from the Third World countries, whose representatives thought too much attention was being devoted to safeguards, which were mainly the concern of the advanced countries, and not enough to technical assistance, of which the principal beneficiaries were the developing countries. And it was indeed true that the IAEA's safeguards and inspection system had become the agency's dominant and most fund-hungry preoccupation.

The Implementation of International Safeguards

By the end of 1973, 20 years had passed since Dwight D. Eisenhower had proposed his "Atoms for Peace" program, 10 since the Soviet Union had rallied to support the IAEA's safeguards system, and 5 since the conclusion of the NPT. International inspection by the IAEA had progressively become accepted, both politically and practically, as a normal part of nuclear exchanges anywhere in the world.

The IAEA regulations did not in principle apply to facilities, but to basic nuclear materials (natural uranium compounds) or special fissile materials (enriched uranium or plutonium) contained in the plants or reactors. The safeguards were based on inspection, accounting systems for the nuclear materials, and analytical methods enabling the presence of these materials and their quantities to be assessed.

The safeguards agreements concluded with the IAEA during the 1960s listed the installations for which nuclear materials were subject to verification. General directions had been worked out concerning the manner in which the control was to be applied and even the frequency of routine inspections, according to the type of installation.

With the NPT signed and in force, the entire nuclear fuel cycle became subject to control within the signatory states. As a result, the methods of application of the IAEA safeguards were amended. The strategic points at which the control should be exercised were defined for each facility, together with systematic methods for the storage and surveillance of materials, which would avoid the need for too frequent stock-takings in large industrial installations.

The countries in which the Western World's principal uranium mines were situated — South Africa, Australia, and Canada — did not wish to have safeguards applied before the refining stage, i.e., when the uranium products reached nuclear purity. The United States accepted this; mines and plants for the chemical treatment of crude ores were accordingly exempt from control, which allowed the uranium-producing countries to protect their commercial data on ore extraction and the production of concentrates from the indiscretions of international inspection.

This was the opposite of what took place with Euratom, where one of the purposes of the control was to verify that declarations made to the supply agency were in true conformity with production data. The absence of such verification represented a serious loophole in the IAEA system, for which countries such as Australia and Canada, among the most devoted supporters of this system, would bear responsibility. Even in the case of NPT countries fully subject to safeguards, their compulsory declarations were the only means the IAEA had of knowing what stockpiles of uranium ores and concentrates were held and what quantities were being exported.

In fact, one of the main features of the IAEA (or any other) nuclear control system must be the "right of pursuit." Substances produced by transmutations of safeguarded materials must in turn be themselves subjected to safeguards. Matters can become most complicated when both nonrestricted and controlled materials are involved in the same process or installation; in such cases it is normal to control the materials on a proportional basis.

In sum, a whole new philosophy and methodology had to be developed for this first application in history of an international control system. The limits to its effectiveness were evident, for the system could neither prevent illicit diversions nor foresee an infraction before either had occurred. Nevertheless, it had dissuasive power since it was designed to detect any failure to comply with undertakings of peaceful utilization. The basic problem was to ensure rapid detection of infractions without incurring extravagant costs and without impeding the legitimate nuclear activities of the country concerned.

The system was, and is, more a matter of politics than of policing, despite its foundation in technical inspection and surveillance. While all countries must operate their own national policing systems as protection against theft or misuse of fissile materials by individuals motivated by terrorism or greed, the international control is intended to detect infractions by the authorities themselves in the countries concerned. To be effective, therefore, the international control must be sufficiently rapid and reliable to persuade a government that, should it decide to break its engagements on peaceful utilization, it will not try to do this secretly (because of the risk of detection by the inspectors), but will prefer to announce publicly its decision to resume the freedom of action it had renounced.

Under present regulations it is always open to an inspected country to reject any inspector, from a list submitted in advance, whose nationality might cause political difficulties. It is nonetheless obvious that the system is completely ineffective in the case of a government that is determined to break its nuclear pledges and close its borders to international inspection.

Some 20 years of experience have now made international safeguards and controls accepted facts of the nuclear scene. During this time no infractions or diversions of materials, which could be described as true breaches of international engagements, have been detected. On the other hand, it is true that a good number of countries accept these inspections without enthusiasm and try as far as possible to avoid them. France, for long hardly considered by the United States and Canada as a full-fledged nuclear weapons power, let alone treated as one, was still heading the rebellious countries in the early 1970s; all the more so because it was hard to see what purpose could be served by such controls on the home territory of a weapons state commonly known to have a stockpile of nonrestricted resources far in excess of military requirements.

The French government had refused, during the London "Cashmere meetings" (see p. 286) to support the establishment of a common policy under which all countries supplying nuclear materials and equipment would undertake to include conditions of peaceful utilization and safeguards in their sales contracts; the French therefore refused to become part of a mechanism designed to prevent the undermining of the control system as a result of commercial competition.

Paris in fact was then reserving the right to consider each case individually, according to its commercial and political context, a philosophy diametrically opposed to the automatic application of general rules as practiced by the Americans, toward whom the French demonstrated — not without some satisfaction at that time — their nuisance value.

The French attitude in commercial sales operations, a sensitive area from the proliferation point of view, therefore varied according to circumstances from ''no trust, no sale'' to unconditional sale, with as the midposition a pledge of peaceful utilization that might be without verification, with verification by French national inspectors, or even by the IAEA.

It was this approach that had initially suggested that peaceful utilization of the Vandellós power station, built in Spain by French industry, could be adequately guaranteed through a system of joint management by Electricité de France and the Catalan Association of Electricity Producers. At a later stage in an exchange of letters, the two governments concerned declared their intention of ensuring that the power station would be operated with the greatest possible economic efficiency, implying a high degree of fuel irradiation, which would ensure that the plutonium formed would have very little military value. Plutonium produced in this plant, moreover, would be extracted in the Commissariat à l'Energie Atomique's (CEA's) installations and could only be used in Spain in a civil program.

In 1968, when the NPT was being signed, France, through her representative at the United Nations, gave a solemn pledge that she would in the future behave exactly as if she had signed the treaty. This pledge was indeed kept during the years that followed, as the French government began to require the application of IAEA safeguards to all nuclear exports.

However France continued to show a certain reticence toward the possible application of the same safeguards to her own activities. This became evident during negotiations on how the NPT controls should operate within countries signatory to the treaty and also members of the European Community. These countries had the right to negotiate jointly their control agreement with the IAEA, and in this they wished the European Commission to act in their names. The commission was very ready to do this, believing it would thus have the best chance of protecting the privileged position of the Euratom control system. To do this, the commission needed French agreement.

But it was no less obvious to France than to other countries that introduction of the IAEA into Euratom's safeguards operations would noticeably modify them, and the French government did not see why it should be subjected to the consequences of a treaty it had not signed. It therefore obtained from the commission, in exchange for agreeing that the latter should negotiate with the IAEA on behalf of the whole community, an undertaking that the future Euratom control, based on and now including the IAEA system, should not apply to nonrestricted materials in France. This advantage obtained by France was not greatly appreciated by her European partners.

The IAEA-Euratom Agreement was accordingly negotiated in 1971 between the European Commission and the IAEA. Euratom wanted the agency's role restricted to verification of the community's own inspections, whereas the IAEA wanted to carry out the controls itself. Euratom finally obtained what it wanted on paper, though without the benefit of a privileged situation compared with other countries not belonging to the community. In fact, the concept of verifying existing national control was substituted for that of independent control in all the NPT agreements concluded by the IAEA, whether with community countries or not.

During the same year of 1971, when IAEA expenditures suddenly increased following the entry of the NPT into practical operation, the French government, opposed to countries that had not signed the treaty being obliged to contribute to the extra costs incurred, decided unilaterally to pay only 90% of its share of the safeguards budget. Some years later, her attitude toward nonproliferation having moved closer to that of the other major powers, France discontinued this symbolic withholding of funds, and even paid the arrears.

In fact, a bilateral control clause, i.e., one involving French inspection, in the CEA's 1970 agreement with the Indian Atomic Energy Commission to provide aid for the construction of an experimental fast-neutron reactor near Madras, marked the last occasion when France accepted an agreement without IAEA safeguards for an export to a nonnuclear weapons partner.

Also in 1970, France had transferred the industrial technology of fuel reprocessing to Japan, when Saint-Gobain sold that country the design for a first plant, which the company was helping to build. Because this plant was intended to treat only irradiated fuels already subject to safeguards, it was transferred without a control clause. However in 1972 the situation was regularized through a trilateral agreement with the IAEA.

France subsequently required IAEA controls in all her major agreements for aid to foreign countries. Despite a period of reticence, she had finally come round to support the application of a policy of controlled assistance that the United States had been practicing for 20 years, culminating in the NPT.

But this policy, at the very moment when it was becoming almost universally accepted, was about to be profoundly modified by its promoters. The Indian atomic explosion was to be the catalyst.

Act Four

CONFUSION
1974 to 1981

This last act, which began in 1974 and whose end still cannot be foreseen seven years later, is being played out against a backdrop of worldwide confrontations over energy, reduced growth in industrial countries plagued with inflation and unemployment, and a doomwatch on armaments.

The Middle East petroleum states, suddenly aware of the immense power at their disposal, have used it twice — following the 1973 Yom Kippur war, and again five years later after the Iranian revolution — to implement huge and savage increases in the world prices of crude oil.

The United States, technologically triumphant in space although turning away from supersonic air transport, administratively handicapped by legal mechanisms and a paralyzing Congress, and politically traumatized by Vietnam and Watergate, appeared overtaken by events and unable to take the lead in the ever more important north-south dialogue. It seemed powerless to remedy galloping energy consumption, its 1973 "Project Independence" (aimed at the total suppression of oil imports by the early 1980s) having proved unrealizable. Nor was the situation helped by the creation in 1974 within the Organization for Economic Cooperation and

Development framework in Paris of an international energy agency that included 16 Western countries but excluded France, who was reluctant to participate in a prearranged plan for international oil sharing.

However, the industrialized countries held a major trump card with which to face the energy crisis: nuclear power. Its basic fuel was to a large degree under their control. Its technical problems had been overcome during 40 years of continual progress, and the threshold of economic electricity production had long been crossed.

In 1954, there had been a single nuclear submarine engine and a single 5-MW(e) nuclear power station in operation in the world. A quarter of a century later, in the early 1980s, there were to be hundreds of reactors powering submarines and warships, and hundreds more in power stations supplying, in the major industrial nations, eight percent of the world's electricity. The maximum unit capacity for these stations was 1,300 MW(e), 25 percent higher than that of the largest conventional coal- or oil-burning units. Many countries were already planning to produce half their electricity from uranium before the end of the century; for France this date was 1985.

The technology of slow-neutron reactors seemed mature and stable, all new units having uranium oxide fuels and being cooled by either pressurized or boiling water. Only the moderators varied: ordinary water, heavy water, or graphite. Within the fuel cycle, two stages were still in the process of evolution: uranium enrichment by new methods and the final conditioning of highly radioactive wastes.

However, despite the discovery of new uranium deposits in Canada, Africa, and Australia, known reserves in the non-Communist world appeared to be insufficient for the programs envisaged beyond the turn of the century. The massive production of usable energy still seemed dependent on the industrial perfection of fast-neutron breeder reactors and of the corresponding plutonium fuel cycle. The first precommercial power station of this type, the 1,200-MW(e) Super-Phénix, was under construction in France with completion expected in 1983.

Controlled fusion, on the other hand, still awaited its technological breakthrough, with its potential contribution to world energy consumption therefore still uncertain and in any case distant.

This brief summary highlights the incredible progress made since the start of laboratory research into the fission of the uranium nucleus in 1939. For any other great modern discovery, once this advanced stage of industrial evolution had been reached, only the laws of economic viability would have affected its future development. But for atomic energy, marked from birth by the nuclear sin, the situation was very different.

At a time when the need to escape from the stranglehold of dependence on oil was reinforcing the arguments of those who saw nuclear power as an available technically reliable and economically viable source of electricity, that same source was suffering from difficulties due mainly to irrational factors, both political and psychological. In fact, the multiplication of atomic installations in the world had led to a double fear: that of the danger to world stability resulting from the potential increase in the number of countries possessing nuclear weapons and that of theoretical health risks for the population due to the radiations inseparable from production of the new energy.

The result was a veritable resurgence of American nonproliferation policies, with all the rules of international nuclear trade again under question, leading inevitably to growing mistrust between suppliers and customers for nuclear materials and services.

Simultaneously, antinuclear opposition was gaining strength in the United States, and taking shape in the rest of the Western industrial world. It was a kind of rejection phenomenon, which also originated in a feeling of mistrust, this time at the national level, between the public on the one hand and the scientific and political authorities on the other.

The Indian nuclear explosion in 1974 and the Israeli attack on the Iraqi research reactor in 1981 symbolically highlighted this period, which was also marked by the 1979 accident at the Three Mile Island power station, by concern for nonproliferation, and by controversy, doubt, and confusion.

In parallel with these at least partially technical developments, the last years of the period were marked by reverberating political events: the detention for more than a year of hostages from the U.S. embassy in Teheran, the Soviet invasion of Afghanistan, the grave political developments in Poland, the tougher policy of the Reagan administration toward the Soviet Union, and finally the ever-increasing terrorist assassination attempts against heads of states and their repercussions on world stability.

The renewed significance of nonproliferation policies was to have a far-reaching effect on international atomic relations, while the antinuclear movement seriously hindered the development of nuclear energy in many parts of the Western World, France alone seeming somewhat less vulnerable to this opposition.

However, the energy crisis must surely dispel the mists of confusion and make the revival of nuclear power inescapable. This book ends with hope for such a fresh start.

1. The Plutonium Conflict

Temptation

The Third World made a significant entry on to the atomic scene in 1974. The staggering rises in oil prices had given a new impetus to the nuclear electricity programs of the industrial countries; while at the same time several Middle East states — Libya, Iran, Iraq, and Saudi Arabia — overflowing with "petrodollars," announced their intentions to acquire nuclear facilities. Iran in particular, with an ambitious nuclear electricity program in view, bought a 10 percent shareholding in Eurodif.

These various projects, perhaps not all entirely devoid of military inclinations, were to become the focus of competition over the next few years between the leading powers of the world.

Germany won the first contracts, for two Iranian power stations. France followed with two more and in addition an order for a large research reactor in Iraq. Libya turned to the Soviet Union for its first nuclear station, and Cuba did likewise.

However, the most significant event at this time was the Indian atomic explosion in May 1974. It disrupted the nuclear policies of the non-

Communist world, which was led to seek substantial changes in the rules of the international nuclear game. The United States, creators 20 years previously of the ''Atoms for Peace''policy, were again behind the new orientation, but this time they were unable to impose their own particular will. Their principal ally in the matter was Canada, whose technical aid had enabled the Indian atomic program to be launched. Indeed, the Ottawa government resented the painful and ironic twist of fate that had attributed to it a share of the responsibility for the Indian blow to nonproliferation, which Canada had always and staunchly supported.

India had systematically refused to sign the Non-Proliferation Treaty (NPT) and reaction to its world-shaking demonstration could have been limited to stricter measures with countries dissenting from the treaty. But nothing of the sort was done, and the Rajasthan explosion was made a pretext for a large-scale operation against all nonnuclear weapons states, whether or not they had joined the NPT. The basis of this new policy was a new importance given to the risk of temptation. Until then, possession of nuclear materials that could readily be used for military purposes — plutonium or very highly enriched uranium — was considered safe as long as these materials came under a pledge of peaceful utilization and were subject to international safeguards. The same applied to irradiated fuel reprocessing facilities producing plutonium and would also apply to any enrichment plant that might be built.

The architects of the new political orientation did not reject international safeguards, which they hoped would continue to spread so as to include the widest range of nuclear activities throughout the world. However they no longer fully relied on them, arguing correctly that such measures were powerless in the face of a country whose government might suddenly decide to refuse access to its territory for international inspectors, at the same time redirecting its civil nuclear program toward the achievement of one or several explosions and subsequently toward an arms production effort.

International safeguards were designed mainly for the timely detection of secret diversions by governments of fissile materials; they are incapable of preventing such diversions, or even flagrant appropriations, by governments determined to use such materials for military purposes. The objective of the new policy was therefore to prevent nonnuclear weapons countries from ever having possession of fissile materials suitable for military purposes or of having facilities allowing their production: the designated facilities did not include reactors (although of course these are the real plutonium producers) on the grounds that ''production'' in the practical sense occurred only at the stage of fuel reprocessing.

This attempt to impose further technical restrictions was at the root of a new form of discrimination, all the more difficult to accept because, besides being an addition to the discrimination inherent in the NPT, it was in complete contradiction to Clause IV of the treaty. This clause guarantees signatory countries the inalienable right to pursue research and development for the peaceful production and use of nuclear energy in exchange for their renunciation of the military nuclear option and their acceptance of generalized controls.

The NPT had never imposed special prohibitions on "sensitive" materials or activities, such as those closest to the explosive utilization of nuclear power. Quite to the contrary, industrial countries such as Germany and Italy had insisted, as a precondition during the negotiation of the treaty, that they should have free access to all the operations of the fuel cycle including reprocessing and uranium enrichment. Germany had even declared when it signed the treaty that "the transfer of information, materials, and equipment to nonnuclear weapons states must not be withheld on the pretext that such a transfer might assist the manufacture of weapons or other explosive nuclear devices."

Moreover, the new orientation was difficult to reconcile with the existing nuclear armaments and balance of terror as well as with the degree of national independence in which the countries of the non-Communist world took such pride.

Indeed, from the moment when the existence of sensitive materials or installations in the hands of a nonnuclear weapons state subjected to safeguards on peaceful utilization becomes a matter for concern, there must be even greater concern over the contrast between the risks of diversion of a few pounds of plutonium or uranium-235 and the general acceptance of the dangers from the stockpiling on the home territory of the same or another nonnuclear weapons country — for example, either of the two Germanys — of tons of these same materials contained in thousands of tactical weapons.

In 1953, Dwight D. Eisenhower's proposals had sought to launch détente in the Cold War by withdrawing some amounts of fissile materials from military stocks and making them available for civil uses, uses subsequently subjected to international controls. A quarter of a century later, détente between the two power blocs having progressed, the presence of plutonium or uranium-235 on the territory of a nonnuclear weapons state and under International Atomic Energy Agency (IAEA) safeguards on peaceful utilization was regarded with more anxiety than if these substances were present in nuclear bombs under the reassuring and watchful eyes of one or the other great lords of the military atom.

Furthermore, the new policy had already been tried out, being applied rigorously and successfully in a large part of the world — the Communist dominated one. Indeed, the Soviet Union had from the start taken care of this risk of temptation by adopting, as part of its nuclear power station export policy, a system of supplying fuel elements on loan and taking them back after irradiation. This system obviously made the countries within the Russian sphere of influence completely dependent on the Soviet Union (sole possessor in the group of enrichment and fuel-reprocessing plants) for the nuclear part of their energy production. Such a dependence would not be readily acceptable to Western industrialized countries or to the more advanced Third World states, who are deeply sensitive over anything reminiscent of colonialism.

The problems of technical restrictions on arms proliferation thus presented different aspects in the East where they were resolved and in the West where they suffered from the contradictions between their specific purpose and the right of any country — as recognized in the NPT — to develop nationally all the stages of the civil production of nuclear energy.

Despite these contradictions, the United States and Canada were determined to impose their restrictions on sensitive stages of the fuel cycle, and more especially on the production of plutonium, the fissile material that the Indian explosion had put in an unfavorable light. They were about to engage in a long war over plutonium, in which they were to make use of every technical, legal, diplomatic, and political weapon at their disposal. Mutual trust, which until then had formed the basis of international nuclear relations, was the main victim of this campaign. It is by no means evident that there was any benefit for the nonproliferation regime.

The Canadian government, which had been the most shocked by the Indian explosion, was the first to enter the battle with concrete action. Its main weapon was insistence on a right of veto over the reprocessing of all irradiated uranium of Canadian origin and of all spent fuel in a reactor built with Canadian aid. In brief, the government insisted that Canadian "prior consent" should be obtained before any extraction of plutonium that had been produced with a Canadian contribution — direct or indirect — could take place. This requirement was obviously in addition to the safeguards on peaceful utilization, to which the materials concerned were in any case subjected.

In late 1974, Ottawa announced a decision to reinforce the political constraints attached to Canadian nuclear exports and — something which had never been done before — to make this policy retroactive. The government therefore made known its intention to "renegotiate" its current bilateral agreements, which in fact meant confronting each importing country with the choice of accepting more stringent conditions or having the supply contract in question cut short.

Lengthy negotiations followed with both India and Pakistan, the two first countries that had adopted Canadian-type nuclear power stations. The negotiations failed and in 1976 Canada cancelled its agreements with both countries. The history of international nuclear relations had seen the beginning of an era of unilateral decisions and broken agreements.

In Pakistan, operation of the country's only nuclear power station, which supplied half of Karachi's electricity, was soon affected by the Canadian action. The situation in India was even more serious.

The breakdown of relations with Canada resulted from Indian refusal to accept IAEA safeguards on any existing or partially finished facilities that included Canadian equipment or technology. Since 1964 the Indian nuclear program had been directed (except for the single American light water power station built at Tarapur) toward the Canadian reactor type. Four power stations of this type were envisaged. One was at Rajasthan, one near Madras, the third in the state of Uttar Pradesh and the fourth in Gujarat; all four were to be comprised of two 220-MW(e) reactors and would provide their respective areas with an important part of their electricity requirements.

The first of the two Rajasthan reactors, which was commissioned in 1976, had been built with direct Canadian aid, both being subject to IAEA safeguards. The others, built nationally, were to be exempt from this control despite the fact that they embodied certain equipment from Canada and were to use Canadian heavy water. The second Rajasthan reactor was just commissioned in 1981, thanks to nationally produced heavy water supplemented by supplies bought from the Soviet Union. The startup of the next three stations was to be spread out over the next 10 years. The breakdown of relations with Canada in fact forced India toward complete nuclear self-sufficiency. This was achieved under the leadership of Homi Sethna, chairman of India's Department of Atomic Energy, which is the only nuclear organization in the world still to possess a complete monopoly of national development in this field. Building and owning fuel cycle installations and heavy water plants, as well as all nuclear power stations, this department has full responsibility for India's nuclear electricity supplies.

But the achievement of complete nuclear self-sufficiency inevitably required time. Meanwhile, and particularly while awaiting the completion of four heavy water plants (one of national design and construction, one German, and two French) and despite a purchase from the Soviet Union, the Indian program was delayed by some four to five years as a result of the break with Canada. On this vast continent where production of electricity is one of the most promising contributions toward conquering poverty and misery, it was distressing to see, as I did in 1979 in Rajasthan, a completed reactor lying idle for lack of heavy water — at a time when the Canadians had overabundant stocks.

If the Indian explosion had taken place, like the Chinese one, before the entry into force of the NPT, it would certainly have created less commotion. For the first time, such an operation had proved counterproductive for a country — at least in the short term — though in the long run India must surely benefit from its "long march" toward autonomy.

In the United States, during the months following the Indian explosion, proliferation risks became a focus of interest for the press and the general public. Professors of political science began to give special attention to the subject, some basing their approach on the calculable quantities of plutonium being produced in currently operating nuclear power stations. In 1974 this production, for all the nonnuclear weapons states, already totalled several tons per year, a figure expected to increase tenfold before the end of the century. It was of course highly irradiated plutonium, of little use for making weapons, but nevertheless capable of producing a tremendous explosion.

Other theoreticians on nonproliferation discussed in great detail various possible acts of nuclear violence, including thefts of fissile materials by members of terrorist or revolutionary groups that would then — it was claimed — manufacture homemade bombs.

Despite the wide diffusion of data relating to the less sophisticated atomic weapons, it remains practically impossible to manufacture a nuclear bomb by makeshift methods, a task "as difficult as building a space rocket at home" according to Andrei Sakharov, father of the Soviet H-bomb and holder of the Nobel Peace Prize.

In any attempt at nuclear blackmail based on theft of fissile materials (a risky operation indeed for the thieves!), it would be extremely unlikely that the criminals would have at their disposal either a suitably equipped clandestine laboratory or the necessary highly specialist technicians. The possibility of such a venture must therefore be extremely low and in any case a matter well within the competence of national police services.

Nonproliferation was also a subject of official preoccupation. While the Canadian government was dealing with the issue bilaterally with its customers, the U.S. State Department, supported by Congress, decided on a multilateral approach. It believed the time had come to resume discussions between nuclear exporting countries, which had been tried without great success 10 years earlier during the "Cashmere meetings" held in

London. France had contributed to the failure of those meetings, and her positive participation would now be an even more indispensable condition for their success should they be restarted. The reasons for France's past reticence had in fact partly disappeared: her status as a nuclear power was no longer disputed, she possessed large resources of nonrestricted uranium, and would soon have equally large quantities of enriched uranium at her disposal. Her nuclear independence was assured.

The London Guidelines

In these new circumstances President Gerald R. Ford, during his first summit meeting with Valéry Giscard d'Estaing at the end of 1974 in Martinique, suggested that in view of the renewed risks of nuclear weapons proliferation a concerted approach to the problem should be arranged between a number of advanced "supplier countries," including the Soviet Union. Aware of the gravity and importance of the matter, the French president accepted.

The negotiations were once again held in London. Their purpose was to establish general rules of conduct, accepted by all the supplier countries and designed to protect the nonproliferation policy from suffering as a result of growing commercial competition. Participation this time was different from that of the 1960s series of meetings: it was comprised of the four leading nuclear powers — the United States, the United Kingdom, the Soviet Union, and France — together with Canada, West Germany, and Japan. Several meetings were held, between 1975 and 1978, and eight other industrial countries from East and West* joined in during this time.

The first result of the negotiations, and doubtless the most important, was official confirmation from France that she would now abide by the NPT rules in the letter as well as in the spirit. Relinquishing the advantages she had previously derived from the freedom of choice over political conditions to be attached to her nuclear sales, France accepted that IAEA safeguards should now apply to all future exports of items on an agreed list of materials, equipment, and technological information that was jointly prepared by the participant countries. This list was broadly similar to that established earlier by the countries participating in the NPT, as an aid to the treaty's enforcement.

The French government, nonsignatory to the treaty, had therefore

---

*Belgium, Czechoslovakia, East Germany, Italy, the Netherlands, Poland, Sweden, and Switzerland.

decided to follow its main "executive instruction." By this act France greatly reinforced the treaty, for there was no longer any excuse for Germany, Italy, or Japan to withhold their ratifications, which they had done on the grounds that French industry was enjoying a privileged position in international competition. Not only, in fact, had France associated herself with concerted international action over nonproliferation, she had added her support to the NPT itself, one of the pillars of the entire structure.

Once French support had been obtained for the NPT rules governing exports to nonnuclear weapons countries, the London negotiations focused on trade in "sensitive" materials and installations.

The United States and Canada, assured in advance of automatic Soviet support, now tried to enforce the requirement of "prior consent," from the original supplier of uranium fuel or of enrichment services or from the supplier of the reactor in which the fuel had been irradiated, before that fuel could be reprocessed in the recipient country. The two North American countries also wanted to impose an embargo forbidding the provision of technical aid to any nonnuclear weapons state where there was any nuclear activity not subject to IAEA safeguards. This was the so-called "full scope safeguards" clause. Finally, they planned to limit the provision of aid in respect to sensitive installations only to such installations that were under multinational management.

These proposals, which bore traces of the philosophy of the earlier Acheson-Lilienthal plan regarding international management for "dangerous" installations, were not agreed upon at first; then, in a spirit of compromise, they were later accepted on an optional rather than obligatory basis.

France, discreetly supported by the Federal Republic of Germany, Italy, and Japan, had opposed their being made obligatory. France thus rejected any form of interference or embargo beyond that called for under the IAEA safeguards, and in particular opposed the use of exports as a means of exerting pressure on the policies or technical programs of purchasing countries.

Finally the supplier countries undertook, from the end of 1975, to respect certain nonretroactive "guidelines" setting out minimal conditions to be required from a purchasing country. These directives, made public only in early 1978, specified that transfers of sensitive technologies and corresponding installations, despite the application of IAEA safeguards, should be undertaken only with the greatest caution.

This was a first step toward refusal to transfer sensitive technologies or to the prohibition of their use. In this respect the London "guidelines" were contrary to the NPT, which guaranteed in Clause IV the inalienable right of a nonnuclear weapons state to carry out research on nuclear energy, to produce it, and to use it for civil purposes. France, not having signed the NPT, was therefore the only one of the participant countries not to have violated the treaty by subscribing to these directives.

A new form of discrimination had appeared. Germany and Japan were prepared to accept it, so long as it was not applied to them, thus confirming their status as "honorary nuclear powers."

The importing countries, particularly those of the Third World, quickly challenged the London meetings, likening them to a cartel intended to perpetuate its members' advanced positions and to undermine the promises given to nonnuclear weapons states in the NPT. Their fears were not unfounded, for what they called the "Suppliers' Club" was now creating new discriminations among countries without nuclear weapons and between the more industrial ones that possessed the sensitive technologies and were determined to exploit them (such as Germany and Japan) and the less-developed ones to which these stages of the nuclear fuel cycle were to be virtually forbidden.

The determination of the American and Canadian delegations to the London talks was not unconnected with internal political pressures at home. In the United States, no less than three congressional commissions were studying ways of reinforcing the existing legislation, and in 1976 an amendment to the Foreign Aid Act was adopted to this effect. Henceforth any country not subject to the IAEA's "full scope safeguards" that undertook the construction of an enrichment or reprocessing plant could be deprived of American economic and military aid. The amendment was due to Senator W. Stuart Symington, who had a quarter of a century earlier, when an adviser to the White House, been in favor of threatening the Soviet Union with nuclear weapons to settle the Korean War.

Contested Agreements

The London guidelines had barely been adopted in 1976 when loud protests were voiced in the United States against three recent nuclear agreements concluded with Third World countries, each relating in part to the supply of "sensitive" installations. All three proposals had in fact been most carefully negotiated to ensure that they conformed exactly to the new directives. In particular, it was laid down that any duplication of the sensitive installation by the importing country would also come, automatically, under IAEA safeguards. The IAEA's board of governors had approved these agreements and had accepted the duties of inspection.

The most spectacular of the three concerned an ambitious program of nuclear power station construction in Brazil [10,000 MW(e) by the 1990s], to be carried out as a collaborative effort between local and German industry. The agreement, which in this period of recession was of vital importance to the German company, Kraftwerk Union, also included a German pledge to equip Brazil with plants covering all the stages of the fuel cycle, from uranium prospection to plutonium extraction, and including uranium enrichment.

Washington, exerting pressure at the highest levels, made every effort to persuade Bonn and Brazilia to renounce this transfer of sensitive technologies, above all that of reprocessing, for which only a small pilot plant was initially planned.

The governments concerned refused to give way, and Germany's prestige in the eyes of the Third World grew accordingly; Germany had made it a matter of honor not to go back on the conditions of a nuclear contract. But although the agreement therefore remained unchanged, the program became considerably more costly and was due to be reduced in size and delayed by at least five years when the recession hit Brazil, a development that went some way toward satisfying American wishes.

The other two offending agreements, financially less important, concerned the construction of irradiated fuel reprocessing installations in South Korea and Pakistan. They were to be built by a French firm, now Saint-Gobain's specialized subsidiary in this field, and were to be of smaller capacity than the plant then being completed by the same firm in Japan.

The Pakistani contract had taken a very long time to conclude. Negotiations had begun in the 1960s, but had suffered many delays because of uncertainty over the size of the plant and its financing, coupled with Islamabad's unwillingness to accept IAEA safeguards, which for the first time had been required by Paris. In 1974 however, following the Indian explosion, these obstacles suddenly disappeared: Pakistan found the needed finance and also accepted the safeguards.

In the case of the second contract, this was abandoned by South Korea in 1976 as a result of external pressure. As is often the case in such an affair, the actual origin of the pressure was not necessarily the officially declared origin. It was the French who claimed paternity while the Americans, whose troops were stationed in Korea, maintained that they were not responsible!

On the other hand, despite an international press campaign and personal representations by American Secretary of State Henry A. Kissinger, the French government refused to cancel the newly signed 1976 agreement with Pakistan. A first step toward its fulfillment was in fact taken with the transfer of detailed plans. Then, as previously mentioned (see p. 205), following discussions between the two governments concerned as to the final purpose of the plant — the construction of which was certainly difficult to justify economically at this stage in Pakistan's nuclear development — the project was shelved in 1978 before any sensitive equipment had left France.

Only a few months later it became clear that the Pakistani threat of proliferation, for which the French government had been pilloried because of the 1976 agreement, came primarily from a centrifuge enrichment project, and that the sensitive technology was coming from Urenco, in particular from the Netherlands.

These trends and developments had not been lost on the French government, which had been drawing its own conclusions on these delicate matters since late 1976. It was then that the president of the Republic had inaugurated a Council on External Nuclear Policy, grouping under his chairmanship all the ministers concerned, together with the Commissariat à l'Energie Atomique's (CEA's) general administrator.

The first meetings of this council defined the government's position. France, convinced of the importance of nuclear power as a competitive and indispensable source of energy for the development of many countries, was ready to help these countries by supplying fuel cycle services, especially enrichment and reprocessing, and would be prepared to meet bona fide requests for the transfer of technology.

The French government's oppostion to proliferation was profound and it was most anxious to reinforce the safeguards attached to its nuclear exports and also to prevent commercial competition from adversely affecting the objective of nonproliferation. It therefore intended to exercise the greatest possible restraint over the transfer of sensitive technologies to other countries.

Going a step further, from caution to refusal, the government next decided that, until further notice, it would no longer authorize bilateral transfers of industrial reprocessing technology. However, the French government was not prepared to exert pressures in the nuclear, economic, or political fields to force the submission of any country's installations to safeguards, nor was it prepared to prevent that country from acquiring or exporting any such sensitive technologies independently. It was over this last attitude that the French and the Americans differed.

Having once adopted these principles of a nonproliferation policy in 1976 and then instituting the necessary system for their application in 1978, France strictly adhered to this policy. From that time onward, her positions in the commercial field and in international discussions were marked by an effort toward coherence and stability, in contrast to the extreme and sometimes incoherent positions of some other countries. In particular, the French government never ceased to point out that uranium enrichment was at least as important a potential cause of proliferation as was the production of plutonium by reprocessing irradiated fuel.

In keeping with these ideas, the CEA, during an important nuclear conference organized by the IAEA in Salzburg in May 1977, revealed the existence of a new process for uranium enrichment that was not suitable for producing high concentrations of uranium-235 and could therefore be communicated and exported without undue risk. The process had been developed in secret by French technicians for over 10 years. It was based on chemical exchange and solvent extraction processes and was economically comparable to gaseous diffusion, while its use involved no more than conventional apparatus of the chemical industry.

The unexpected announcement of this new process coincided with a summit meeting of the seven most advanced Western nations, whose agenda included precisely the problems of nonproliferation. But the French announcement was in no way a technical/political bluff, and Paris immediately proposed that other countries should join in the industrial development of the process and thus verify its nonproliferant characteristics. Negotiations with a view to such collaboration were undertaken in 1978 and 1979, notably with the Americans and Germans. But these negotiations were blocked by the impossibiity of revealing the details of the process, which were essentially very simple, before obtaining a firm commitment from the future partners. Conversely, the latter were unwilling to commit themselves blindly. At the end of 1979, the Americans were the first to conclude an agreement with the CEA, enabling their Department of Energy (DOE) to evaluate the process. They had, in accordance with the rules of poker, ''paid to see.''

Following this evaluation, the DOE judged the process to be interesting, less proliferant than the ultracentrifuge, and capable of development to improve its economic viability. However, such development would also be likely to reduce — though not to eliminate — the nonproliferation advantages vis-à-vis other processes.

Meanwhile the Australians, anxious to market their uranium in the most developed form possible, were again considering ''home enrichment,'' as they had done in the early 1970s. So in 1981 they began a detailed study to enable them to choose between the two processes they thought the most attractive, centrifugation and chemical enrichment. In the case of the centrifuge process, not only were Urenco and the Japanese competing but also American industry — with more advanced centrifuges of greater capacity. This is proof of the liberal attitude of the Reagan administration, now ready to export such sensitive equipment to a friendly nonnuclear country. Should the French process ultimately be selected, France would most probably participate in financing the project, which in any case would also be of interest to Australia's important uranium clients, the Japanese.

The French problem of interesting potential partners in the chemical enrichment process without disclosing too much technical detail was also encountered by the South Africans during their fruitless search for foreign participation in the projected industrial exploitation of their own secret enrichment process (see p. 374). In this particular case, general political factors added to the complications.

The problems of enrichment, and even more those of reprocessing, had assumed worldwide public importance as early as 1976. The issues had become, for the first time, important subjects in a U.S. presidential election campaign.

The Carter Policy

On December 12, 1952, 10 years and 10 days after the first criticality of the Fermi pile in Chicago, the first serious accident with a nuclear reactor occurred in Canada, luckily involving only material damage and no human injury. Following a sudden, too-rapid unplanned power increase, the large Chalk River heavy water research reactor, the result of the tripartite collaboration at the 1943 Quebec Conference, was severely damaged. The delicate operations of dismantling and repairing the reactor lasted two years, and the United States contributed to the work by sending one of their two teams then preparing to run their first two nuclear submarines on which construction work had recently begun. Taking part in this operation during his brief 11-month atomic career as a land-based nuclear submariner was Lieutenant Jimmy Carter. In 1953 he returned to civilian life on his peanut farm in Georgia.

A quarter of a century later, U.S. President Jimmy Carter became the second American leader, after Eisenhower, to attempt to orient world nuclear development in a new direction. Unlike the Eisenhower ''Atoms for Peace'' initiative, however, this time the reorientation was to strike the development of nuclear energy the heaviest blow it had received throughout its history.

As leader of the great country that had been at the head of all the scientific and industrial stages of this great adventure, that had discovered plutonium and the principle of the breeder reactor, Carter — determined to reinforce the new nonproliferation policy to its utmost extreme — prohibited the use of plutonium for civil purposes.

In his electoral speeches as Democratic candidate, he had advocated giving nuclear power the lowest priority compared with other energy sources; he had also proposed a moratorium on sales of uranium enrichment and reprocessing facilities, to have a retroactive effect on the German-Brazilian and French-Pakistani contracts. On the eve of the November 1976 vote, his rival Gerald R. Ford publicly adopted the same stand.

Once elected, Carter — whose entourage was by no means free from hostility toward nuclear energy — decided to practice what he had preached, inspired by a group of professors of political science and economics. These had recently set out their ideas in "Nuclear Power Issues and Choices," a report commissioned by the Ford Foundation. This report focused on the proliferation risks that would be presented by a world expansion of the "plutonium economy," by recycling plutonium in light water reactors, or by using it in fast-neutron breeder reactors. According to the Ford Foundation study, resort to these operations was premature, if not unnecessary, since world uranium resources would alone allow a sufficient development of nuclear energy, using slow-neutron power reactors, while awaiting the perfection of better technical and political solutions to the nonproliferation problem.

During an important speech in 1977, the president endorsed these ideas and, in order as he said to set a good example to the world, he had no hesitation in postponing indefinitely the commercial reprocessing of used fuels in the United States. To this end he prohibited the completion, then scheduled for three years hence, of the only remaining private reprocessing plant in the country, being built by Allied Chemical at Barnwell, South Carolina. He also proposed to stop further work on the country's prototype breeder reactor at Clinch River, near Oak Ridge in Tennessee, on which over half a billion dollars had already been spent. He declared this reactor, which had already suffered many delays, to be already obsolescent; as a result of this new misfortune, the American breeder program itself became even more obsolescent!

By this extraordinary and unique act of self-mutilation, an already declining American industry was to become paralyzed in two key sectors of future development, fuel reprocessing and breeder reactors, precisely the sectors in which the United States was already between 5 and 10 years behind the Soviet Union and western Europe, in particular, France. The president, in thus giving satisfaction to American opponents of nuclear energy, attracted fierce hostility from his own industrialists as well as a vigorous opposition in Congress.

However, the president's zealous efforts to set a good example had limits. Hence in 1976, when the long controversy over the transfer of U.S. enrichment plants to private industry was finally resolved (by a narrow congressional vote to maintain public ownership) and the construction of new enrichment capacity had then to be decided, the choice in 1977 was for the centrifuge method rather than the current gaseous diffusion process. The reason for the choice was energy economy; the president did not oppose this despite the far more ''proliferant'' characteristics of the centrifuge method. Production of weapons-grade uranium-235 by building a clandestine plant or by adapting an existing plant is in fact much easier (and much harder to detect) when using the centrifuge method. American policy, obsessed by the plutonium issue, paid far less attention to enrichment, the other sensitive stage in the fuel cycle. Paris made many attempts to convince Washington of the importance of controlling enrichment processes. Development of the South African enrichment process and the later Pakistani project justified the French concern.

In his April 1977 speech, President Carter, hoping to encourage countries to continue using their light water reactors for as long as possible before considering extracting and recycling the plutonium produced, had declared that he wished the United States to resume the role of a completely reliable and trustworthy supplier of enrichment services. He even unearthed Eisenhower's old idea of an international fuel bank, the purpose of which would now be to guarantee customers against any breach of contract for motives other than nonproliferation (including therefore erratic changes of opinion in the U.S. Congress!). In the same speech, Carter announced his government's intention to ''renegotiate'' current bilateral agreements so as to bring them in line with his new policy. He even appointed an ambassador-at-large for this purpose, Gerard Smith, a veteran of nuclear diplomacy who had negotiated the SALT I agreements. The announcement of this renegotiation of past contracts could hardly be expected to encourage confidence in the inviolability of those to come.

The first country to come under pressure to renegotiate a current agreement was Yugoslavia, a country toward which the United States should have had the most consideration since it was the only socialist state to have ordered an American nuclear power plant. Washington had in fact stopped the issue of export licenses since 1976 for the component parts of the 600-MW(e) Yugoslav installation being built by Westinghouse. In addition, the Americans were insisting on IAEA safeguards and on the right to veto the reprocessing of any fuels irradiated in the reactor, even if not of American origin. The Yugoslav government refused these conditions, and early in 1977 referred the affair to the IAEA board of governors.

An official complaint against another country to the agency was unheard of and made the worst possible impression on the Third World. Without being finally settled, the affair lost some of its drama following Marshal Josip Broz Tito's official visit to Washington a few months later. The disagreement was not resolved for a long time, and construction of the power station suffered a costly delay. It was not completed until 1981.*

President Carter was also determined to convince the rest of the world that his policy was well founded, and he had decided to invite all interested countries to take part in a technical program designed to evaluate the fuel cycle from the viewpoint of nonproliferation. This program, the International Nuclear Fuel Cycle Evaluation (INFCE), was one of the main elements of his action at the multilateral level.

In May 1977, he officially put this proposal before the summit meeting in London of the seven most industrialized Western countries. He also put his name to the communiqué following this meeting that confirmed the world's need for increased reliance on nuclear energy. This did not prevent him, this time in front of experts from 40 countries gathered in Washington that October for the inaugural meeting of the INFCE, from telling us that he believed the world's requirements for nuclear energy had been greatly exaggerated.

At the suggestion of President Giscard d'Estaing of France, the INFCE studies were preceded by a preliminary analysis carried out by experts from the seven countries that had taken part in the London summit. The main objective of the analysis was to clarify conditions of participation in the evaluation. France played an important part in the definition of these conditions: the research program was to cover two years; its results must not be prejudged; therefore current energy programs should be pursued and no decisions likely to affect them should be taken during this ''period of truce'' — which the U.S. Congress hastened to break.

The INFCE program was reminiscent, on a considerably larger scale, of work undertaken 30 years earlier at the United Nations by scientists making a first attempt at controlling atomic energy. This time, however, the technology to be safeguarded was industrially developed and practically free from political secrecy.

Several hundred experts from over 50 interested countries, including some from the Third World, surrendered themselves, with logistic help from the IAEA, to the delights of seeking a technical-diplomatic com-

---

*A similar affair occurred in 1979 with the Philippines, when Washington refused Westinghouse authorization to build the first nuclear power station ordered by this country on the pretext that the site chosen by the Phillipino government was unsuitable from seismic aspects.

promise in a sort of giant scientific "happening." The various issues in question had already been debated at length in international conferences and seminars over many years, and on matters of disagreement all the experts had well-established (and well-known) beliefs that sometimes even had a degree of backing from their respective governments. They were not, therefore, likely to alter their attitudes in the face of the already known arguments of their colleagues. Hence, the task of the INFCE consisted mainly in presenting the issues as a compromise between the most diametrically opposed concepts and, more exceptionally, in drawing attention to disagreements.

At a cost of around $100 million (including salaries and traveling expenses), a 20,000-page collection of studies was produced, which contained nothing new and from which eight reports were written, each under the responsibility of a working group assigned to one of the stages in the fuel cycle. The conclusions were approved at a final conference held in Vienna, in a somewhat disillusioned atmosphere, in February 1980. While governments ought to have been able to derive from these conclusions the technical arguments for their future nonproliferation policies, because all differences of opinion had been attenuated in the effort to reach compromise, every government could in fact find — with help from its own experts — arguments both in favor of its own actions and opposed to any actions it did not support. Nevertheless, on the whole there was a predominance of conclusions that did not support the American policy of President Carter.

The Russians, who had solved the problems of proliferation within their sphere of influence in their own way, participated only with great caution in the INFCE exercise, the purpose of which they had initially found difficult to understand, while they observed with curiosity and interest the differences between the capitalist countries. The comparative isolation of the American experts, whose only regular support was from the Canadians and the Australians, seemed quite remarkable to those familiar with the international nuclear conferences of the past when the pace was set in Washington by a State Department often indifferent to contradiction, even from allied states.

Despite some real differences over estimates of the quantities of uranium available until the turn of the century, the INFCE honorary jury of experts clearly acquitted those principally accused in the affair, namely plutonium, reprocessing of spent fuel, and breeder reactors. As generally expected, although contrary to the U.S. president's hopes, the experts were unable to suggest a miracle way out from the fundamental contradiction between independent national civil nuclear energy development and the elimination of factors favorable to the production of nuclear weapons. Most of the experts were in fact quite convinced that all options should be kept open and that no technical solution should be forbidden because of proliferation risks — even if it were a solution among those methods more obviously susceptible to being used for clandestine production of nuclear explosives, such as centrifuge enrichment.

The INFCE exercise nevertheless emphasized the advantages, recognized long ago, of the internationalization of sensitive operations, and particularly of plutonium stocks, together with the value of IAEA safeguards in any nonproliferation policy and the importance of further strengthening these two factors.

Despite its disproportionate size compared with the results obtained, the INFCE nevertheless contributed toward a better awareness of the problems of nonproliferation and better understanding by the governments concerned. This was done within a framework from which — in contrast to that of the Suppliers' Club — the importing and Third World countries were not excluded. On the other hand, it did not produce the result sought by the United States: the adoption of their point of view by the other countries.

As for the U.S. Congress, it neither heeded nor even waited to hear the INFCE results. One year before the exercise was due to be completed, it adopted new legislation, the Nuclear Non-Proliferation Act, that was both strict and constraining. By applying this act immediately, Congress contravened the pledge of a 2-year standstill that had been given by the American president as a precondition to the launching of the INFCE.

The New American Legislation

The Nuclear Non-Proliferation Act was promulgated in April 1978. A monument to the complexity of American legislation, it included all the provisions of embargo and veto rights that the United States and Canada had proposed during the London meetings, and that had been rejected by the main non-Anglo-Saxon Western suppliers. Everything in the act was designed to oblige nonnuclear weapons states to submit their entire nuclear programs to IAEA safeguards and to dissuade them from any attempts at reprocessing or enrichment. By outlawing reprocessing, the act also condemned the principle of breeder reactor development.

The act provided for an embargo on nuclear aid or services to any country refusing either the NPT or ''full scope safeguards,'' or developing any technology considered by the United States to be proliferant; it also established the right of prior consent, or of veto on the reprocessing of fuels made from U.S. enriched uranium or irradiated in a reactor of U.S. origin.

The president was legally obliged to open negotiations to bring existing bilateral agreements into conformity with the new rules and there was a "guillotine" clause under which supplies would cease within 30 days to any country refusing to renegotiate.

Finally, as a way of escape from the otherwise inextricable situations that it was bound to provoke, the act included a number of provisions for its application to be waived on the president's initiative and with the agreement of Congress. Having tied themselves down in the name of nonproliferation, the Americans were to use this method of presidential exception to untie themselves in the name of their national interests. Despite its apparent inflexibility, the act thus allowed a case-by-case approach to its application.

This case-by-case approach was later adopted, rather than a vain attempt to amend the new act, in the policies from 1981 onward of the Reagan administration, less sectarian and more pragmatic than its predecessor over questions of nonproliferation.

Prior to this, however, American policy was very much determined by the constraints of the act. This made it imperative, if the U.S. national industry was not to be severely handicapped in international competition, that the other main supplier countries be persuaded to adopt the same export conditions imposed by the new legislation, which went well beyond the "London guidelines." This of course was in parallel with, and independent of, the INFCE conclusions, of which Congress took very little notice since they did not conform to its position.

The scenario for a generalized control of peaceful utilization in the "Atoms for Peace" program was thus repeated a quarter of a century later, though this time success proved far more elusive. Of much greater concern and seriousness was the threat of American policy to the energy independence of countries that, like France, were particularly short of conventional energy resources and were less lucky than the United States in having such enormous coal deposits. The attempted moratorium on uranium reprocessing and breeder reactors was clearly an attack on freedom of choice among available technical options, as well as affecting key stages of the fuel cycle in which several industrial countries were already engaged. It came at a time when no nonnuclear weapons state seemed willing to accept any further discriminations, either factual or theoretical, that went beyond those of the NPT.

Prior Consent

In the past, the main U.S. weapon for imposing their control policies, beyond their technological advance, had been their monopoly of uranium enrichment. This had been gradually eroded since the early 1970s and by 1979 it had suffered to the extent that half the enriched uranium entering the European community had been produced in a Soviet isotopic separation plant. Moreover the Eurodif gaseous diffusion plant, once in operation, would make several West European countries, in particular France, independent of the need for any outside enrichment services by 1982 at the latest.*

Nevertheless the Americans still had some trump cards to play in their crusade against plutonium. In particular, several important agreements for the supply of enriched uranium were still in operation, giving Washington rights over transfers from country to country and sometimes over the actual reprocessing of this enriched uranium after irradiation.

On the other hand, the American antireprocessing policy was considerably strengthened by events in two of the countries richest in uranium, Australia and Canada, the former being as yet only a potential producer, since the powerful Australian Miners Union had since 1973 opposed the opening up of the recently discovered extensive ore deposits. The risk of proliferation was one of their favorite arguments to justify this paralyzing action. It was not until 1981 that exploitation of two major Australian deposits began, and this not without initial difficulty, for the railway men temporarily refused to transport the uranium concentrates, which had to be sent by private air freight.

This tempted the Americans to try to replace their failing stranglehold on enrichment with a stranglehold on uranium, their two Anglo-Saxon partners being willing also to insist on a right of prior consent for the reprocessing of uranium they had supplied.

In fact, Ottawa had anticipated Washington on this issue by being the first to call for renegotiation of nuclear contracts. The American secretary of state had even acknowledged, in a letter to his northern colleague, Canada's "leadership" in the prevention of nuclear proliferation. This must have been one of the very rare occasions the United States conceded leadership in a political field to another country: the "good conduct certificate" was all the more surprising for being awarded to the power that had facilitated the Indian explosion!

The American policies of prior consent and plutonium prohibition were strongly opposed by the governments of the leading members of the

*During 1981 the uranium enrichment purchases in the European community were: Eurodif, 61 percent; Soviet Union, 16 percent; the United States, 13 percent; and Urenco, 10 percent.

European community and of Japan, which had made an investment in uranium reprocessing. They found it unacceptable to give a right of intervention in the technical orientations of their energy policies to the government of another country that happened to produce and supply uranium or concentrated fissile materials, whether it was a Third World country with no atomic program, such as a future independent Namibia, a major uranium producer; or an advanced country such as Canada, the United States, or Australia; not to mention the Soviet Union which, though an important supplier of enriched uranium, had never claimed such a right.

The importing countries did not deny that, from the viewpoint of nonproliferation, the reprocessing stage was a sensitive step, because it enabled plutonium to be extracted. But this stage was also part of the route toward the final treatment of radioactive wastes, and of the process of "breeding" which, by reducing uranium consumption to a minimum, was the basis of long-term nuclear development.

The early skirmishes in this reprocessing conflict left neither side with a decisive victory.

Canada was the first to enter the fray. In early 1977 its government decided to impose an embargo on uranium exports to the European community, exports that concerned Germany and the United Kingdom in particular. The action was basically a protest by the Canadians at the slowness with which their agreement with the European Atomic Energy Community (Euratom) was being renegotiated. They finally lifted their embargo at the end of the year, having temporarily relinquished their claim to the right of prior consent for the reprocessing of all uranium delivered over the next three years, declaring themselves satisfied instead with consultations involving no obligations. In this way, the problem was being shelved until the end of the INFCE "truce," in accordance with one of the preconditions to the start of that exercise. Australia was also to adopt this solution, theoretically in its case, since the new mines were not scheduled to begin production until 1981.

The truce was also maintained on the renegotiation of the USA-Euratom Agreement, though not without some serious skirmishes. The agreement, which was valid until 1955, had been designed to facilitate the entry into the community of the light water reactor. In exchange, the member states as a whole were to be treated as a single territory for the purposes of transfers of nuclear materials among themselves and for reprocessing. Under the agreement, therefore, Washington's approval was not required for the reprocessing of fuels of American origin, irradiated in another Euratom country, to be carried out in the French plant at La Hague. The new American legislation clearly obliged Washington to start renegotiations within 30 days for the suppression of this arrangement.

Despite the threat of an embargo on U.S. deliveries of enriched uranium, particularly awkward in the case of highly enriched material not available from the Soviet Union, the European commission, on instructions from the governments concerned and especially thanks to the French stand in this matter, held its ground, refused to negotiate, and in spring 1978 the 30-day deadline passed without concession to the American "ultimatum." Washington, unwilling at this time to provoke a major crisis, did not carry out its threats.

Two months later when the affair had blown over, the community, wishing to avoid any real breakdown in relations with Washington, authorized the commission to undertake "discussions," it being understood that renegotiatons could only begin at the end of the INFCE operation. The U.S. government had modified its demands, and it was good for both sides to save face.

Countries other than those of the European community also suffered from Washington's attempts to impose its views on irradiated fuel reprocessing. The first victim had been Yugoslavia, in circumstances already described. The most encumbered by American "attentions" was probably Japan, though in that case it was not a question of renegotiation but simply the strict application of clauses in current contracts.

During the summer of 1977, the first Japanese irradiated fuel reprocessing plant was nearing completion at Tokai-Mura northeast of Tokyo. The construction contract had been won by French industry against competition from British and American firms. The cost was $200 million and the work had never given rise to the slightest protest. The plant was capable of producing some 3,000 pounds of plutonium per year from enriched fuels used in light water reactors. The Japanese nuclear power stations of this type were at the time running exclusively on fuel enriched in the United States. In 1978 the Japanese, in an attempt to boost the falling dollar, had even purchased $1 billion worth of enrichment in advance.

However, by virtue of their bilateral agreement with Japan, the Americans had a right of prior consent over the reprocessing of fuels enriched in U.S. plants (Canada, through its own bilateral agreement, had the same right as a result ot its sales to Japan of natural uranium). The purpose of the clauses had been to ensure that the plutonium extraction installation was built so as to facilitate effective safeguards; Washington decided to use them to prevent the plant being put into operation.

The attempt failed because of the violent reaction of the Japanese government, which went so far as to draw comparisons with the American embargoes on strategic materials that had been one of the causes of the attack on Pearl Harbor and of Japan's entering the war in 1941. Washington gave way, and granted a limited 2-year authorization to reprocess American fuels in the Japanese plant, on the condition that this was also used for research that would be of interest to the INFCE experts, and more especially on the condition that the plutonium produced would be kept in solution, making its possible diversion more difficult.

The affair left deep scars and reinforced Japan's intention to behave in the civil field exactly as if it were a full-scale nuclear power, pursuing the construction of a pilot uranium enrichment installation as well as preparatory work toward a breeder reactor and a second, larger, reprocessing plant. Two years later, in 1981, the Reagan administration authorized the Japanese to run their reprocessing facility normally until the end of 1984 while American industry became a strong contender for the order to construct the larger one — again in competition with French industry.

Disagreements between Tokyo and Washington did not stop here, however, for most of the irradiated Japanese fuels were due to be reprocessed in France or in England under contracts signed between 1975 and 1978 with the fuel cycle subsidiaries of the atomic energy organizations in the two countries. These contracts represented for each country a revenue well over $1 billion, an important fraction of which had been paid in advance to finance the necessary extensions of the respective units. Here again the Japanese government had the greatest difficulty in gradually extracting from the Americans their permission to send the fuels to France, in spite of a Franco-Japanese agreement not to return the corresponding plutonium to Japan before 1990, and then only on conditions mutually agreed upon between Paris and Tokyo.

The Japanese contract with British Nuclear Fuels Limited was eventually shelved following a long public inquiry into the necessary extension of the plant at Windscale. This inquiry procedure, to which we shall return later, had a favorable outcome: The judge responsible, referring to the new practice of renegotiating contracts, declared that he did not believe the best way to conclude a new contract was first to break an existing one.

Washington, in fact, had not hestitated to intervene at a high level to persuade the British government to defer its decision on the proposed extension. The decision was eventually made, following a debate and vote in the House of Commons that went very much in the government's favor. During this debate, David Owen, the then foreign secretary, had made an especially forceful criticism of the American policy on reprocessing, particularly since it came from Washington's most conciliatory ally. In Owen's view, the recycling of plutonium was the best way of usefully eliminating the nuclear explosive plutonium, whereas the storage of irradiated fuels represented both a loss of more than 90 percent of the contained energy and a danger from the viewpoint of proliferation. Indeed, with the gradual decrease in radioactivity of the fission products, stored irradiated fuels would become progressively more easily accessible as sources of plutonium. This was a heavy blow for the Carter policy.

The INFCE conclusions, the resistance of European countries and Japan, a new overabundance of uranium accompanied by an inevitable drop in price, competition in the enrichment market, and above all the change to a Republican administration in the United States combined in 1981 to trigger a weakening of the position of the Anglo-Saxon suppliers.

While U.S. industry was authorized to become involved in the second and larger reprocessing installation in Japan, there was a complete change of attitude when negotiations began on a Franco-Australian agreement that would give France some access to the uranium resources of the Australian continent. In exchange the Australians could, if they so wished, have access to French enrichment technology.

The conclusion of this agreement at the end of 1980 led to a compromise between the protagonists and opponents of prior consent — a compromise that was later embodied in a general agreement between Euratom and Australia. It consisted of giving advance consent for any fuel cycle operations (enrichment or reprocessing) needed for an agreed program of nuclear electricity production. It was going to be difficult for the United States and Canada, in their turn, to ignore this compromise.

Dealing with the Rebels

The countries that had refused to sign the NPT, and whose nuclear activities were therefore not subject to IAEA safeguards, also underwent pressure from the American administration of President Carter. His new legislation would ultimately have entailed an embargo on all U.S. nuclear supplies to them.

There were five countries in this position: South Africa, India, Argentina, Brazil, and Israel. Spain should really have been in the same category but French participation in the operation of the only Spanish power station not subject to IAEA safeguards, the graphite-moderated, gas-cooled Vandellós plant, no doubt combined with the essential role played by the country in American military structures and strategies to give protection from Washington's wrath. (In 1981 the power station was finally placed under IAEA safeguards).

In India's case, the Americans' only means of exerting nuclear pressure was through the supply of enriched uranium for the country's first power station, the boiling water plant at Tarapur. Obliged by contract to obtain their supplies from the United States, the Indians were free to approach other suppliers if the Americans defaulted. In such circumstances they also considered themselves free to extract the important quantities of plutonium produced in the reactor since its entry into service in the late 1960s because the Indians believed that if the Americans broke their supply contract, the whole agreement would become null and void including the clause giving Washington the right of prior consent for the reprocessing of the irradiated fuel.

President Carter first decided to stop all supplies in 1979 because the previous year, during a meeting with Prime Minister Morarji R. Desai, he had been unable to persuade the latter to adopt "full scope safeguards" for the Indian civil program, particularly at a time when there were rumors from the west of Pakistani centrifuge plants.

The first embargo on American enriched uranium was thus to be directed at the first nuclear power station supplied to a Third World country under the "Atoms for Peace" program; although the peaceful purpose of this plant, subject to IAEA safeguards, had never been in question. American industry rather than India would have been the main sufferer.

However, following the events in Afghanistan and the supply to India of substantial quantities of Soviet arms, President Carter was obliged to reassess the situation and, not without difficulty, he secured a vote in the Senate in October 1980 exempting Tarapur from the new U.S. legislation. A first delivery of enriched fuel was made, with a second foreseen approximately a year later, provided India "behaved correctly." The final decision on this was to fall to President Reagan — and once more in 1981 those for and against the strict application of the law were to find themselves in conflict in Congress, in the administration, and in the eye of public opinion, while negotiating teams met fruitlessly, alternately in Washington and New Delhi.

The problem of South Africa was even more complex, as the new American president was to find out when dealing with it in 1981. In 1977 South Africa had been voted out of its seat on the IAEA's board of governors, a seat that it had held under the agency's statute since IAEA's creation because it was the most advanced African country in the atomic field. The expulsion had become possible as a result of the enlarging of the IAEA's board of governors in favor of the Third World countries, which are naturally hostile toward South Africa. These developing nations were also determined to eliminate, in the agency's management, the predominance given by the IAEA's statute to advanced countries in possession of both technology and essential raw materials.

One year earlier in 1976, the IAEA had agreed to apply its safeguards to two 900-MW(e) South African reactors in the Cape region, construction of which had recently been started by the French company, Framatome, following a close and eventful competition with bids from American and German industry. To allay any concern over plutonium, the South Africans had voluntarily agreed not to reprocess fuels from these reactors within their own territory.

A double threat nevertheless hung over the future of this facility, which was due to be completed in 1983. This first light water station to be exported by French industry was being supplied to the country that in 1963 had refused to discriminate between France and the two Anglo-Saxon nuclear powers over the supply of nonrestricted uranium (see p. 287).

In the first place, there was a threat that the hostility of the United Nations, particularly the Black African and Third World countries, over the acute issues of apartheid and Namibia, might lead to a decision by the U.N. Security Council prohibiting general trade, or perhaps simply nuclear trade, with South Africa, and that this might happen before the completion of the 2-reactor power station.

Secondly, the enriched uranium required for the first two fuel charges had to come from the United States according to a bilateral American-South African agreement to this effect. There was a clear risk that the contract would not be honored by Washington if Pretoria, suspected as we have seen of preparing a nuclear explosion, maintained its refusal to sign the NPT and to accept IAEA safeguards on its pilot enrichment plant that embodied an original and secret process. This likely American action could delay the commissioning of the new power station by two to three years, the time the South Africans themselves would need to produce the necessary enriched uranium.

Since President Reagan was certainly better disposed toward Pretoria than his predecessor, it was probable that the United States would veto any sanctions resolution in the Security Council, but it looked as if his administration would not find it easy to proceed with the nuclear fuel supply contract in the face of the dual hostility of the opponents of apartheid and of those who insisted that American nonproliferation legislation should be strictly enforced.

Finally, with the tacit consent of Washington, the South Africans were able to solve their problem in 1981 by turning toward the European market and especially to some of the Eurodif clients oversupplied with enriched uranium because of the slowing down of their national electro-nuclear projects.

The French Position

From the early 1970s, France's role on the international nuclear scene began to take on new dimensions. Her leadership in the multinational exploitation of uranium deposits in the Niger; the successful construction and commissioning of Eurodif, which was expected in 1982 to contribute to nearly half of Western enriched uranium production; the international role of the fuel-reprocessing plant at La Hague; the sales by French industry of the large but ''stillborn'' research reactor to Iraq and of light water power stations to South Africa and Iran (those for Iran were abandoned in the early stages of construction, due to the revolution); hopes of further sales elsewhere, unfulfilled in the case of China but becoming more concrete in the cases of Korea and Egypt; and European participation in the construction of Super-Phénix had together by the beginning of the 1980s made France the most advanced nuclear country in the Western World.

The French government's attitude toward nonproliferation was an important element in the picture. Its moderating influence had already made itself felt in 1975 during the negotiations on the London guidelines, and also during the establishment of the INFCE's mandate. It had initiated action to underline the proliferation risks of enrichment, which the Washington policy — obsessed by plutonium and breeder reactors — tended to underestimate.

Supporting the development of a comparatively nonproliferant chemical enrichment process and multilateral gaseous diffusion plants, France had sought to slow down the spread of proliferant technologies and processes, such as today's ultracentrifugation and perhaps tomorrow's laser method, which involves delicate engineering but is more dangerous since it will allow high degrees of enrichment to be achieved directly. This method is also being studied in the United States for upgrading the content in plutonium-239 (while diminishing the proportion of plutonium-240) in civil plutonium so as to render it more suitable for military use.

It was clear that during the 1980s France would remain the principal supplier of reprocessing services and the leading producer of plutonium for civil uses. She therefore had an essential contribution to make toward the outcome of the plutonium conflict. In this field, devoid of any nonproliferant (or only slightly so) process, the French government since the late 1970s has been against the spread of small-size plants and in favor of large and even multinational facilities, built in politically stable areas and using the most advanced nuclear technologies.

Once plutonium has been separated from the unused uranium and the fission products in irradiated fuel, the most delicate political problem concerns the storage of the material before it is recycled in new reactor fuels. Without a reasonable and effective solution to this problem (which must also avoid giving the suppliers of natural or enriched uranium any excuse for seeking veto rights in this stage of the fuel cycle), normal development of international reprocessing services cannot be expected to take place.

Fortunately, the IAEA statutes provide for the storage, under international supervision and safeguards, of plutonium produced in reprocessing plants. This provision had in fact been the object of serious contention during the adoption of the statutes in 1956 (see p. 283). In 1978, the French government was in favor of such storage of plutonium on the sites of the reprocessing plants concerned, it being agreed that return of the material to its country of origin would only take place as and when it was required for fueling reactors, thus avoiding the accumulation of national stocks that might develop into worrying temptations.

Difficulties with Euratom

The application of any such restriction to the return of plutonium to a nonnuclear weapons member of the European Community in the reprocessing contracts concluded with France was to create difficulties with the Euratom commission and with the member countries it could concern, in particular with the Germans, for whom the Euratom treaty had always provided a means of limiting or avoiding the discriminations resulting from their renunciation of the military option.

Backed by the commission, these countries maintained the thesis that there should be free circulation of nuclear materials within the community and that France had no right to impose conditions on the return to them of their own plutonium.

The commission cited both its theoretical ownership under the Euratom treaty of all concentrated fissile materials on community territory other than those used for armaments, and the application to the Euratom treaty of regulations imposed by the separate Common Market treaty. In this way the commission deduced that a so-called "nuclear common market"' must exist, which should overrule the regulations of the Euratom treaty alone, which only prohibited quantitative restrictions and tariff barriers in intercommunity exchanges, and not political restrictions connected with potential military usage.

Referring to its prerogatives in matters of external relations and supplies of nuclear materials, the commission also claimed that problems of nonproliferation, as well as those concerning physical security and involving protection against nuclear terrorism, came under its authority. However, at the time the treaty was drafted, the first of these problems was far less acute than it had since become, while the second had not even been considered. At that time, in fact, no one foresaw the possibility of nuclear terrorism.

As early as 1959, however, the French foreign minister, in connection with the sale of nonrestricted uranium to Denmark and Sweden, had rejected the commission's right to intervene in the political arrangements linked with such nuclear exports, maintaining that these were matters of foreign policy and defense, and were therefore beyond the field of the Euratom treaty.

In view of the importance of these questions, particularly those concerning nonproliferation, the commission decided to have the matter referred to the European Court of Justice. For this purpose, in 1978 Belgium submitted a test case to the court, involving the problem of physical protection, with the object of obtaining legal confirmation of its (and the commission's) thesis by apparently challenging it. As the Belgian government had hoped, the court disagreed with the viewpoint provocatively advanced by the Belgians and developed arguments implying that not only physical protection but also nonproliferation came within the competence of the community, in the framework of the "nuclear common market" mentioned above.

This Court of Justice ruling was not compatible with the position of the French government, nor with that of several other member countries who wished to retain control of their external nuclear policies, notably those concerning nonproliferation.

Pursuing this guerrilla warfare, in 1978 the commission refused to approve an agreement under which the United Kingdom was to obtain uranium supplies from Australia. Such an agreement, between two Anglo-Saxon countries, gave Australia the right to forbid subsequent reexportation from the United Kingdom. The restriction was included for reasons of nonproliferation and was, in the eyes of the commission, incompatible with the "nuclear common market" concept. The nuclear powers replied to this action by refusing the commission a mandate under which it could have concluded a Euratom-Australian agreement, and so obtained the monopoly it sought over the supply of uranium from that continent to the community.

The affair was finally resolved in July 1979. The British signed their agreement, and the Australians gave up their right to veto the reexportation of uranium, under certain conditions, to other community countries. Two months later, the Council of European Ministers authorized the commission to conclude its own agreement with Australia, thus complementing those negotiated bilaterally.

Disagreements of this sort with the European commission were additional proof that the Euratom treaty required updating to take account of changes in the industrial and political situation and in the scale of development of nuclear power — all unforeseeable at the time of the treaty's original signature. The rules relating to the supply of uranium were particularly unsuited to the new world situation.

A sharp exchange of words on the subject of the Court of Justice's consultation and verdict, which took place in the French National Assembly in 1979 between the foreign minister, Jean Francois-Poncet and the former prime minister, Michel Debré, revealed that neither had any illusions over the position. Francois-Poncet, under attack from Debré, asked the latter why he had, when in office, " . . . never suggested the denunciation or even the revision of the Euratom Treaty?" Debré's answer was plain: "Because, sir, it was not being enforced"; the other's rejoinder was pithy: "It's not being enforced today either."

That was the nub of the matter: Whether or not the Court of Justice's ruling still allowed the nonapplication of the controversial aspects of the treaty.

Wishing to clarify the situation, in July 1979 France submitted a memorandum to her community partners and to the commission calling for revision of the chapter of the Euratom treaty relating to the supply of fissile materials, such modification being explicitly provided for in the treaty itself. The memorandum specifically called into question the nonenforced articles, and especially those relating to the supply agency's monopoly of transactions and to the principle of "equal access" to nuclear materials. The particular purpose was to ensure both freedom and security for those industries that had invested in the nuclear fuel cycle and to prevent member states that had made no financial contribution from nevertheless claiming access to the materials.

In September 1979 the Council of Foreign Ministers accepted in principle the updating of this part of the treaty and set up an expert committee to study the matter. It remained to be seen whether, this time at last, a solution could be found to this important and irritating problem that had been under discussion for so many years. It still seemed doubtful at the beginning of 1982.

During the foreign ministers' council meeting, the French minister also raised the question of free circulation in the community of plutonium and highly enriched uranium. He restated that this matter was entirely within the competence of the member states, and that the practical removal of restrictions on such movements could only result from an agreement between the Nine on joint regulations concerning nonproliferation.

The worldwide conflict over nonproliferation and plutonium, which thus reanimated the 20-year-old debate on the faults of the Euratom treaty, was not over. The struggle had certainly slowed down the nuclear energy programs of some Third World countries, notably India, while no real obstacles had been created for those determined to keep open a military option, or even to force their way into the lobby of the Atomic Club. The conflict had certainly led to deteriorations in Washington's nuclear relations with non-NPT Third World countries, as well as with some of the most faithful past advocates and supporters of U.S. policies, such as Germany and Japan — countries for whom the struggle has provided a reinforced determination not to allow their nuclear programs to be dictated by others. Finally, the conflict confirmed facts that were already known: that the risks of proliferation result from political intentions far more than from the civil nuclear capacities of the countries concerned; and that measures designed to check the slow and virtually inevitable proliferation of the number of nuclear weapon states must therefore be essentially political rather than technical.

Only time will tell whether the nonproliferation issue has, in the end, benefited or suffered from this plutonium conflict, which ultimately had no noticeable effect on the rate of peaceful development of nuclear energy, at least on the world scale. The effects on that development of a determined antinuclear movement, however, were soon to prove far more severe.

2. The Nuclear Debate

Of all the difficulties encountered or created throughout the history of the exploitation of the benefits of fission, the psychological and emotional obstacle, the basis of antinuclear opposition, seems today to be the most difficult to overcome.

In the era of computer science and space exploration, but also of growing awareness of environmental problems, this opposition has the appearance of a highly organized rejection phenomenon that is peculiar to the Western industrial world. The phenomenon bears little relation to the hostility shown toward the introduction of certain great inventions of the past, such as railways some one and one-half centuries ago.

The absence of any nuclear accident resulting in human casualties in the neighborhood of an atomic installation and the progressive education of the public at large, enabling it to evaluate the real and minimal importance of the risks involved compared with those in other branches of industry, should have led over a period of time to general public reassurance. Nothing of the sort happened; on the contrary, public anxiety over the dangers of radioactivity increased in step with the growth of programs for world nuclear electricity production.

For the first time on such a scale, a fraction of the general public (more women than men) began to show real mistrust of the statements and claims of scientists and technicians, many of whom had accepted, during and since the war, the constraints of governmental secrecy: it could perhaps be questioned whether such specialists could ever free themselves completely from these constraints. Further, because they had necessarily been involved and interested in the development of atomic energy, the experts could be accused of partiality or even suspected of being, consciously or unconsciously, the champions of the immense industrial edifice to the creation of which they had contributed.

The answer to antinuclear attacks might appear to lie in providing better information for the people, but in this case the extreme complexity of nuclear technology, emphasized in the title of this book, intervenes. Here, newspaper articles and television programs will not familiarize the man in the street with such complex notions as the stages of the nuclear fuel cycle or the mysteries of the nuclear chain reaction. For many people, the easy way out must consist in no longer attempting to follow the expert in his difficult demonstrations (though often complaining later of insufficient information or consultation) but rather in trusting the nonspecialist and his usually pessimistic vision of the risks involved.

In the early development of this situation, lack of understanding progressively grew on either side of the argument, until a substantial gap appeared which then continued to widen. Questions that should have involved no more than technical differences over the accurate evaluation of the risks of the new technology, or discussions of the criteria for determining an acceptable level of risk for the populations concerned, gradually degenerated in several Western countries into a profound emotional conflict, the issue at stake being whether to accept or reject the production of electricity by nuclear means.

The antinuclear protest movement found a fertile breeding ground in the public's irrational fear of atomic radiation, but it did not confine itself to this one motivation. Parts of the movement took on a political character, being used as a weapon against today's industrial-type society; other parts had a technical aspect, expressing rational doubts over the risks involved, and such concerns were thus positive and useful; others again sought philosophical inspiration in nostalgia for a world "close to nature," with neither industry nor commerce to spoil it.

This heterogeneity of the protest movement is no doubt a major reason why it is so difficult to discuss matters or reach any agreement with it.

The controversy began in the United States where, as we have seen, it became widespread in the early 1970s; in the middle of the decade it spread

to Europe. It reached a climax in 1979 over the dramatic accident at the American Three Mile Island nuclear power station, the consequences of which we shall examine before ending this book with a survey of the state of advance of nuclear energy in the world in the early 1980s.

American Antinuclear Maneuvers

In 1975 the American antinuclear movement won a first major victory by contributing to the collapse of the country's nerve center, pillar, and monument of atomic energy, the U.S. Atomic Energy Commission (U.S. AEC). The U.S. AEC was no longer to be allowed to exercise the roles of both sponsor of the new form of energy and watchdog over its use. It was replaced in early 1975 by two new organizations: the Nuclear Regulatory Commission (NRC), whose task was to control and regulate nuclear activities; and the Energy Research and Development Administration (ERDA), responsible for research into and development of all forms of energy. Atomic energy promotion, in the country that had led its development for so long, ceased to be the responsibility of a single organization with this exclusive purpose; the new regulatory body was to some extent divorced from nuclear techniques.

This disorganization of atomic energy in the United States, based on its treatment as just another form of energy, was to continue. No sooner had the ERDA overcome its formative problems than it also disappeared, being absorbed in 1977 by a new government department, the Department of Energy (DOE).* At the same time, and in token of the new trivialization of atomic power, the great champion and defender of the nation's nuclear progress, the congressional Joint Committee for Atomic Energy, was also abolished.

One of the last acts of the U.S. AEC was to publish a report on the safety of reactors, the "Rasmussen Report," which concluded that the risk of a catastrophic accident was negligible and that the likelihood of being killed by a nuclear power plant was, for someone living in the vicinity, comparable with that of being killed by a falling meteorite.

The pressure of the opposition against nuclear power stations could now afford to relax. Orders for new plants were becoming more and more scarce as the electric utilities became discouraged by the longer administrative delays due to two official organizations now operating in the field, as

---

*The DOE was initially headed, until mid-1979, by James R. Schlesinger, the next to the last chairman of the U.S. AEC.

well as the increasing severity of NRC norms and regulations.

The protest movement therefore focused its attention on the later stages of the fuel cycle, which were also the least advanced industrially. Indeed, since 1973 these stages had provided a new issue that had been growing in importance for the opposition, namely the management of the highly radioactive wastes resulting from nuclear power plant operation, and the alleged absence of a "sufficiently proven" method for conditioning them to withstand the passage of hundreds — some suggested even thousands — of years. The wastes were of course considered to be a threat to the environment.

In the period 1974 to 1975, while Ralph Nader was organizing annual meetings in Washington for all the nuclear protest movements, plutonium appeared on the scene, having become a main focus of public attention following the Indian explosion: it was quite erroneously denounced as "the most toxic substance known to mankind." (Plutonium is in fact less toxic than radium and considerably less toxic than many biological poisons.)

As an additional sensation a young Princeton student appeared on television and explained that he had collected, from the published scientific literature, adequate information to produce a homemade nuclear bomb: all he needed was some plutonium!

Some months earlier, an industrial firm, Kerr McGee, dealing with significant quantities of this element had been accused of taking insufficient precautions against the risks of its theft. Shortly afterward the accuser, a woman called Karen Silkwood, an ex-employee whose apartment was found to be slightly contaminated with plutonium, was killed in a car accident, in allegedly mysterious circumstances. She was said to have been on her way, while under the influence of sedatives, to hand over incriminatory documents to a journalist. The documents were never found. The affair became a cause célèbre that four years later led to the accused firm being found guilty of negligence and ordered to pay $10 million in damages to the young woman's family. But in 1981 this decision was dismissed in appeal.

Early in 1975, a fire completely destroyed the important and delicate control panel of a large public-owned nuclear power station in Browns Ferry, Tennessee, causing some very expensive damage, although there was no radioactive escape since it had been possible to shut down the reactor by direct manual control. Ironically, this first fire in a nuclear power plant had been caused by an ordinary candle, carelessly used by a workman. As usual, the media headlined the story, exaggerating the risks involved and making the most of the absurdity of the affair.

The first "occupation" of a nuclear power station building site took place in 1976, at Seabrook, near a popular summer tourist area on the New

Hampshire coast, where construction work had recently begun. For several years this site remained one of the hot spots of American protest. During a demonstration there in 1977, nearly 1,500 people were arrested and the New Hampshire prisons were filled to capacity. More recently, in 1981, the Diablo Canyon power plant on the California coast became a regular rendezvous for nuclear opponents and police after it had been given a startup authorization following a lengthy official inquiry caused by the proximity of a seismic fault. However, it later became the example of a "comedy of errors" as a number of minor but unforgivable construction mistakes prevented the startup again for many months.

A different form of action comprised the so-called "initiatives" in six states, including California, by which the opposition tried to mobilize the electorate through a legal process requiring any "petition" that received sufficient public support to be transformed into a draft law and submitted to popular vote. All six draft laws were designed to prevent the issue of contruction permits for further nuclear power plants until the state legislature had assured itself that the federal government had — inter alia — approved the installation's safety systems, a reprocessing program for its used fuels, including waste storage facilities, and accident liability guarantees; in short, the electorate was to vote for or against a nuclear moratorium. For the first time, the future of the expansion of a new industry was to depend on a public referendum.

It is interesting to point out that in each of the six states in question there was at least one of the following military installations: a nuclear weapons assembly plant, an intercontinental missile silo, or a nuclear bomber base; any one of which should surely have caused more concern than a power station, yet toward which both public opinion and the anti-nuclear opposition seemed relatively unmoved.

Many voters had relatives in the nuclear navy, but this had never caused any special concern, even in the case of the country's largest nuclear-powered aircraft carrier, aboard which over 5,000 people were living, perfectly normally, on top of eight powerful atomic reactors. Here again, the contrast between civil and military activities was surprising: it was only necessary to install an atomic power plant in the hold of a gray-painted ship, and submit it to service discipline, and all worries and fears disappeared!

The first vote was in California and was an unqualified success for the nuclear supporters; the others, held some months later, showed similar results. In all, six million electors rejected the moratorium proposals by two votes to one.

Once again it was difficult for the man in the street to understand and appreciate the many technical and economic arguments put forward in the pre-poll campaigns. Each campaign therefore appeared rather as a test of confidence, the outcome of which was of capital importance, particularly for the American nuclear industry which was by no means indifferent — for the votes would affect the future of some very considerable financial investments.

In the end the electorate showed its faith in the experts, despite the fact that once again the opposition accused them of partiality, and to be distrusted because they were spokesmen in their own industry.

The Arguments

During the "initiatives" campaigns, the experts had constantly stressed that a nuclear moratorium would inevitably lead to increased economic and political dependence on Middle East oil. No new source of energy, other than that resulting from the fission of uranium, could be expected to make a significant contribution to the national (or world) energy balance before the start of the next century. In particular, solar and geothermal energy would be unable to play any great part until technical and economic breakthroughs — as yet uncertain — had been achieved.

Controlled fusion was even less sure technologically; moreover even if eventually successful, it would also give rise to radioactive wastes and proliferation risks since the intense neutron fluxes produced in the thermonuclear reaction must inevitably create radioactive elements in the structural materials of the reactor and could also be used to make nuclear explosives.

From the environmental veiwpoint it was easy to show the terrible effects on marine biology of the "black tides" from oil tanker accidents and leaks from off-shore oil rigs. Greater dependence on coal or the possible processing of the abundant national reserves of bituminous shales to extract fuel oil would present considerable problems of waste disposal or stabilization of residual materials after processing.

The specialists pointed out that during normal service a nuclear power station, which does not consume oxygen and does not produce carbon dioxide, is infinitely less polluting than a plant burning fossil fuels; and they stressed the dangers, comparatively recently recognized, of microashes and harmful gases resulting from fossil fuel combustion, which incidentally always contain a certain amount of natural radioactivity.

In addition, there was a risk of modifying the world's climate due to the constant and irreversible increase in the atmosphere's carbon dioxide content due to fossil fuel combustion. This increase in turn resulting, by a form of "greenhouse effect" in a slow rise in the temperature of the globe, which could in the long run lead to unpredictable ecological consequences, notably an as-yet hypothetical and far-distant risk of the melting of the arctic polar icecap.

Safety had been the constant concern of nuclear power station constructors from the very beginning. Nobody living near such a power plant had ever suffered any ill effect whatsoever to health, and risks that were specifically related to atomic energy had always been kept to a minimal level, perfectly acceptable compared with the risks from the exploitation of conventional energy sources. The maximum "imaginable accident" — a sudden and complete breakdown of a reactor's cooling system leading to a meltdown of its core — is an accident that could only occur if all the multiple safety systems were to fail simultaneously, but which could then take on catastrophic proportions causing many fatalities and requiring temporary evacuation of the neighborhood. The likelihood of this type of accident occurring, according to the U.S. AEC and other estimates, was so small as to be scarcely credible. It was certainly very much less than the likelihood of the tragic fire-damp explosions in coal mines or the bursting of hydroelectric dams.*

As for nuclear radiation, which seems so mysterious because it cannot be detected directly by the human senses, although it is readily measured with appropiate instruments, it was easy to explain that man is constantly subjected to bombardment by rays from interstellar space (cosmic rays) as well as from radioactive matter in the earth's crust. Current norms and regulations meant that the maximum extra doses of radioactivity received in the neighborhood of a nuclear power plant over a year were comparable to those of cosmic origin received during a single Atlantic crossing by jet aircraft: these were very low doses compared with those of natural or medical origin.

The experts also explained that the risks of biological mutations due to the operations of the nuclear industry, which were often highlighted as severe dangers, are minimal compared with those caused by natural radiations, and doubtless much less significant than those owing to the chemical industry or to tobacco smoking.

As for the problems of long-term management of radioactive wastes, the specialists explained that technically satisfactory solutions were available, from which choices would be made according to economic and national political criteria. The transformation of highly radioactive wastes

*From 1975 to 1980, dam bursts caused several tens of thousands of deaths in the world.

into a stable form of glass, which was then poured into steel containers (a process developed and now used successfully in France), was mentioned as the most satisfactory solution. The containers may be stored indefinitely in stable geological formations such as granite* or rock salt deposits.

Whatever method of storage might be chosen, after several centuries the residual radioactivity in the wastes would be comparable to that contained in the original uranium ores, the presence of which in various parts of the world, usually in geological formations that are not particularly stable, has evidently never had deleterious effects.

But this problem, like many others, was once again one that had never given rise to any anxiety when the defense services had been responsible for the presence in the Nevada Desert, actually in the soil, of radioactive wastes resulting from hundreds of underground explosions, equivalent to those produced by the prolonged operation of a few nuclear power plants. Indeed, it would not be surprising if the management of all the nuclear wastes arising in the United States, whether from civil or military installations, was ultimately entrusted to the Department of Defense. The Nevada site, among others, could surely be used for this waste storage.

The opponents of nuclear energy also stressed the dangers of nuclear terrorism which, according to them, would make necessary reinforced physical security measures and so lead to a form of fascism, ending in a police state. The answer to this argument was simple: It was terrorism that should be eliminated, not nuclear power stations, and in fact it was a great deal easier to stop the theft of radioactive materials than to prevent the hijacking of an airliner. Moreover, a would-be terrorist had means at his disposal, far less dangerous to himself than the theft of plutonium, with which to terrorize a population.

Broadly speaking, these were the arguments put forward in answer to the opposition during the ''initiatives'' campaigns in the United States, which had nationwide repercussions despite the fact that they concerned only a small number of states. Most of these nuclear votes took place in November 1976, at the same moment when the national polls bore Carter to the presidency.

The supporters of atomic energy had won an important victory, but the new president was soon to give substantial satisfaction and help to the opponents by his changed attitude toward nuclear power and by his antiproliferant policy of blacklisting plutonium and blocking both the civil reproc-

---

*The extraordinary degree of conservation, after 4,000 years, of the tombs of the Pharaohs in the Valley of the Kings is proof that the long-term storage of wastes in granitic desert regions would be an excellent solution.

essing of irradiated fuels and the building of a breeder reactor.

On the other hand, the opponents did not give up their campaigns and were making every effort to persuade states to adopt various regulations concerning the construction and operation of nuclear power stations, all of which were aimed at delaying or even stopping the development of atomic power. These attempts to harass and impede nuclear progress were not without success. With the help of inflation, the recession, the rising price of uranium, administrative hurdles, and constantly changing safety standards, the campaign led to electric utilities not only virtually abandoning placing new orders, but also canceling most of those for which construction permits were still needed. Thus, from 1975 onward the United States was virtually faced with a moratorium on new nuclear power stations.

The situation had yet to affect western Europe, where the antinuclear controversy was beginning to make itself known.

Europe's Turn

The antinuclear opposition spread like an epidemic from the United States to Europe in 1974, at the moment when the oil shock of autumn 1973 had given fresh impetus to nuclear electricity programs, an impetus that in many cases subsequently petered out due to the economic recession, itself partly a result of the rise in oil prices.

The opposition was comprised of minority protest groups and ecological movements, which appealed to the sensitivities of a public fearful of the notion of radioactivity but as yet unable to think of nuclear power stations in the same way as other industrial plants presenting similar levels of risks.

Parts of the protest movement reflected profound doubts about the organization of modern society and the relationships between the quality of life and industrial production, thought by some to be mutually exclusive. But paradoxically, these reactions were in general mistakenly oriented, for the special concern in the nuclear field for safety, together with the minimization of pollution, were particularly evident in comparison with other energy forms.

All the West European countries concerned were infected to varying degrees by the epidemic, which however seemed not to have spread to the Eastern bloc countries or to the Third World countries interested in using nuclear power.

The Soviets had, in fact, long ago let it be known that they had the answer to problems of public uncertainty and fear: all that was needed was to educate, then convince the opponents, and later accept them as members of pronuclear committees. Nevertheless, an article in 1979 in the *Central Committee Review* acknowledged the existence of the same preoccupations over safety and ecology that were troubling Western countries; it suggested that nuclear power stations and fuel-reprocessing plants should be sited well away from the highly populated areas in the west of the country and it recommended devoting more effort and investment to safety.

But radioactivity certainly does not excite in the Eastern countries the same fears as in the West. The proof is the existence in Czechoslovakia, at Jachimov (formally Joachimsthal,* an area of rich mineral deposits in which uranium and radium were discovered in 1789 and 1896 respectively) of a thermal spa of repute where even today patients from many Eastern countries bathe in the waters, which come from the mines and whose radioactivity is believed to be a cure for many illnesses.

In the Western democracies, as in America a few years earlier, the nuclear opposition first took root at the local level. Then, in countries where this local action had succeeded in delaying or stopping a project, the success, amplified by the media, had nationwide repercussions, the antinuclear theme becoming a factor of political competition, which had not been the case in the United States. However, the dividing line between supporters and opponents did not always run between political parties, groups, or trade unions, but often right through them.

This involvement of the opposition in national politics was to make itself particularly felt in countries where the governmental balance could be altered by only small swings at the polls. It was also to prove more effective in the northern countries than in those of the south, in decentralized countries with a federal structure rather than centralized ones, and in those where electricity production was not in the hands of a national monopoly. Finally, nuclear opposition became most developed in countries where the energy supply seemed most secure, particularly when a substantial recourse to nuclear power was announced. Sweden, Austria, Switzerland, and the Federal Republic of Germany were the most affected. On the other hand, the United Kingdom, France, Belgium, and to a lesser extent Spain, were able to confine the attacks to local levels. This also applied to Italy; however, the impossibility of finding suitable sites for new power stations (whether nuclear or conventional) was an additional factor leading to paralysis of the proposed program.

In another part of the world, Japan's first antinuclear opposition was limited to local actions, particularly by fishermen's associations, which

---

*In the 16th century, Joachimsthal was famous for its silver mines and gave its name to the German silver coin, the "thaler," from which the word "dollar" is derived.

claimed and obtained substantial financial compensation since all the proposed nuclear plants were to be sited on the coast.

Finally, countries such as Denmark (where a nuclear referendum was planned for the early 1980s), Ireland, and Norway were awaiting the outcome of nuclear contestations in the rest of Europe before deciding whether or not to build atomic power stations. The Netherlands, comparatively uncommitted, also prudently awaited developments in other countries before constructing new stations.

In general, the opposition chose to concentrate its activities on proposals for new nuclear power plants, although in the most advanced countries it was also especially concerned with problems of wastes, reprocessing plants, and breeder reactors — all facilities where large quantities of plutonium would be involved. This opposition drew much strength and inspiration from U.S. nonproliferation policies; a report presented in 1978 to Congress on the subject of antinuclear protest in Europe removes any doubts as to the link. Without going so far as to welcome the development of the protest movement in Europe, the report emphasized the importance for American nonproliferation policies of an effective opposition to the commercialization of breeder reactors and to the use of plutonium as a nuclear fuel.

The opposition therefore carried out its campaign on three fronts: on the local level, over the choice of sites; on the national political level; and on the level of advanced techniques and waste disposal problems.

The Sites

In the 1960s and earlier, ecologists in France, Sweden, and Switzerland made use of the advent of nuclear energy in their opposition to the building of hydroelectric dams. During the early part of the following decade, their attitudes completely changed; and one of the reasons for their hostility toward nuclear was the very large number of the new power stations planned in the valley of the Rhine — in France, Germany, and Switzerland.*

The first antinuclear marches took place in 1970, at Fessenheim in Alsace, the future site for the first large French light water power station. The year before, there had been some opposition to the building of a storage

---

*In 1969, when an accident occurred at a small experimental heavy water nuclear plant built in a cave in the Canton of Vaud, damaging the installation beyond repair, the general public showed no particular concern and certainly no fear.

depot for low activity radioactive wastes at the nuclear center of La Hague in Normandy, where however the previous construction of France's second fuel-reprocessing plant had caused no difficulties. Again in France, the publication by Electricité de France (EDF) in 1974 of a list of 36 possible sites for nine future nuclear power stations then envisaged predictably and unnecessarily provoked unenthusiastic declarations in most of the local communties that found themselves involved.

The French demonstrations had very little influence on the development of the substantial projected nuclear electricity production program, the *Conseil d'Etat* having granted, after carrying out the appropriate local inquiries, the necessary ''declarations of public utility'' for construction on the sites concerned.

On the other hand, areas of violent protest appeared in 1975 in various West European countries: in Sweden, over a proposed nuclear power plant to be built too close to Stockholm; in Germany, over a station planned in a wine-producing area; in Switzerland, over the construction of an installation near Basel; and finally in Spain, over a nuclear power plant in the Basque country, an area particularly sensitive in principle toward decisions made in Madrid.

The first European site occupations took place at Wyhl in Baden-Württemberg and at Kaiseraugst in the Basel Canton. No doubt this must have been the first instance of civil disobedience in modern Switzerland — Kaiseraugst became a center of ferment and a focal point for the Swiss protest movement.

In these early days, the ''site occupation forces,'' mainly coming from environmental protection organizations, which between them had set up a veritable international ecological network, were generally well-behaved multinational groups. Soon however, a different class of people became involved and the occupations took a violent turn, as in France at Creys-Malville,* the site between Lyons and Geneva of the Super-Phénix breeder reactor; or they became quasi-insurrectional, as in Germany at Brokdorf and Kalkar, another breeder reactor site; or even worse, they degenerated to armed terrorist attacks, as at Lemoniz in the Spanish Basque country.

These violent demonstrations became less frequent from 1978 onward, particularly in Germany where instead the opposition turned to exploiting the network of administrative and legal procedures available at federal, provincial, and city levels in attempts to delay, stop, or question the validity of construction permits, even for projects already being built.

---

*This demonstration, which involved 20,000 protestors and left one dead and several wounded, marked the climax of the protest movement in France during the 1970s.

On the whole, these local actions had no profound or lasting effects, except where the problem of antinuclear opposition became a factor affecting national and electoral policies in the country concerned.

National Politics

The European countries that were the least affected by the protest movement were those where no important political party had adopted the antinuclear theme as part of its policy. This was the case in Belgium, France, Spain, and the United Kingdom.

In France, a comprehensive parliamentary debate took place in 1975 when all the political parties declared themselves in favor of the nuclear electricity production program. There were a few expressions of reticence, mainly from Socialists, who had doubts as to the speed of the development which they thought was too fast, and from the Communists, who regretted the recourse to a reactor type built under American license. During the parliamentary general elections in spring 1978, the antinuclear pressure movement was disowned; the Ecologist party obtained less than 2 percent of the vote; and the Socialists, who had finally proposed a 3-year moratorium on building any further nuclear power stations, found themselves in the defeated opposition. This position they shared with the Communist party, which was in favor of an important nuclear program and had the support of the Confédération Générale du Travail, a Marxist-inspired trade union with very strong representation in EDF.

Sweden was the first country where the nuclear equation upset the political balance, contributing to the downfall of the Social-Democratic party, which had been in power for almost half a century, and whose then prime minister, Olaf Palme, had decided in 1975 to increase the national nuclear program to 13 power stations designed to produce nearly two-thirds of the country's electricity by 1985.

The small Centre party adopted an antinuclear stand as the main plank of its electoral campaign in 1976. Its leader, Thorbjörn Fälldin, undertook to stop further work being carried out on any of the nuclear plants then under construction, and eventually to close down the five that were already completed. His election success enabled him to become prime minister, but he was obliged to abandon his electoral promises since the two other right-wing parties in the government wanted the nuclear program to continue.

This disagreement led, two years later, to the breakup of the coalition. Although the minority interim government, formed to cover the period before the new elections in late 1979, was in favor of nuclear power, following the Three Mile Island accident it decided to hold a national referendum in 1980, thus excluding nuclear questions — except by implication — from the 1979 electoral campaign. The elections in fact went, by a very small margin, in favor of the "bourgeois coalition," carrying Fälldin back to power. The referendum, held in March 1980, produced public support for the operational startup of the country's four complete or near-complete power reactors, as well as for the construction of two others previously decided upon. Fifty-eight percent of the votes were against and 38 percent were in favor of the antinuclear lobby, which called for the progressive closing down of the six power plants already operating and producing 25 percent of the country's electricity.

The first referendum to be won by the antinuclear lobby took place in Austria in November 1978. Its history is a veritable "comedy of errors." The issue at stake was whether to accept or refuse the operational startup of Austria's first nuclear power station, which had been built by the German firm Kraftwerk Union and completed in 1977. The plant was to produce 7 percent of the country's electricity.

The decision in 1971 to build the installation, and the choice of a site at Zwentendorf, some 30 miles west of Vienna on the Danube, had caused no complaints or difficulties among the political parties, the trade unions' federation, the employers, or the local population.

The first protests appeared over the choice of a site for a second unit, the proposals for which, when public reaction became evident, were rapidly withdrawn. However, the protests had been noted by the Populist opposition party, which in 1977 had realized, following the downfall of the Social Democrats in Sweden, that nuclear power could offer them a way of making problems for their political rivals. The Populists therefore adopted the antinuclear theme, despite the fact that their electorate was more in favor of atomic energy than was that of the Socialists. The latter, despite their absolute majority in Parliament, dared not authorize the operational startup of the completed Zwentendorf station in 1977, so Chancellor Bruno Kreisky decided, with the agreement of the opposition, to hold a referendum. The campaign was an impassioned one, in which the chancellor's own son and daughter-in-law joined the antinuclear ranks.

Just a month before the vote Kreisky had confided to Michel Pecqueur, the Commissariat à l'Energie Atomique's general administrator, and myself that he was certain of winning the referendum by a small margin, but that he was afraid his party would lose the general elections to be held in spring 1979, as much of his socialist electorate would reproach him for his pronuclear attitude.

Exactly the opposite happened for, at the last moment, Kreisky made a major tactical error and completely confused the situation by announcing that he would consider the vote not only for nuclear but also as one of confidence in his party and himself. Instead of rallying the hesitant Socialists, he incited the pronuclear electors of the Populist opposition to vote against him, and finally his opponents won by just 50.5 to 49.5 percent.

However he did not resign, as he had let it be thought he would, and his party easily won the general election in May 1979. All this time, however, the power station that had become so much a political symbol was lying idle, a symbol also of the nuclear deadlock in the country and of lost electricity production at a time of worldwide energy and eonomic difficulties. The Zwentendorf plant had cost over half a billion dollars.

The Austrian referendum was followed three months later, in February 1979, by a national vote in Switzerland. Trans-frontier contagion was no doubt partially nullified by the threat of reduced oil supplies following the Iranian revolution. This time in fact, the balance went just in favor of the pronuclear lobby, by 51 to 49 percent of the votes. The opposition had campaigned for the ''democratic development'' of the new energy form, whereby construction of any nuclear plant would be conditional on its acceptance by a majority of the electorate in each township within a radius of 20 miles from the proposed site. Such a condition would have automatically blocked any new project, since in this type of local poll there is often less than a 50-percent turnout.

A second national referendum took place in Switzerland in May of the same year. It strengthened Parliament's power in nuclear matters by giving it the responsibility of approving any new project for a nuclear station, the need for which must have been demonstrated on the national scale, as well as the corresponding long-term management program for the wastes. These powers would no doubt lead to increased delays affecting the rest of the power station construction program (including the controversial unit at Kaiseraugst). Nevertheless, by the end of 1979, with four stations operational and one under construction, Switzerland was producing 30 percent of its electricity from nuclear sources and in this was leading the world, followed by Sweden and Belgium with some 25 percent each.

In West Germany, the political parties took up the antinuclear theme in 1976. The country's power reactor construction program and the establishment of a national nuclear industry had been proceeding smoothly for over 10 years.

From 1975 onward, however, local moves led to a flood of legal appeals for the suspension of projects. The Social Democrats in power were divided over the nuclear issue: Chancellor Helmut Schmidt was in favor, but the small Liberal party, whose support he needed in the governmental coalition, was somewhat hostile. On the other hand, the Christian opposition parties were strongly in favor.

In 1976 the federal government decided to make all further authorizations for nuclear stations conditional on the existence of a "satisfactory solution" for the final management of the unit's radioactive wastes. This entailed a quasi-moratorium on all new projects. The following year, Parliament adopted an energy policy that gave priority to the use of national coal and lignite resources, nuclear energy to be limited to the strict minimum needed to avoid a shortage of electricity. This was the result of a compromise between the parties; it left the problem of wastes unresolved.

Waste Disposal

The problem of "wastes" from nuclear power stations is the area in which opponents of nuclear power have had the most success, having in fact succeeded in convincing the public, and even some governments, that it is a question of capital importance for which no solution is in sight. Technically, this is not the case.

Nevertheless, the opposition has succeeded in persuading a number of governments — notably those of Sweden and West Germany, as well as the state government of California — that they should insist on the "demonstration" of a "final solution" to this problem before issuing construction permits for any further nuclear plants, or in some cases allowing those already built to be commissioned. But at the same time the opposition has sought by every possible means to block the building of installations necessary for the management of these wastes: especially reprocessing plants for irradiated fuels but also the waste storage facilities whose nonexistence is one of the causes of complaint. Even preliminary geological surveys, to confirm the suitability of potential storage sites, have been resisted. This double game demonstrates the opposition's fundamental hostility to the very development of nuclear energy.

In recent years proposals for reprocessing plants in Britain and Germany have been attacked. The two encounters, with only a year between them, ended respectively in a victory and a defeat for the supporters of nuclear energy.

The British campaign concerned the construction, at the existing Windscale facility in the northwest, of additional reprocessing capacity for oxide fuels from the British advanced gas-cooled reactor power stations and from foreign light water stations for which contracts had already been signed, in particular with Japan. A very thorough public inquiry, symbol of British democratic processes, was entrusted to a leading judge, Justice Parker, who until then had known nothing of the mysteries of fission. His task was to decide whether or not such a plant would constitute a danger to the population or danger from the weapons proliferation viewpoint.

The inquiry lasted several months. Representatives of all attitudes and "lobbies" took part: advocates of "soft energies," American professors of political science, a fisherman who strangely ate nothing but lobsters caught near the plant's disposal point for residual cooling water, rubbed shoulders with the cream (and some of the dregs) of British science. The judge's conclusion was unambiguously in favor of the project, to the disgust of the opposition and displeasure of the mandarins of American antiplutonium policies: As mentioned earlier (see p. 421), the judge's decision was swiftly followed, in spring 1978, by a parliamentary vote authorizing the start of work on the new plant.

One year later and some 600 miles to the east, near the frontier between Lower Saxony and East Germany, the same question received the opposite answer. The federal government had proposed building, over a disused salt mine at Gorleben, a giant nuclear complex that would include a reprocessing plant, a fabrication plant for plutonium fuels for future breeder reactors, and a permanent underground depository for radioactive wastes in the salt domes, which experts had pronounced geologically stable over the past hundred million years. The project, which was to have cost $6 billion and taken at least 10 years to complete, quickly became a new major issue for the German protest movement. Again a public inquiry was held, during which numerous representatives of the two factions gave evidence. Despite the importance of the project for the federal government, which would have thus acquired independence in this stage of the fuel cycle (hitherto the privileged prerogative of the nuclear weapons states alone), in May 1979 the Christian Democrat government of Hanover State refused its authorization, not for technical reasons but essentially because of the difficulty in convincing a still reticent population. The Christian Democrats, who in fact were in favor of nuclear energy, wanted to avoid — in view of the forthcoming general election in 1980 — the apparently two-faced attitude of the Social Democrats, who defended the federal nuclear program in Bonn while seemingly opposing it at the provincial level.

In October 1979, an agreement was reached between Ernst Albrecht, president of the Lower Saxony government, and Chancellor Schmidt that recognized the long-term need for the Gorleben complex and accepted that geological studies of the salt deposits should go ahead. Meanwhile, to maintain the industrial capacity concerned, a number of irradiated fuel depositories, and possibly a small reprocessing plant, would be built on other sites.

During this period, the French reprocessing plant at La Hague had a virtual monopoly for the treatment of irradiated fuels from both home and abroad, and Compagnie Générale des Matières Nucléaires (COGEMA) soon had some important foreign contracts. Although every one of these specified that the radioactive wastes should in due course be returned, after conditioning, to their country of origin, the opposition dubbed the La Hague facility "the radioactive dustbin of the world." It became a target of criticism both from outside, by ecology groups who demonstrated when ships containing irradiated fuels arrived at Cherbourg, and from inside by the French Democratic Workers' Confederation, the trade union most opposed to the national nuclear program.

Generally speaking, however, the Western World's unions supported the development of nuclear power, though with one notable exception: the Miners' Union of Australia, a country endowed with vast coal resources. As mentioned earlier, during the few years from 1973 onward when the Australian Labour party was in power, the union paralyzed exploitation of the country's uranium deposits, the richest in the world, on the grounds of awaiting a solution to the problem of radioactive waste management. One can hardly imagine vineyard workers, even in Australia, making their participation in the grape harvest conditional on the worldwide eradication of alcoholism, which is a real scourge, whereas properly conditioned wastes are not dangerous at all. Even as late as 1981 when some Australian mines were in production, some transport and dockers unions were still opposing the handling of the uranium concentrates that were ready to be shipped to foreign customers.

The protest movement had thus struck many times in various parts of the Western World and at several stages of the nuclear electric cycle, even uranium mining. Yet no accident at a nuclear power station having serious outside consequences had ever happened. It was generally feared that, if one day such an accident did occur, it would surely be in a country that had only recently launched into nuclear technology.

In fact, it was to be in the United States.

Three Mile Island

In March 1979 the film "The China Syndrome" was being shown throughout the United States. It gave a detailed account of a fictitious atomic accident at a nuclear power station in which, despite the confusion of the operating team, there were in the end no serious consequences. The film's dramatic story, however, was concerned with the plant owner's reaction, which was that the affair should be kept quiet, while a group of journalists, who happened to have been present at the time, felt it their duty to warn the public of the dangers should the reactor be prematurely restarted. The leading part in the film was played by Jane Fonda, a well-known actress renowned for her impassioned participation in the campaign against the Vietnam War and in the more recent one against nuclear energy.

The timing was particularly effective. President Jimmy Carter would soon be obliged to declare his intentions concerning the country's energy problems, and he was being pressed from all sides to give his public support, this time unequivocally, to the nuclear option. When he made his decision on April 2, there was no mention of the atomic program: five days earlier, in the state of Pennsylvania, an accident had occurred which at least on the surface could seem technically similar to that in the film.

Ownership of the damaged reactor, a 900-MW(e) pressurized water unit built by Babcock and Wilcox, was shared between several electricity producers together known as General Public Utilities, which included the one whose order for the Oyster Creek power station had triggered the American nuclear boom in 1963.

The accident was due to a combination of human errors and material faults, together with unsatisfactory control instrumentation. The reactor's automatic shutdown system had functioned correctly within seconds of the appearance of an operational abnormality, but due to a human error in the control room, the residual heat in the core was not satisfactorily removed. In addition, a hydrogen bubble had formed in the dome of the reactor pressure vessel and for several days it was feared, quite incorrectly, that this might cause a dangerous explosion.

Nevertheless, the automatic safety systems had functioned correctly, and there had been neither extra health hazards for the operators nor any harmful external contamination: The maximum dose received by the neighboring population was on the same order as that received in a year from natural sources. On the other hand, the reactor itself, which had cost some billion dollars, was put out of service for a considerable number of years, if not forever, its core having suffered considerable damage.

The consequences of the accident were not only material and financial, but above all psychological and sociological. Both the public authorities and public opinion were profoundly disturbed by a whole series of statements issued and by panic measures taken immediately when a slight increase in radioactivity was detected in the atmosphere outside the installation. The statements included a vast amount of alarming and often contradictory information, which was given extraordinary attention by the media, by then represented by 300 American and foreign journalists. The special measures included advice to the local population to stay behind closed doors and windows and listen to radio instructions; the recommendations made by the Nuclear Regulatory Commission (NRC) that the governor of Pennsylvania should consider evacuating the population within a radius of 10, then 20 miles (650,000 persons, nine hospitals, and one prison); and finally the recommended evacuation of school children and pregnant women within a radius of 5 miles and the closure of all schools within the same area. The schools reopened eight days later, the crisis having ended and having no further acutely dramatic character after the first four days. In all, over 150,000 people spontaneously left the area during the alert.*

The accident was of course officially brought to the attention of President Carter, who was kept constantly informed of developments: accompanied by his wife, he visited the power station on the fourth day of the crisis. Neither did Congress remain indifferent to the affair, ordering the publication of verbatim conversations between the five NRC commissioners recorded during the crisis period, such recording being required by law whenever three or more of them are together. This publication cruelly highlights the uncertainties faced by those responsible for nuclear safety, both in the technical evaluation of the accident and its risks, and over the best way to inform the public, the authorities, and President Carter himself.

The NRC did not admit officially until a month after the accident that there had not been the slightest danger of the hydrogen bubble inside the reactor exploding, and consequently there had never been any reason for calling the affair a catastrophe.

Meanwhile, the box office count in the United States for "The China Syndrome" reached $100 million, which was more than 10 percent of the cost of the power station. In France, the public gave the film a much more reserved reception.

The Repercussions

The public in the United States could no longer complain of being deprived of nuclear information; it was saturated with it. No event since the

---

* A few months later, when a train carrying chemicals in Ontario in Canada was derailed and caught fire, 250,000 people had to be evacuated, but the international press hardly mentioned the affair.

last days of Richard M. Nixon's presidency had been the subject of national debate on such a scale.

The opposition had been admirably served; despite the absence of any dangerous radiation dose, the media had constantly dwelt on the possibility that, for the population in the vicinity of the power station, there would be increased incidence of leukemia and cancer in 20 to 30 years' time.* Once the affair died down a T-shirt appeared in the area and quickly became popular. It bore the legend "I think I have survived Three Mile Island!"

In any case, the worst accident in the history of commercial nuclear power had happened, and led to the partial evacuation of a population by no means prepared for such an eventuality.

Without doubt the technical lessons learned from the accident would in due course lead to improved safety systems for protection against the consequences of inevitable human and technical failures: from this point of view, the warning was beneficial. The political implications were much more serious.

Following the accident, four commissions of inquiry were set up, the first by the president and the others by the two houses of Congress and the NRC. The presidential commission was the first to report, seven months later, with conclusions that were reassuring both *a posteriori* over the accident and because they confirmed that there had been no risk of the reactor exploding and a negligible probability of health risks for the population.

However the report was severely critical of all those who, however remotely, had been concerned in the accident: the constructor for the poorly designed instrumentation system and reactor control panel (100 alarm signals had been simultaneously activated); the operators for lack of competence; those responsible in the company owning the reactor for the lack of credibility in their initial statements; the NRC experts for wrongly evaluating the risks and hence contributing to the unnecessary evacuation of the local population; the media for their incompetence and irresponsibility, and the local authorities for their total confusion.

Although this was in no way a matter to be taken lightly, it might be noted that all the failings criticized had the same origin: the complete lack of experience with such a potentially serious accident — which in one sense was a tribute to the safe operation of many other nuclear plants over a quarter of a century.

---

* A report by the French Academy of Sciences evaluated the likely effects of the accident as causing less than one (0.3) additional cancer case to add to the 400,000 "natural" cases that would inevitably affect the two million people closest to the power station.

Without going so far as to recommend a moratorium on construction permits for further power stations and on operating licenses for completed units, the report implied this in practice by calling for a revision of the regulations and a reorganization of the NRC whose task it was to enforce them. In fact, the NRC itself was immediately to enforce such a moratorium at the expense, among others, of seven nuclear stations that were ready to be commissioned and had cost some $8 billion. Such a hypothesis was no longer an unlikely figment of the imagination.

The possibility could even be foreseen that a fresh start for nuclear power in the United States might become conditional on the establishment of a system of mutual insurance against financial risks of this nature, guaranteed by all the electric utilities together or even by the federal government. Such a system could carry the seeds of eventual nationalization of all nuclear electricity production.

On the other hand, closure of the 80 existing nuclear power plants that were already producing 13 percent of the country's electricity (40 percent for the city of Chicago), or 4 percent of the total world energy consumption, was unthinkable. It was even out of the question to delay the construction of a similar number of further units on which work was already under way. President Carter himself, in a speech in July, had had to admit that resort to nuclear power was unavoidable if the U.S. energy balance was to be maintained.

There were similar repercussions throughout the Western World, with new orders continuing to fall off and *de facto* moratoriums on further construction. But at the same time, existing plants continued to function and new construction work already begun was pursued.

In May 1979 the European heads of state, meeting in Strasbourg, reaffirmed their belief in nuclear power, the ineluctability of which was unanimously accepted a month later during a summit meeting of Western industrial countries in Tokyo.

However, the opposition had certainly produced some severe difficulties. For many world leaders, nuclear energy, which today is indispensable, seemed at best a controversial and complex benefit and more probably a necessary evil, to be used only within the limits imposed by shortage of conventional energy resources or while awaiting the availability of new energy forms.

In the United States, the Three Mile Island facility was to remain a persistent and paralyzing monument on the road to nuclear development. The opponents tried to stop the clean-up operations (decommissioning), which involved pumping out a million gallons of contaminated water from the reactor containment, concentrating its radioactive content into solid

residues, and discharging the decontaminated water into the great Sus-
quehanna River, which incidentally would still be less radioactive than
many of the mineral waters drunk throughout the world. General Public
Utilities, the group responsible for the unfortunate power station,
threatened with bankruptcy, could obtain neither the necessary finance nor
the authorization for such an operation.

At the same time the administration's policies on waste management,
an important problem in the pursuit of national nuclear development, were
facing a whole series of difficulties in 1980; in particular, it was becoming
more and more unlikely that the four or five possible sites for a "nuclear
cemetery" could be "evaluated" by 1985. Such delays were in sad contrast
to the wartime practices of the Manhattan Engineer District.

In the same year, the American nuclear opponents tried for the first
time to close down an operating power station, the only one in the state of
Maine, which had been commissioned in 1972. At the same time they
attempted to have the building of any other such plant in the state prohibited. The vote was in September, and pronuclear good sense won by 59
to 41 percent.

Nuclear questions remained relatively absent from the 1980 U.S.
presidential election, though the winner was known to be much more
favorably disposed to atomic energy than his unhappy opponent and predecessor. President Ronald W. Reagan had in fact pledged himself, in his first
year of office, to revitalize the American nuclear power industry and to
reduce the tangled administrative procedures that made nuclear power
station construction in the United States take nearly twice as long as in
France or Japan. The president was also determined to give new impetus to
reprocessing breeder reactors and waste disposal, activities that the previous administration had practically paralyzed. But the whole American
nuclear machine had lost its past momentum and its recovery could only be
slow, and in any case would depend on the nation's economic recovery.

During this same period, while the nuclear controversy was becoming
more political in several European countries, Sweden moved in the same
direction as the United States. Following the referendum of March 1980
Prime Minister Fälldin, who had come to power in 1976 on an antinuclear
platform, respected the later verdict of the people, and the country resumed
once more its place among the world's leading civil nuclear powers. The
planned program for 12 power stations continued with the commissioning
of units already finished and the construction of the last ones planned.

In Spain, the extreme wing of the Basque Separatists continued to call
for the destruction of two 1,000-MW(e) units at the Lemoniz power station
near Bilbao, which at the beginning of 1981 were 90 and 25 percent

complete, with commissioning expected in late 1981 and 1984, respectively. Terrorists kidnapped the station's chief engineer and, when the Madrid authorities refused to agree to their demands, killed him a week after he had been seized, so creating the first martyr to civil nuclear power.

In the early 1980s, as already mentioned, a wave of demonstrations in favor of nuclear disarmament, similar to those that in the 1950s had preceded the Moscow treaty banning aerial testing, swept through the Scandinavian countries as well as the nations that had accepted the December 1979 North Atlantic Treaty Organization decision to deploy modernized medium-range missiles on their territories as a riposte to the Soviet SS20 rockets.

Contrary to what had happened in the 1950s, this later neutralist and antinuclear armament wave spread to the civil field so that the two themes, antiweapon and antipower station, were often pursued together by the main protest groups. In the Netherlands, for instance, where the Socialist party was totally opposed to nuclear power, the installation of either new missiles or power stations seemed practically impossible.

In Germany, a planned power station at Brokdorf near Hamburg, whose construction had been blocked for five years, continued to be the main focus of opposition hatred, symbolized by the struggle between the government of Schleswig-Holstein, who wanted the work restarted and the Senate of the city of Hamburg, whose opposition was expressed through the city's mayor, shortly to resign his post. There were huge street demonstrations, involving 100,000 people in support of the nuclear opponents.

In France, where the military program had long been accepted by the four main parties, a section of the Socialist party supported the ecologists who had concentrated their actions on the Plogoff power station site in Brittany, chosen by EDF in agreement with the regional authorities concerned if not with all the local population; on the reprocessing and breeder programs; and finally on the overall size of the national electricity program, dubbed "all nuclear."

The winner of the French presidential election of May 10, 1981, Francois Mitterrand, had promised during his campaign to suspend the part of EDF's program corresponding to the projected stations for which construction had not yet started and submit this to a "vast national debate." The antinuclear campaign in France seemed, for the first time, to have taken a threatening turn.

On the scale of the Western World as a whole, the nuclear controversy had, by delaying or stopping the completion of new plants, dealt a severe blow to nuclear energy development in the industrialized countries, and ultimately to those countries' energy balances. But its offensive undoubtedly came some 10 to 15 years too late, for it had not prevented the construction of a considerable number of nuclear power plants, which will certainly continue to increase.

Without these hundreds of power-producing units, the outcome of the campaign for nuclear energy could have seemed uncertain. However, these atomic stations are due to provide an increasing proportion of the world's electricity requirements releasing corresponding quantities of oil for other purposes. But perhaps even more important, their satisfactory performance will give the general public, which is becoming more and more alarmed over the energy crisis, the only demonstration likely to restore confidence that the risks linked with the production of electricity from nuclear sources are minimal.

3. The 1980s — Nuclear Energy's Challenging Years

The Energy Shortage

The exceptionally rapid growth of the world economy during the 30 years from 1950 to 1980 was accompanied by a fourfold increase in energy consumption. It is certain, unless some dramatic worldwide crisis intervenes, that this growth will continue, albeit at a somewhat slower pace. Even the most pessimistic forecasts made in 1980 foresaw an increase approaching 50 percent, and possibly even more, between now and the end of the century.

Energy shortage, in the industrial world, must inevitably result in reduced activity, increased unemployment and political instability; while in the Third World, already victim of continuing and unavoidable population expansion, the miseries of poverty and malnutrition will sound the death-knell of any plans for improving living standards.

Lack of energy is bound to breed social problems, political tensions, and increasingly bitter competition for shares of the world's main fuel resources which, if worse came to worse, could well end in armed conflict and even the risk of a new world war.

In 1955 the Indian scientist Homi J. Bhabha, chairman of the first United Nations "Atoms for Peace" Conference in Geneva, declared that "no energy is more expensive than no energy." A quarter of a century later we may paraphrase his words with the assertion that no energy can be more dangerous than no energy.

In 1980, a little over 40 percent of all the world's energy supplies came from oil, nearly one-third of this being consumed in the United States alone. Oil production between now and the end of the century will at best increase very slightly or remain stable; so that essential increases in energy consumption can only come from gas, from coal with its ecological constraints and transportation difficulties, from nuclear power with its political and psychological problems, or to a certain extent from as-yet unexploited hydropower reserves mostly in Third World countries. Solar and geothermal energy and biomass as well as thermonuclear fusion are for the moment no more than "maybes" for the 21st century.

Nuclear power is the only new energy source to have been discovered and developed industrially in the present century. It has taken four decades to reach the stage of world utilization, yet even today that utilization is for the moment limited to electricity generation, although we can also foresee, in the not too distant future perhaps, the production of usable heat. Furthermore, and in contrast to the case of oil, the industrial countries of the West are richly endowed with supplies of the basic nuclear fuel, uranium. In 1980 their known economically recoverable reserves were estimated at 5 million tons. Used in light water reactors, these reserves would have an energy equivalent of about one-half of all currently known reserves of oil.

But light water reactors are relatively inefficient. The advent of the fast-neutron breeder reactor will multiply by a factor of at least 50 the usable energy from these uranium reserves, hence greatly prolonging the period over which nuclear fission can provide a sure and reasonably priced energy supply. Moreover, as the costs of alternative energy supplies increase, the exploitation of vast additional uranium resources — at the present uneconomical — will become viable.

During 1980 there were some 250 nuclear power reactors generating electricity in over 20 countries. Their combined power output of 140,000 MW(e) was equivalent to about three times the French electrical capacity at that date and a third of the capacity in the United States. Altogether, more than 2,000 reactor-years of experience had already been accumulated.

In that same year a further 230 more powerful nuclear stations were under construction, representing slightly over 200,000 MW(e) of new generating capacity.

In 1980 nuclear power supplied a little over 2 percent of the world's total energy consumption from all sources, nearly 8 percent of all electricity. By 1985 the nuclear contribution to global energy demand is expected to reach 6 percent, and by the end of the century it could even have attained 15 percent.

Of the total world nuclear electricity production in 1980, the United States consumed 40 percent, the rest of the Western World's industrial countries 45 percent, the Soviet Union and its partners 12 percent, and the developing countries 3 percent.

In Sweden, France, Belgium, and Switzerland, between 20 and 30 percent of electricity production was already from uranium, while a larger group of nations — including Bulgaria, Finland, Taiwan, Canada, the United States, the United Kingdom, Japan, and West Germany — had a production between 10 and 13 percent. Spain and South Korea, with a relatively large number of stations under construction, were well on the way to joining these leading producers of nuclear power. By 1981 it seemed that Italy too, after many years of nuclear uncertainty and polemics, was on the point of launching a sizable effort in the field with backing from government and parliament: even the local authorities, which had been responsible for many past difficulties, seemed more ready to accept the construction of stations on the various sites under consideration.

For the future, it is clear that the United States will be the largest bulk producer of nuclear electricity for many years. Second place in this league is at present held by France, who displaced Japan from that position early in 1981, a year during which French electricity consumption was 37 percent nuclear.

By the beginning of the 1980s, the cost of the nuclear kilowatt-hour was generally less than that of a kilowatt-hour from coal or oil, the ratios depending on the country concerned and other specific conditions, such as costs of conventional fuels, plant construction lead-times, licensing procedures, and interest rates. In France at the beginning of 1982, the cost of a kilowatt-hour produced in a pressurized water reactor station was two-thirds and two-fifths, respectively, of the costs from coal-burning and oil-burning power stations.

In addition to this civil atomic development and expansion, by 1980 there were some 300 nuclear-powered missile-launching or attack submarines, principally Russian or American, as well as some two dozen American nuclear-powered cruisers and aircraft carriers and three Soviet ice breakers patrolling the oceans of the world.

Thus by 1980, a very considerable operational experience was being provided by some 600 power reactors, on land or at sea, mostly of the light water type whose safety characteristics could only be further improved following the Three Mile Island accident. An immense nuclear capital had been acquired, to which should be added industrial experience of the associated fuel cycle and the advanced work on breeders.

In the face of an ever-menacing world energy situation, with its threats of formidable geopolitical consequences — not excluding war — it is indispensable that this most precious capital be protected against all attacks, and helped to develop and expand under the most favorable conditions.

The 1980s must therefore be the decisive decade for nuclear energy, and this is the challenge we must face.

North-South Tensions

The small part currently played by nuclear power in most developing countries is due to the fact that a nuclear power plant, to be economically viable, must be above a certain minimum size — at present from 700 to 1,300-MW(e) capacity — and few developing countries as yet have a sufficiently advanced electricity distribution system or sufficient total demand to absorb so large an output. It is this criterion, and not lack of interest in the new source of energy, which limits current Third World participation in the nuclear race to a very few rapidly developing nations: Argentina, Brazil, India, Taiwan, and South Korea. Iran would have been in this category, had not the Islamic revolution crushed so many of the country's ambitious plans.

However, this does not mean that all the developing countries cannot already draw benefit from nuclear power. Every nuclear generating plant in an advanced industrial country reduces that country's requirements for other energy supplies, notably oil and coal, which therefore become available to the less-developed countries and so can contribute to easing North-South tensions.

These tensions, however, have become evident in the global nuclear politics since the late 1970s, and a new factor has appeared: the increasing role the developing countries are determined to play in the main decisions to come. These nations feel, rightly or wrongly, that they have been unfairly treated in the implementation of the Non-Proliferation Treaty (NPT); they refer to the London guidelines, the Carter policy, and even the International Nuclear Fuel Cycle Evaluation exercise. This dissatisfaction was clearly demonstrated by the failure of the second review conference of the treaty in 1980 as well as in a U.N. decision to organize in 1983 a more political than technical conference on the transfer of nuclear technology.

The advanced countries were more and more in a defensive position at the International Atomic Energy Agency (IAEA) in the debates of the board of governors and the general conference. One of the causes of disagreement was the comparison of rapidly increasing costs of safeguards, which are paid from the regular budget, and the funds for technical assistance, which are raised by voluntary contribution. During my chairmanship of the board of governors in 1980, I secured an increase in the target for these funds, to be decided in advance for three years; however, this only partially alleviated the discontent of the Third World nations who still requested that this major item should be part of the regular budget.

But these North-South tensions reached their peak with the difficulties raised by the election of the successor to Sigvard Eklund, the IAEA's director general for 20 years. For five months, from May until September 1981, the board of governors was bogged down, perhaps not without some complacency, in an unbelievable electoral psychodrama. The advanced countries were incapable of agreeing upon the name of a single candidate among six presented, while the developing nations, who had a common choice that the other side refused, made the maximum of the fact that they had a blocking third of the voting power in this contest, which had to be resolved by a two-thirds majority. Finally after more than 35 votes spread out during the whole period, on the last night of the general conference the "southern" countries agreed to the choice of a Nordic compromise with the last candidate: a Swede like his predecessor, a lawyer, and former foreign minister — Hans Blix. But in exchange, they had been given promises that the board would reconsider the level of their participation in the main posts in the secretariat, the size of their membership in the future board of governors, and naturally the transfer to the regular budget of technical assistance.

This kind of electoral bargaining was not a good omen for the future of the agency. Far off were the days when experts and technicians were in the majority on the board of governors. At no moment had the serious plight of the development of nuclear energy in the world been a factor in this electoral fight that in many ways had irrational overtones.

The Irrational Factors

The urgent need to meet growing world energy requirements will force a relaunching of nuclear programs throughout the Western World. This will not happen everywhere simultaneously, nor everywhere with the same impetus; it will, as must always be the case, be influenced by individual circumstances in different countries, in particular their degree of dependence on imported fuels.

A concerted effort will be necessary to overcome the irrational obstacles in the way of nuclear development: those connected with the risks of radiation as well as those involving the dangers of proliferation. A profound evolution will have to take place in public attitudes toward nuclear radiation. Increased attention to safety, the evident absence of any significant pollution, and the satisfactory operation both of existing nuclear power stations and of those which will soon be completed must in the end convince the people that the main risk to be faced is not that of radiation (which is already well taken care of) or of proliferation (which does not depend to any significant degree on the expansion of civil nuclear power) but the severe and growing danger of a global energy famine.

We have clearly seen that nonproliferation is a never-ending conflict between the inevitable spread of scientific and technical know-how and measures that will always be deemed insufficient. The last years of the 1970s have clearly shown that in the complex domain of nonproliferation the best is the enemy of the good.

By 1982 about 98 percent of the publicly known atomic facilities in the world's nonnuclear weapons countries were under international safeguards, and 111 of those countries were party to the NPT. This world acceptance of the IAEA's safeguards, as well as such an impressive number of adhesions to the NPT, are precious assets that should be protected against all excesses such as undue denial of materials or technologies, or brutal action as in the case of the destruction of the Iraqi reactor, an action that unfortunately must surely encourage the Arab countries to seek their own military nuclear capacity.

In the late 1970s, the vain pursuit of an absolute and permanent guarantee of peaceful pledges created strong feelings of mistrust between various groups of nations: suppliers and importers, advanced and developing. It had very little restrictive effect on the activities of countries determined to keep open the nuclear option.

This is why additional nonproliferation restraints in internationally safeguarded exchanges, contrary to the spirit and the letter of the NPT, should be applied only exceptionally; for like delicate drugs, if used too frequently or in the wrong way, they can be more harmful than beneficial — in this case by encouraging countries toward nuclear autarchy or by increasing the risk of unbalance in the world's energy supply.

The most serious menace to peace certainly remains the nuclear arms race, especially the continually growing American and Soviet arsenals. In this context we may fear that the dangers of geopolitical destabilization, resulting from a worldwide (or even regional) energy shortage partly brought about by a halt in civil nuclear power production, could well be greater than the dangers resulting from a new country acquiring nuclear explosive capacity. We may note that these latter dangers are viewed differently in each case and by each country; for example, the majority of the countries of the United Nations showed relative indifference over the questionable nuclear activities of Pakistan, but far greater concern over unconfirmed suspicions that South Africa might possibly have carried out a test explosion.

European opponents of civil nuclear power have well understood how to exploit public sensitivity over nuclear armaments in Europe. Since 1980, by fighting the proposed North Atlantic Treaty Organization (NATO) plan for the deployment of modern Euromissiles, they have added a powerful though illogical dimension to their propaganda against the peaceful atom. This is a further factor that could impede nuclear energy expansion in some European countries.

Future Prospects

The dynamic and highly successful French nuclear power program; the pursuit of nuclear electricity production (even at a somewhat slower pace) in Japan, Spain, Belgium, and Switzerland; the favorable referendum in Sweden; the election of a U.S. president well disposed toward nuclear energy; the reelection of Chancellor Helmut Schmidt and his coalition government in Germany; and the firmly expressed intention of Prime Minister Margaret Thatcher to put an end to nuclear decline in Britain were all promising signs in 1981 that a general revival of nuclear progress could be expected. The promise was fulfilled in Sweden, but in the other countries mentioned it gave way to less happy reality.

In Britain, a 1979 decision to adopt the pressurized water reactor and to start building a first unit within three years was by 1981 already running into difficulties and delays due mainly to the never-ending problems of reorganization in the battered nuclear industry. Two public enquiries were planned, one concerning this new (for the United Kingdom) type of reactor, the other to open the way to building a powerful breeder station in collaboration either with the United States or with France and Germany. But electrical overcapacity due to the deepening depression and the discovery of new and rich coal deposits for which existing markets were insufficient, together with growing production of gas and oil from the North Sea, made the need for more nuclear power less obvious and certainly less urgent. Indecision in Britain could still be justified, and could still justify inaction.

In Germany, two nuclear issues were contributing to a weakening of the political coalition: the installation of modern Euromissiles and the building of new power stations. Chancellor Schmidt, challenged by his own party over his stand in favor of the missiles, was less firm on the civil uses of nuclear energy, which he felt unable to impose on his country in the face of substantial opposition. So by 1981, six years of quasi-moratorium on new nuclear power stations showed no signs of ending, and two prestigious advanced reactor prototypes — a breeder and a high temperature reactor — were facing ever-increasing costs and delays.

In the United States, cancellations exceeded new orders for nuclear units in every year since 1975; during 1979 and 1980, 31 units were cancelled and none ordered. Electric utilities throughout the country, and above all their financial backers, still live in the shadow of the Three Mile Island accident, especially the indecision over decommissioning the reactor and the financial repercussions for its owner, General Public Utilities, which by mid-1981 was on the verge of bankruptcy. The combination of reduced growth in electricity demand, the considerable difficulties over raising money for any kind of power plant, the bewildering maze of administrative rules and public hearings, and the ever-present hostility of a vociferous — if minority — section of the general public, together have seemed, at least temporarily, to have proved too much even for the benevolent attitude toward nuclear power of the Reagan administration.

In October 1981 the new administration indeed lifted the ban on reprocessing of spent civil fuel, gave orders to proceed again with the Clinch River Breeder Reactor, and to accelerate the search for permanent disposal sites for highly radioactive wastes. It decided to speed up the licensing process to cut down by a factor of nearly two the time in between the planning and the operation of new stations and to deliver in 1982 and 1983 the operating authorization to 33 nuclear plants finished or in an advanced state of construction. It was also aiming, Congress permitting, at more flexibility toward the most severe aspects of the Nuclear Non-Proliferation Act with a view toward regaining the past American position as a reliable world-wide supplier.

But at the same time that these indispensable measures had been taken, the administration was proposing for economic reasons to dismantle the four-year-old Department of Energy, risking once more throwing into confusion the nation's nuclear nerve center at a time when new orders for nuclear power stations seemed still some time away.

However, sooner or later it will become essential for the independence of the country, and for its overall defense, that a greater fraction of its energy balance be nuclear-based. This will probably have to be achieved through some kind of government financing, including guarantees in case of an accident. Any such governmental intervention must evidently depend on changes in the attitude of the electric utility industry, changes unthinkable in the past but possible in the coming years.

Meanwhile, the first years of the 1980 decade are unlikely to prove particularly encouraging for the American nuclear industry, which will have to be satisfied with the completion of nuclear power station construction already begun at home and a limited number of foreign orders.

The picture is very different in the socialist world. The Soviets and their partners, free from any public nuclear opposition, relaunched their own atomic programs in the late 1970s. In June 1979 the heads of government of the seven main European Socialist countries, meeting in Moscow, declared themselves in favor of nuclear energy (as did the participants in the Tokyo summit meeting of the Western industrialized countries at about the same time) and set themselves the somewhat overambitious target of 150,000 nuclear MW(e), or 30 percent of their electricity supplies, by 1990 — independently of the Soviet Union's breeder program, which also is being pursued with great vigor. Furthermore, the Soviet Union has acquired a notable foothold in the Western enrichment market. In 1981 there was a first importation of Soviet enriched uranium into the United States for fabrication into fuel elements.

This 1979 Comecon decision will probably contribute to delaying the moment when the energy resources of the Soviet Union could become insufficient for itself and its associated countries, at which point Russia could be led to demand a share of Middle East oil production, thus creating a conflict with the United States and other Western countries.

Industrial Competition

The overall drop in the rate of ordering new nuclear power stations in the Western World could have the gravest consequences for the future of the nuclear industry, one of the most important elements of the capital accumulated in this field over the last decades.

Fortunately, in the United States as well as in Germany and Japan (probably a future competitor in the nuclear power market), the main industrial companies responsible for building nuclear plants are specialized subsidiaries of much larger groups whose survival is not dependent on the rate of development of atomic energy either at home or in the rest of the world. Nevertheless, these nuclear subsidiaries have been faced with serious difficulties in keeping together their teams of specialists, and they have found their financial resources running dangerously low precisely at the time when forecasts, earlier in the past decade, had led them to expect — at last — substantial profits.

This situation has naturally led to increased world competition for any new contracts, the principal contenders being American, German, Canadian, and also French industry, the last having had the advantage of a steady flow of national orders.

American industry, eager to regain in the 1980s the leadership it had previously enjoyed in the nuclear trade, was anxious to show itself once again a reliable supplier. Although the Reagan administration basically supported this renewed initiative, it was unfortunately hampered by the country's Nuclear Non-Proliferation Act of 1978, which continued to prevent the honoring of long-standing enrichment commitments toward Brazil, India, and South Africa, countries that had not accepted full scope safeguards. That act also prevented adherence to the clauses of the USA-Euratom Agreement regarding free transfer and reprocessing of American enriched fuel in the community.

In the cases of Brazil and India for their Angra dos Reis and Tarapur reactors, respectively, the matter had taken on a "Kafkaesque" aspect as the U.S. enrichment contracts with those countries forbade them, during the lifetime of the reactor, to purchase the necessary enriched uranium anywhere else than the United States, which by its own new legislation was now forbidden to supply such material to its two customers.

In late 1981 Brazil was "allowed," by a "special case exemption" announced officially by U.S. Vice President George Bush, to purchase in Europe the fuel necessary for its reactor's first core, the multimillion dollar fine for such a "breach of contract" having been waived. In the meantime, the administration was hoping to convince Congress to amend the retroactive application of the clause of the Nuclear Non-Proliferation Act, thus resolving all of the above unsolvable problems. In any case there would be no more attempts to inhibit fast breeder reactor development or fuel reprocessing in other countries with advanced power programs where there was no proliferation risk. The Carter policy was ended and the United States had recognized that they could not dictate the shape and content of nuclear commerce.

Despite the resulting increased competition, the French Framatome Group — basically responsible for all the country's pressurized water reactor power plants — reinforced its links with Westinghouse in early 1981 through an agreement on equal footing covering exchanges in the whole field of this type of reactor technology. This unique agreement replaced an earlier and as-yet unexpired 10-year license agreement, in existence since

1972, which gave the U.S. government a political right to veto before Framatome could transfer any information received from Westinghouse to a third country. In the new agreement, since the information would flow equally in both directions, each partner would have freedom of final decision, following a (nonetheless obligatory) consultation between the French and U.S. governments. Thus Framatome and Westinghouse were now on a completely equal basis.

Outside Europe, Taiwan (a "hunting ground" for American nuclear industry) and South Korea had remained the principal clients for new orders, while China, Egypt, and Mexico were potential ones. For the first time, the Soviet Union had appeared in the market beyond its European neighbors and partners, obtaining nuclear power plant orders from Cuba and Libya.

With the sole exception of South Korea, where a vigorous nuclear program had been maintained resulting in two orders for Framatome in 1980 following a long series of Westinghouse successes, Asia had become a continent of disappointments. The Iranian jackpot — some 20 nuclear stations by 1990 — vanished in 1979 as the Islamic revolution threw out Kraftwerk Union (KWU) and Framatome, leaving two German-designed plants three-quarters completed and two French units scarcely begun.

Framatome's hopes of selling two reactors to China also faded in 1979 when the Chinese were obliged to revise an overambitious modernization program. But the following year these hopes were revived by a project for two units, this time jointly owned by Hong Kong and China and situated on the Chinese mainland. By late 1981 the American government also became eager to be the first to sell the Chinese a nuclear power station, even though the fact that in all probability China would not accept international safeguards went against U.S. legislation.

In Latin America, ambitious German projects in Brazil were being pursued, although at a somewhat slower pace because of the recession. It also seemed possible that Brazil and Argentina, being less tied than the Federal Republic by nonproliferation pledges, such as those of the NPT and the London guidelines, might at a later stage become more liberal exporters of German technology.

In 1979 KWU had in fact strengthened its position in Argentina by winning the third game in a match with Canada for the sale of power stations using natural uranium and heavy water. Germany had won the first game in 1968 and lost the second in 1973. The whole enterprise, to be placed under IAEA safeguards, also included the purchase of a heavy water plant, the contract for which was secured by Switzerland.

Canada had bid for both installations and its lack of success was no doubt due to its demands for full scope safeguards, which neither Germany nor Switzerland had insisted upon, each assuming that this was the concern of the other.

This setback was greatly resented in Ottawa. In fact, beyond Ontario province — the homeland of the Canada deuterium uranium (CANDU) reactor, where a third of all electricity production in the early 1980s was nuclear — the Canadian technology had only been adopted (on a limited scale) in two other provinces, Quebec and New Brunswick; despite great efforts in Japan and Mexico, no export orders had been obtained since 1973 apart from the sale, following an agreement with Rumania in 1979, of two CANDU power reactors and the corresponding license to that country. This last would enable the Rumanians to build future units on their own. But by the end of 1981, the project had scarcely advanced beyond the planning and site preparation stages, probably because of the considerable economic difficulties facing the country. At the same time, the main Canadian commercial thrust was again directed toward Mexico where it faced eager American, French, German, and Swedish competitors attracted by a program calling for 20,000 MW(e) of nuclear capacity by the end of the century — an overambitious aim for an oil-producing developing country.

Meanwhile, Argentina had purchased heavy water from the Soviet Union, a transaction that followed its supplying the Russians with much-needed grain during the Carter-imposed American embargo following the Afghanistan invasion.

The world recession in nuclear activity was also bound to have a detrimental effect on the uranium market, precisely at a time when Australia was at last beginning to exploit its large resources, and when South Africa, Canada, and Niger were increasing their production. Once again, an overabundance inevitably led to a fall in price, from $45 per pound of uranium oxide in 1977 to $30 per pound in 1981, a fall of more than one-half in real value if depreciation of the American currency is taken into account.

Such a drop in world price was particularly serious for the economy of Niger, whose uranium sales comprised its main source of foreign exchange. The French government, Niger's principal customer, was obliged to help its former African colony by making large purchases at above world prices, thus increasing French uranium stocks to the equivalent of about five years of national consumption. On a worldwide scale, the available stocks in 1981 also correspond to about five years of consumption.

Fortunately, and in contrast to what happened in the early 1970s that finally led to the formation of the non-American producers' cartel, the uranium purchasers avoided taking advantage of this temporary glut by trying to obtain still lower prices, for they were conscious that a healthy mining industry is in the long run more important than the transient gains that might be obtained through a few cutthroat contracts.

Thus oversupply of uranium, as well as increased production of enriched uranium resulting from the operation at full capacity of the Eurodif enrichment plant in France, however superfluous they might seem momentarily, will guarantee that the necessary fuel supplies are available for the essential nuclear revival later in the 1980s. Although under these circumstances there is not likely to be a need for further enrichment capacity during the decade, it may well be needed in the 1990s.

The French Example

In a horse race it commonly happens that the competing jockeys and mounts, disappearing into the last turn, reappear in the final straight in a totally different order. In the same way, in the early 1980s, France, following the confusion of the second half of the 1970s, suddenly found herself leading the field in the Western World's nuclear race, not only in scientific research but also in industrial achievement.

Whenever a country's nuclear effort has been able to profit from continuity, with a technical and political consensus giving support to competent technical and executive teams, it has reaped benefits — as can be seen in the Canadian and French examples.

Whenever administrative structures and technical choices have been continually in question, nuclear decline has followed, as in the case of the United Kingdom, as well as in the United States following the dismantling of the U.S. Atomic Energy Commission and the imposition of the Carter policies.

France herself went through a trying period, in the late 1960s, when her nuclear program seemed to be heading for similar difficulties and disappointments. The decline was avoided just in time, thanks to resolution of the conflict between Electricité de France (EDF) and the Commissariat à l'Energie Atomique (CEA), to the adaptation of the CEA to changing circumstances through the creation of specialized subsidiaries, and above all to constant concern for the maintenance of continuity. This was under-lined in 1978 when responsibility for the CEA was entrusted for the first time to an engineer, Michel Pecqueur, whose entire career had been spent there, and who had already made his mark in every major sector of that organization.

Because France is more dependent that most other industrialized countries on imported energy resources, her reaction to the energy crises of the 1970s had to be more positive than that of her neighbors: from that moment her nuclear power program had received and retained national priority.

The choice of the pressurized water reactor was justified in 1969 because it would allow the fullest benefit to be drawn from experience with this reactor type that was being widely adopted throughout the world. Paradoxically it now seems probable that, in the 1980s, it will be neither American nor German industry, both victims of moratoriums on the building of new stations, but rather the French program, which has been able to keep its industry in full employment, that will provide the bulk of new operational experience with pressurized water reactors. Indeed, in the 15 months since March 1980, EDF commissioned fourteen 900-MW(e) units.

Futhermore, France's important efforts with the fast-neutron breeder have led to an important technical advance with a financial investment that has been moderate compared with similar projects elsewhere. It is hoped that this will be followed before too long by economic improvements.

The creation of a complete fuel cycle industry, a result of the natural uranium policy of the 1960s followed by the policy of enrichment and reprocessing since the start of the following decade, has proved a trump card in guaranteeing French nuclear independence, also giving the country a competitive lead as supplier of the most complete fuel cycle services that other countries could require.

But of all continuities, the most important has been the political continuity that France has enjoyed, for every succeeding government for a third of a century has had the same national objectives in both the civil and military fields, namely nuclear capacity and nuclear independence. At the same time, attitudes to the serious problems of nonproliferation have gradually evolved since 1976 to reach a realistic and reasonable midposition between laxity and a rigidity that would have cut relations with the Third World.

Thus, while the United States was trying to impose on the rest of the world their phobic aversion to plutonium, reprocessing, and breeders, as well as propagating — though this time unwillingly — their problems of nuclear controversy, both the public and the political parties in France have remained — in contrast to other Western countries — relatively unaffected by these external influences. As a result, the French program for nuclear electricity has hardly been affected by the crisis of confusion.

This relative immunity would seem to have been due to a variety of reasons: public awareness of a nuclear past that is already part of history — the Curies, the Joliot-Curies, the epic of heavy water; more recent successes, including those in the military field; a certain complacency at now being independent of the United States and even having an advance over the Americans, who have not always been entirely helpful toward French nuclear development; and finally, above all, the maturity and solid good sense of the French people — aware of the grave implications of the energy crisis, influenced very little by nuclear sensationalism in the media, which incidentally has been relatively more moderate in France than elsewhere, and having a lot more confidence than might sometimes be thought in the competence and responsibility of public bodies.

This calmness and confidence were well demonstrated in 1979, in the ''cracks'' episode when metallurgical microfaults were found in certain thick parts of the pressure vessels of power stations under construction. In assessing these faults, the public in general accepted the advice of the specialists rather than the alarming predictions of certain trade union groups. It was accepted that, with responsible surveillance, even the quite unlikely further development of these faults could not do worse than cause a long-term economic risk by somewhat reducing the power stations' lives.

Two years later, however, other more serious cracks appeared in the French nuclear structure. The civil program was attacked for the first time in political elections by the only one of the four major parties to have an antinuclear faction. During the 1981 presidential and parliamentary campaigns, the Socialists in fact accused EDF of having pursued an exclusive ''all nuclear'' policy without giving consideration to the so-called renewable energies and also without taking sufficient account of the wishes of the population over the siting of proposed nuclear stations.

However, it was clearly stated that all nuclear stations already under construction would be completed, thus enabling the nuclear fraction of national electricity production to reach 55 to 60 percent by 1986.* The plan of the previous administration was for 70 percent of the electricity (30 percent of all energy) to be nuclear by 1990, with 56 pressurized water reactors at 16 different sites.

These clouds in a hitherto clear French nuclear sky did not develop into a serious atomic storm. As promised by the new president, Francois Mitterrand, whose Socialist party won an overwhelming majority, a national debate on energy and especially nuclear power policy took place at the National Assembly in October 1981, five months after the elections.

The real debate was not on the floor of the assembly but in a tense closed-door meeting between the divided factions of the Socialist party. Prime Minister Pierre Mauroy, favorable to nuclear energy, won the day

---

*Construction time for these reactors at the end of the 1970s was six to seven years, about half the time then necessary in the United States.

after having proposed to cut down from nine to six the last units to be built in the current program of 56 reactors. Such a cut was compatible with the slow down in the predicted electrical consumption due to the economic recession. Mauroy also announced that the plans to extend the reprocessing plant at La Hague would go ahead because reprocessing is "the best solution" for disposing of irradiated fuels and France intended to respect all her obligations to treat foreign spent fuel. The costs of this extension, about $350 million, were to be paid by the existing contracts with Germany, Japan, and Sweden. As to the next round of breeder stations, a decision would take place in 1984 after the operation at full power of the precommercial Super-Phénix.

At the conclusion of this parliamentary debate, it was also decided to proceed with more consultations than in the past with the local and regional authorities for each new site; the final decision, however, staying in the hands of the government.

In other words, to the great dissatisfaction of the trade unions hostile to nuclear expansion and of the ecologists who spoke of broken promises and treason, the new majority had readopted the previous program, even though it still retained in its midst some strong enemies of extensive nuclear development that a single debate had surely not disarmed.

But we must have no illusions: France's position at the head of the world nuclear race will certainly make her the chosen target of the world's nuclear oppositions. This will make it difficult for France to maintain her leading position as an isolated advocate of nuclear energy in the Western World, particularly if the relaunching of atomic expansion takes too long to materialize in the other main countries such as the United States and Germany.

It is to be hoped that France, having in the past profited from the successes and difficulties of her competitors, will be able to continue to serve as an example and in turn play an important part in helping her competitors in the renewed and essential further expansion of nuclear power in the 1980s.

4. Conclusion

From infinitesimal particles of matter to infinitely destructive potentialities and concentrated power, this has been the story of the nuclear adventure since 1939; often rewarding, at times unpleasant, but always fascinating.

The adventure from the beginning has been constantly and inextricably intertwined with contemporary history, on which it has left some indelible marks and by which it has itself been profoundly influenced.

The rivalry between the Great Powers, their relationships with the other nations, as well as the awakening of the overpopulated Third World to its rights and to the power and influence conferred on it by its mineral riches, have all been major factors affecting the world nuclear scene.

Starting from limited but sufficient uranium resources, the technology has progressed with astonishing speed to the achievement of the most horrendous weapons, gigantic submarines, and the most powerful and economic thermal power stations.

Inevitably, future science and technology will lead to other weapons as abominable or perhaps even more so, but the economic weapons were the first, and it is on their account that man must face the test of reason or folly.

In the coming decades, science and technology will develop other sources of energy, but controlled nuclear combustion is the only one available today to overcome the looming and dangerous deficit of classical energy resources. Also and above all, it offers the only real chance of reducing the danger of atomic annihilation of all that our civilization has achieved, following a generalized East-West or North-South conflict to take possession of the world's main fossil energy resources.

Everything still remains to be done to bridge the great gap between the riches of the industrialized countries and the poverty of the Third World, as a precondition for equilibrium between countries possessing technology and those possessing raw materials.

Everything still remains to be done to achieve nuclear disarmament, for which the fragile East-West détente and the first SALT agreements are no more than insufficient but indispensable preliminaries.

There is nothing that can be done to halt for good the multiplication of nuclear weapons countries without a real start to nuclear disarmament.

There is nothing that can effectively and progressively reduce public fears of radiation without continuing and satisfactory operation of increasing numbers of nuclear power stations.

And there is no energy more dangerous than no energy. So we must pursue nuclear development, insisting as before — but ever more forcefully — on the minimization of risks of proliferation and of accidents. This is the conviction that ends these lines, as the atomic saga continues.

As the reader peruses these concluding words, the world nuclear scene will already have changed. So the reader may enjoy a sort of revenge on the narrator of the past, for he will now see the beginnings of solutions to some of the many unanswered questions raised in these pages. Perhaps the book will have enabled him to understand better, in their present configuration, the many facets of the atomic complex.

Name Index